Vasily E. Tarasov (Ed.)
Handbook of Fractional Calculus with Applications

Handbook of Fractional Calculus with Applications

Edited by *J. A. Tenreiro Machado*

Volume 1: Theory
Anatoly Kochubei, Yuri Luchko (Eds.), 2019
ISBN 978-3-11-057081-6, e-ISBN (PDF) 978-3-11-057162-2,
e-ISBN (EPUB) 978-3-11-057063-2

Volume 2: Fractional Differential Equations
Anatoly Kochubei, Yuri Luchko (Eds.), 2019
ISBN 978-3-11-057082-3, e-ISBN (PDF) 978-3-11-057166-0,
e-ISBN (EPUB) 978-3-11-057105-9

Volume 3: Numerical Methods
George Em Karniadakis (Ed.), 2019
ISBN 978-3-11-057083-0, e-ISBN (PDF) 978-3-11-057168-4,
e-ISBN (EPUB) 978-3-11-057106-6

Volume 5: Applications in Physics, Part B
Vasily E. Tarasov (Ed.), 2019
ISBN 978-3-11-057089-2, e-ISBN (PDF) 978-3-11-057172-1,
e-ISBN (EPUB) 978-3-11-057099-1

Volume 6: Applications in Control
Ivo Petráš (Ed.), 2019
ISBN 978-3-11-057090-8, e-ISBN (PDF) 978-3-11-057174-5,
e-ISBN (EPUB) 978-3-11-057093-9

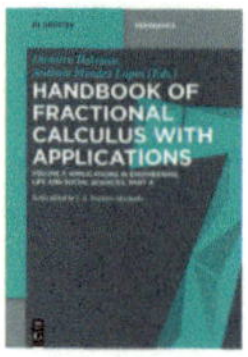

Volume 7: Applications in Engineering, Life and Social Sciences, Part A
Dumitru Băleanu, António Mendes Lopes (Eds.), 2019
ISBN 978-3-11-057091-5, e-ISBN (PDF) 978-3-11-057190-5,
e-ISBN (EPUB) 978-3-11-057096-0

Volume 8: Applications in Engineering, Life and Social Sciences, Part B
Dumitru Băleanu, António Mendes Lopes (Eds.), 2019
ISBN 978-3-11-057092-2, e-ISBN (PDF) 978-3-11-057192-9,
e-ISBN (EPUB) 978-3-11-057107-3

Preface

Fractional calculus (FC) was initially developed in 1695, nearly at the same time as the conventional calculus. However, FC attracted limited attention and remained a pure mathematical exercise, in spite of the contributions of important mathematicians, physicists, and engineers. FC had rapid further development during the last few decades, both in mathematics and applied sciences, being nowadays recognized as an excellent tool for describing complex systems, phenomena involving long range memory effects, and non-locality. A huge number of research papers and books devoted to this subject have been published, and presently several specialized conferences and workshops are organized each year. The FC popularity in all fields of science is due to its successful application in mathematical models, namely in the form of FC operators and fractional integral and differential equations. Presently, we are witnessing considerable progress both in regard to theoretical aspects and applications of FC in areas such as physics, engineering, biology, medicine, economy, or finance.

The popularity of FC has attracted many researchers from all over the world, and there is a demand for works covering all areas of science in a systematic and rigorous form. In fact, the literature devoted to FC and its applications is huge, but readers are confronted with a high heterogeneity and, in some cases, with misleading and inaccurate information. The Handbook of Fractional Calculus with Applications (HFCA) intends to fill that gap and provides the readers with a solid and systematic treatment of the main aspects and applications of FC. Motivated by these ideas, the editors of the volumes involved a team of internationally recognized experts for a joint publishing project, offering a survey of their own and other important results in their fields of research. As a result of these joint efforts, a modern encyclopedia of FC and its applications, reflecting present day scientific knowledge, is now available with the HFCA. This work is distributed by several distinct volumes, each one developed under the supervision of its editors.

The fourth and fifth volumes of HFCA are devoted to the application of fractional calculus (FC) and fractional differential equations in different areas of physics. These volumes describe the fundamental physical effects and, first of all, those that belong to fractional relaxation-oscillation or diffusion-wave phenomena. FC allows describing spatial non-locality and fading memory of power-law type, the openness of physical systems and dissipation, long-range interactions, and other physical phenomena. The most well-known physical phenomena and processes, which are described by fractional differential equations, include fractional viscoelasticity, spatial and frequency dispersion of power type, nonexponential relaxation, anomalous diffusion, and many others. The fourth volume of HFCA focuses on the application of FC in various aspects of classical mechanics and continuum mechanics. The most important basic models and phenomena include the fractional oscillator and Van der Pol oscillator, fractional relaxation and fractional oscillation phenomena, discrete long-range

https://doi.org/10.1515/9783110571707-201

interactions, and nonlinear systems with power-law memory. In the 13 chapters, the non-locality and memory of the power-law type are described in phenomena, such as viscoelasticity, thermodynamics, thermodiffusion in deformable solid, in gradient mechanics, and mechanics of fractal media.

My special thanks go to the authors of individual chapters that are excellent surveys of selected classical and new results in several important fields of FC. The editors believe that the HFCA will represent a valuable and reliable reference work for all scholars and professionals willing to develop research in the challenging, relevant, and timely scientific area.

Vasily E. Tarasov

Contents

J. A. Tenreiro Machado and António M. Lopes

Fractional Van der Pol oscillator

Abstract: This chapter presents the Van der Pol oscillator (VPO). Several formulations of the VPO are analyzed, namely the classical integer-order and the real and complex fractional-order models. The autonomous and forced versions are investigated in time and frequency domains, using phase portraits, spectral analysis, and bifurcation diagrams.

Keywords: Fractional calculus, Van der Pol oscillator, nonlinear dynamics, chaos

PACS: 05.45.Pq, 05.45.Xt, 05.45.-a

1 Introduction

The Van der Pol oscillator (VPO) was introduced by the Dutch electrical engineer Balthazar van der Pol (1889–1959) to describe triode oscillations in electrical circuits [37, 38]. The VPO dynamics is modeled by a second-order nonlinear differential equation, and—in the scope of mechanical systems—can be interpreted as describing a nonlinear mass-spring-damper system with a position-dependent damping coefficient, or analogously, as a resistance-inductance-capacitance electrical nonlinear circuit with a negative-nonlinear resistance. The VPO has been used to describe phenomena in many areas, namely electronics, biology, and acoustics, since it exhibits a behavior that is ubiquitous in several natural and artificial systems [12, 13, 16, 17, 21, 26, 31, 34, 39, 42].

This chapter studies several formulations of the VPO, both in the time and frequency domains, using phase portraits, spectral analysis, and bifurcation diagrams. In the first phase, the autonomous and forced integer-order VPO is analyzed. In the second phase, the influence of a fractional order time derivative is investigated.

In this line of thought, the chapter is organized as follows. Section 2 introduces the autonomous and forced VPO. Sections 3 and 4 study the real and complex fractional-order models, respectively. Finally, Section 4 presents the main conclusions.

J. A. Tenreiro Machado, Institute of Engineering, Polytechnic of Porto, Department of Electrical Engineering, Rua Dr. António Bernardino de Almeida, 431, 4249-015 Porto, Portugal, e-mail: jtm@isep.ipp.pt
António M. Lopes, UISPA–LAETA/INEGI, Faculty of Engineering, University of Porto, Rua Dr. Roberto Frias, 4200-465 Porto, Portugal, e-mail: aml@fe.up.pt

https://doi.org/10.1515/9783110571707-001

2 Integer-Order Van der Pol oscillator

The standard autonomous integer-order VPO model is given by the second-order nonlinear differential equation:

$$\ddot{x} + \lambda(x^2 - 1)\dot{x} + x = 0, \tag{1}$$

where $\lambda > 0$ is a control parameter, or damping, that reflects the degree of nonlinearity of the system.

The state-space model of (1), with $x_1 = x$ and $x_2 = \dot{x}$, is given by

$$\begin{bmatrix} \dot{x}_1 \\ \dot{x}_2 \end{bmatrix} = \begin{bmatrix} 0 & 1 \\ -1 & -\lambda(x_1^2 - 1) \end{bmatrix} \begin{bmatrix} x_1 \\ x_2 \end{bmatrix}. \tag{2}$$

Equation (1) has a limit cycle that attracts other solutions, with exception of the unique equilibrium point $(x, \dot{x}) = (0, 0)$.

Figure 1 depicts the phase portraits and the time domain trajectories of the autonomous VPO for the initial conditions $(x(0), \dot{x}(0)) = (0, 1)$ and different values of the parameter λ. It is seen that as λ varies, the unstable focus and the limit cycle remain the same. In other words, the topology of the phase space does not change and, as λ increases, all solutions tend to approach the limit cycle in a shorter time interval. Therefore, for $\lambda = 0$, the autonomous VPO behaves as a harmonic oscillator, whereas for $\lambda > 0$, it exhibits a stable limit cycle, with amplitude $x_{\max} \approx 2$.

Figure 2 represents the angular frequency of the limit cycle, ω (period $T = \frac{2\pi}{\omega}$), versus the parameter λ. For small values of λ, the frequency is approximately $\omega = 1\ \mathrm{rad\ s^{-1}}$, whereas for increasing values of λ, we verify that ω decreases. Figure 3 depicts the Fourier spectra of x for the autonomous VPO as a function of λ, for $\lambda \in [0, 10]$, confirming the variation of ω with the parameter λ and showing the corresponding harmonics.

We now consider the sinusoidal forced integer-order VPO model:

$$\ddot{x} + \lambda(x^2 - 1)\dot{x} + x = A\cos(\omega_f t). \tag{3}$$

The forced VPO behavior depends not only on the damping λ, but also on the amplitude and frequency of the excitation function $\{A, \omega_f\}$ (period $T_f = \frac{2\pi}{\omega_f}$).

Figure 4 depicts the phase portraits and the time-domain trajectories of the forced VPO for the initial conditions $[x(0), \dot{x}(0)] = (0, 1)$. The damping is set to $\lambda = 5$, and the frequency and amplitudes of the exciting function are $\omega_f = 7\ \mathrm{rad\ s^{-1}}$ and $A = \{10, 28, 48, 53\}$, respectively. Figure 5 represents the corresponding Poincaré maps. For $A = \{10, 48\}$, the forced VPO exhibits quasi-periodic motion: (i) the trajectories

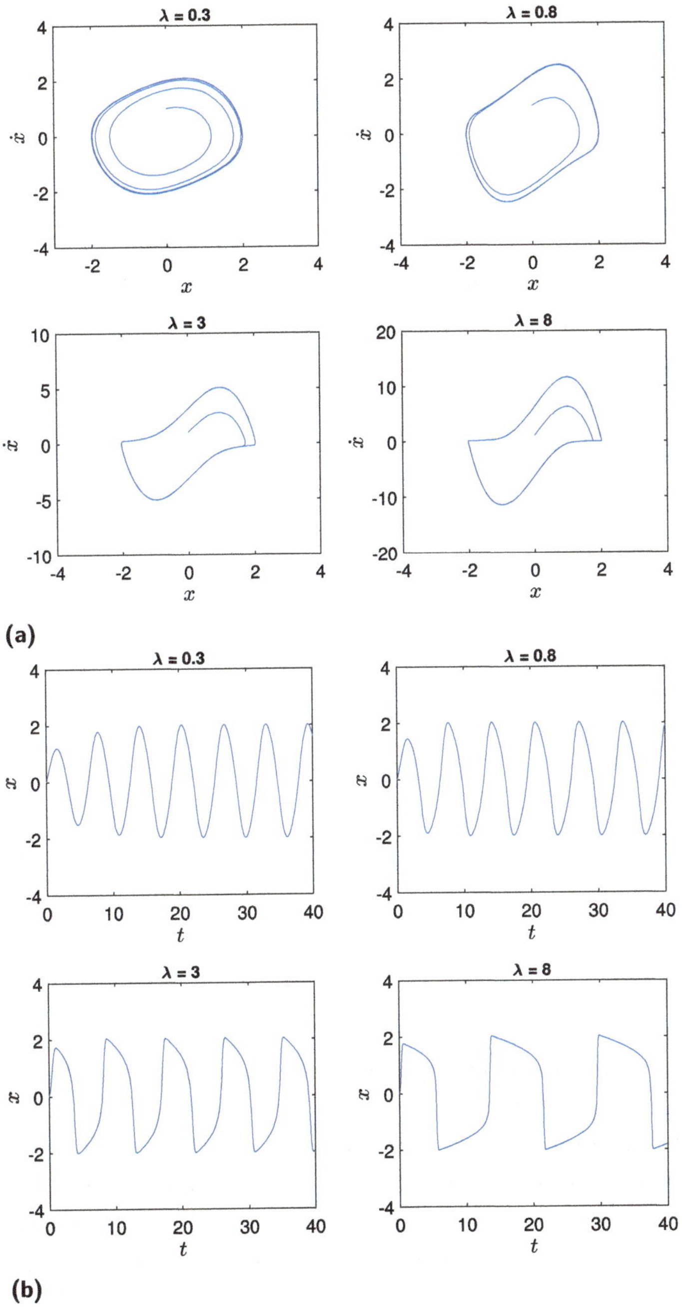

Figure 1: The autonomous VPO dynamics for $\lambda = \{0.3, 0.8, 3, 8\}$: (a) phase portraits $(x, \dot{x})$; (b) time-domain trajectories x.

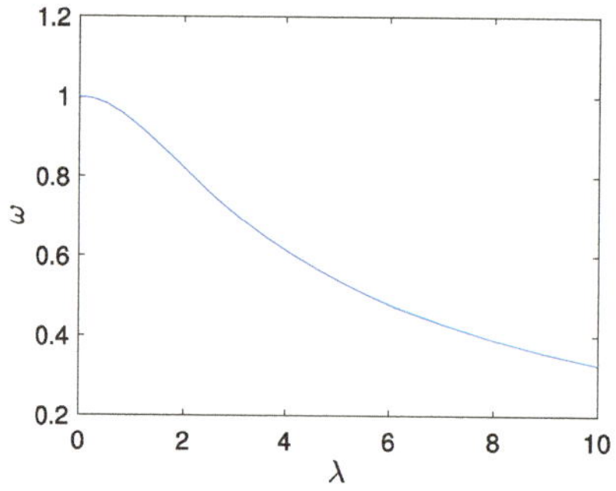

Figure 2: The autonomous VPO angular frequency of the limit cycle ω versus the parameter λ.

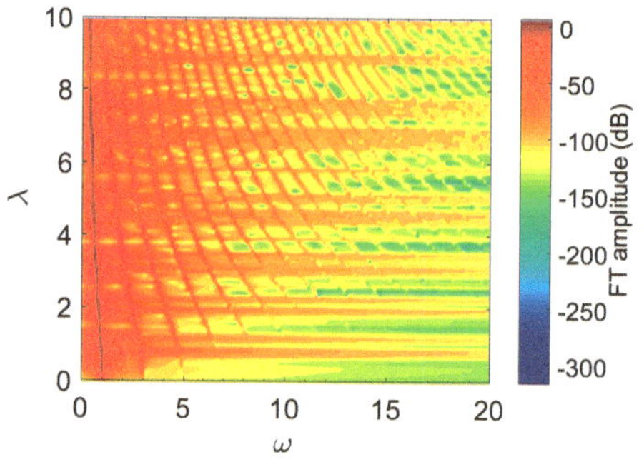

Figure 3: The Fourier spectra of the signal x of the autonomous VPO, as a function of $\lambda \in [0, 10]$.

cover densely the region in the interior of the phase space; (ii) the ratio between the two frequency components visible on the time-domain trajectories is an irrational value; (iii) the orbits in the Poincaré maps are almost closed connected curves, meaning that at the time instants $t = nT_f$, $n \in \mathbb{N}$, the trajectories never return exactly to the same points. For $A = \{28, 53\}$, the forced VPO reveals periodic motion: (i) the phase trajectories do not cover densely the interior of the phase portraits; (ii) the ratio between the two frequency components observed on the time trajectories is a rational number; (iii) the Poincaré maps are sets of points.

Figure 6 shows different bifurcation diagrams of the forced VPO, obtained by applying the method of Poincaré sections. These diagrams represent the loci of the sampled output, $x(nT_f)$, versus the parameters A, λ, and ω_f. When varying the parameter A (Figures 6(a) and 6(b)), regions of periodic, quasi-periodic, and period-locked motion are observed. In the periodic regions, the period of the limit cycle is an odd multiple of T_f. Period-locked motion occurs for the critical driving amplitudes $A = 16.9$ and $A = 73.3$ rad s^{-1}, respectively, and the period of the solutions equals T_f. When varying the parameters λ and w_f (Figure 6(c) and Figure 6(d), respectively) it is verified that periodic and quasi-periodic regions alternate.

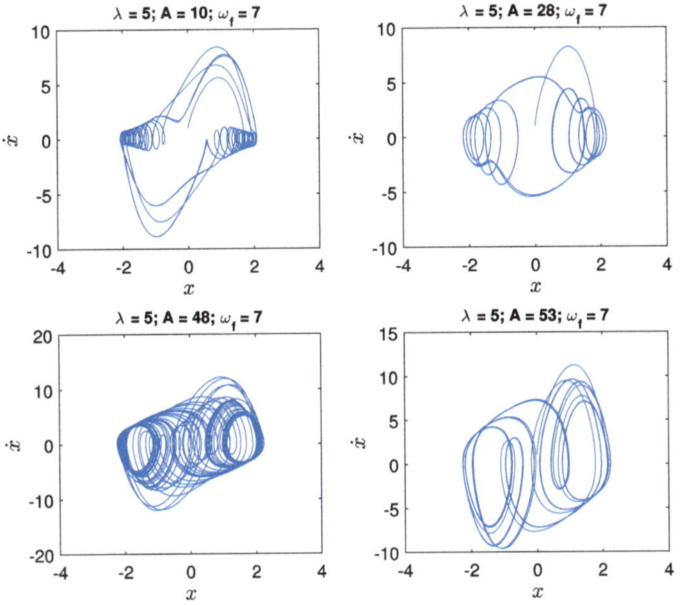

(a)

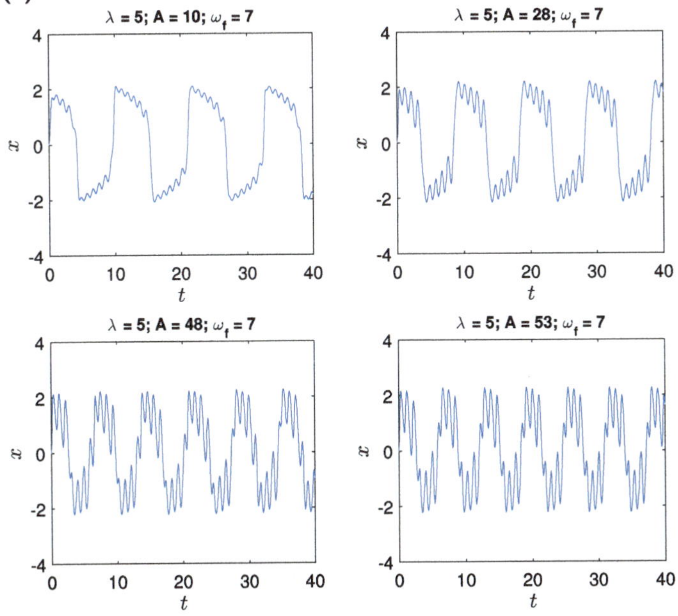

(b)

Figure 4: The forced VPO dynamics for $\lambda = 5$, $\omega_f = 7$, and $A = \{10, 28, 48, 53\}$: (a) phase portraits $(x, \dot{x})$; (b) time-domain trajectories x.

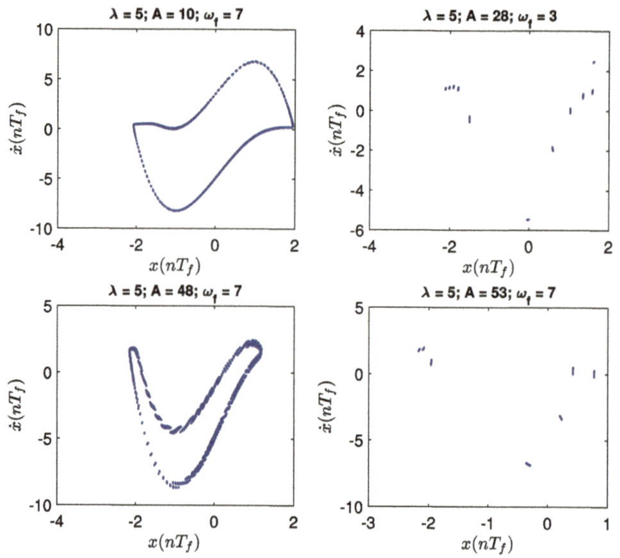

(a)

Figure 5: The Poincaré maps of the forced VPO for $\lambda = 5$, $\omega_f = 7$, and $A = \{10, 28, 48, 53\}$.

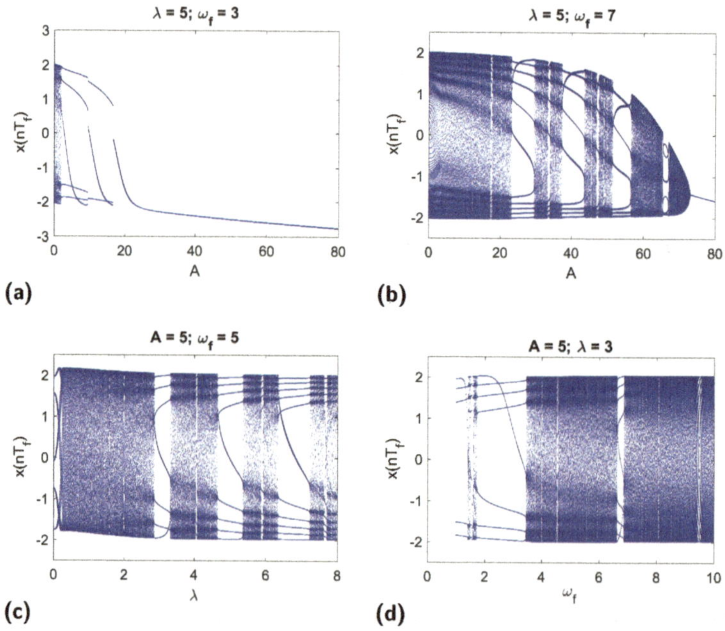

Figure 6: The bifurcation diagrams of the forced VPO: (a) $\lambda = 5$, $\omega_f = 3$, $A \in [0, 80]$; (b) $\lambda = 5$, $\omega_f = 7$, $A \in [0, 80]$; (c) $A = 5$, $\omega_f = 5$, $\lambda \in [0, 8]$; (d) $\lambda = 3$, $A = 5$, $\omega_f \in [1, 10]$.

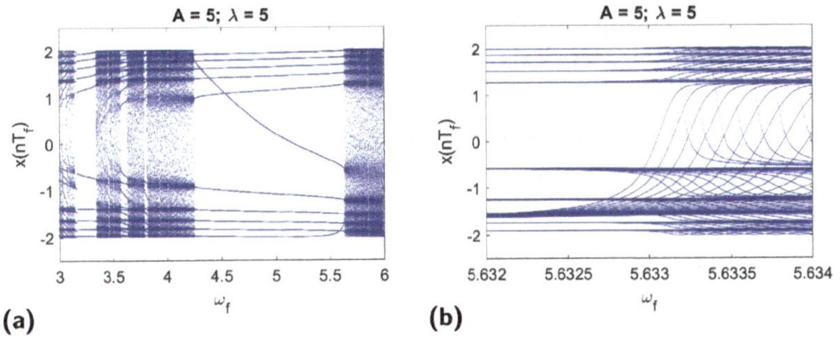

Figure 7: The bifurcation diagram of the forced VPO dynamics for $A = 5$ and $\lambda = 5$: (a) normal view ($\omega_f \in [3,6]$); (b) magnified view ($\omega_f \in [5.632, 5.634]$).

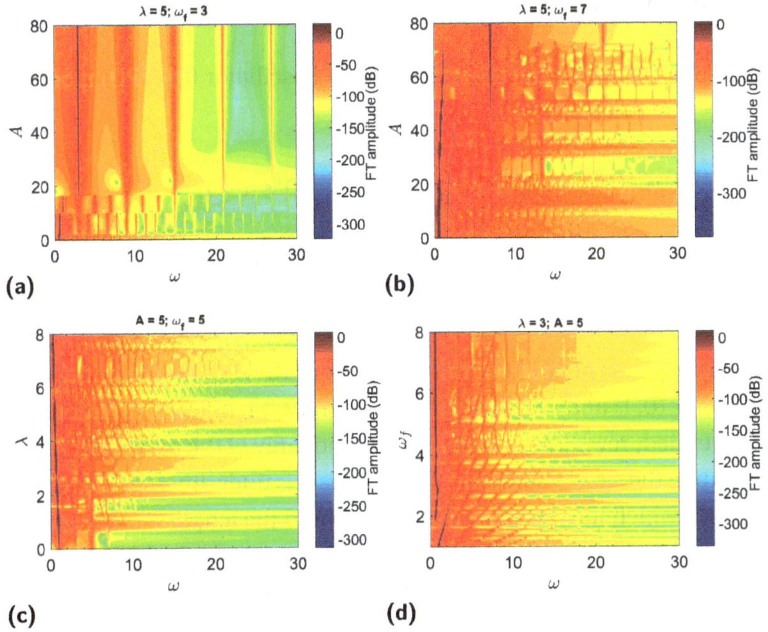

Figure 8: The Fourier spectra of x for the forced VPO: (a) $\lambda = 5$, $\omega_f = 3$, $A \in [0, 80]$; (b) $\lambda = 5$, $\omega_f = 7$, $A \in [0, 80]$; (c) $A = 5$, $\omega_f = 5$, $\lambda \in [0, 8]$; (d) $\lambda = 3$, $A = 5$, $\omega_f \in [1, 10]$.

Figure 7 illustrates chaotic behavior, revealing the existence of period-doubling-cascades phenomena [19, 25, 40]. The bifurcation diagram looks self-similar under scaling, meaning that the structure of the bifurcation curves repeats successively in the parameter space [33].

Figure 8 depicts the Fourier spectra of x for the same parameters used in Figure 6. The periodic, quasi-periodic, and period-locked motions are confirmed.

3 Real fractional-order Van der Pol oscillator

The standard VPO model (1) can be generalized by replacing the integer-order time derivatives with real-order ones, yielding:

$$x^{(\alpha_1)} + \lambda(x^2 - 1)x^{(\alpha_2)} + x = A\cos(\omega_f t), \quad 0 < \alpha_1, \alpha_2 < 2, \tag{4}$$

where $x^{(\alpha_i)}$, $i = \{1, 2\}$, denotes the α_i-order time derivative of x, that is, $x^{(\alpha_i)} = D^{\alpha_i}x$.

Different versions of Equation (4) were studied, differing on the restrictions applied to the derivative orders α_1 and α_2 [2–4, 6, 7, 9, 11, 20, 32]. Recently, the variable-order VPO was proposed [8], based on the concept of variable-order fractional dynamics, which extends the fractional calculus to derivatives and integrals of variable-order [1, 22, 29, 35, 41].

In this section, the real-order fractional Van der Pol oscillator (RF-VPO) model is studied, so that $\alpha_2 = \alpha$ and $\alpha_1 = \alpha_2 + 1$:

$$x^{(1+\alpha)} + \lambda(x^2 - 1)x^{(\alpha)} + x = A\cos(\omega_f t), \quad 0 < \alpha < 1. \tag{5}$$

The integrator $s^{-\alpha}$ is approximated in the Fourier domain by means of the CRONE's recursive method [23, 24, 36]. The order is set to $N = 5$, and the bandwidth is $[\omega_l, \omega_h] = [10^{-2}, 10^2]$ rad s^{-1}.

Figure 9 shows the phase portraits of the autonomous RF-VPO, for the initial conditions $(x(0), x^{(\alpha)}(0)) = (0, 1)$ and different values of α and λ. Significant variations on the limit cycle are verified, revealing a large influence not only of the parameter λ, but also of the order α upon the system dynamics. Such impact is underlined by the results depicted in Figure 10, which illustrate the frequency ω and the amplitude x_{max} of the output versus α and λ.

Figure 11 shows the amplitude Fourier spectra of the output x for various λ and α. It can be noticed that the energy of x is not only concentrated at the fundamental and integer-odd harmonics, but also distributed along the entire frequency domain [6, 18].

In what concerns the forced RF-VPO, Figure 12 illustrates the phase portraits and the time-domain trajectories for several distinct values of the parameters. The initial conditions are $(x(0), x^{(\alpha)}(0)) = (0, 1)$, the derivative order is $\alpha = 0.8$, the damping is set to $\lambda = 5$, and the frequency and amplitudes of the exciting function are $\omega_f = 7$ rad s^{-1} and $A = \{5, 15, 22, 35\}$, respectively. Figure 13 represents the corresponding Poincaré maps. For $A = \{15, 22\}$, the forced RF-VPO exhibits periodic motion, whereas for $A = 5$ and $A = 35$, it reveals quasi-periodic and chaotic motions.

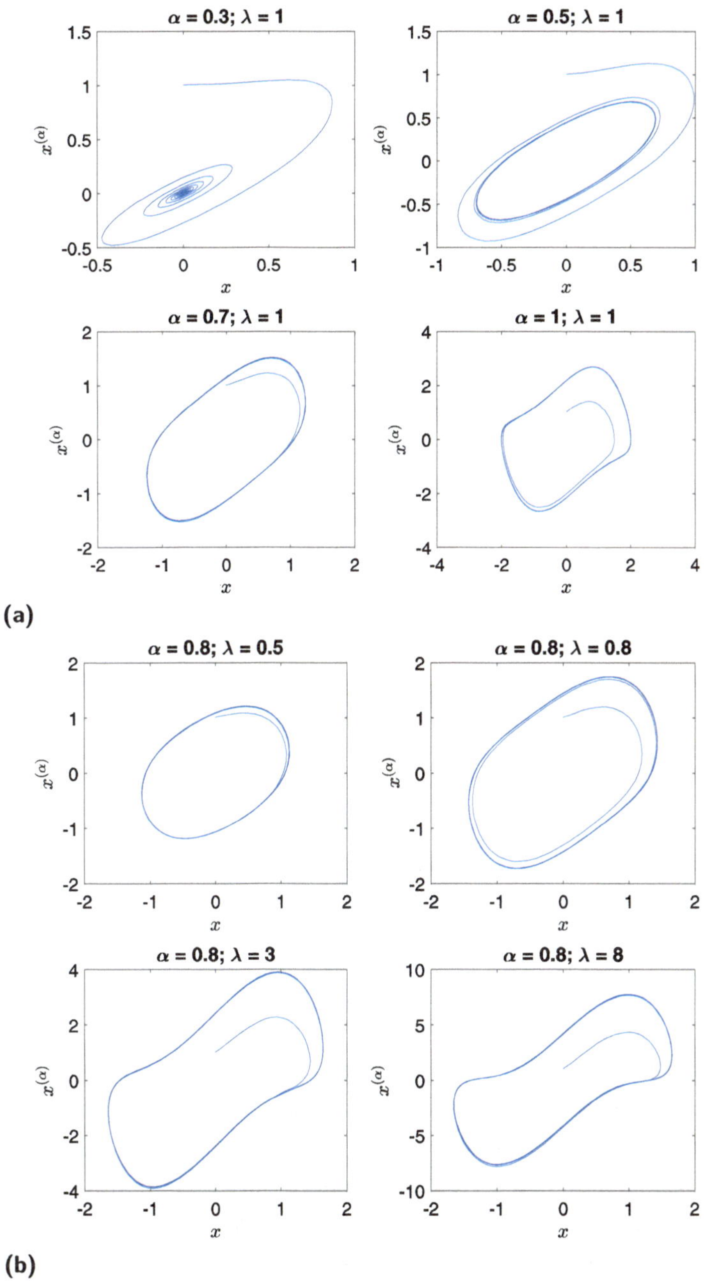

Figure 9: The autonomous RF-VPO phase portraits $(x, x^{(\alpha)})$: (a) $\lambda = 1$ and $\alpha = \{0.3, 0.5, 0.7, 1\}$; (b) $\lambda = \{0.5, 0.8, 3, 8\}$ and $\alpha = 0.8$.

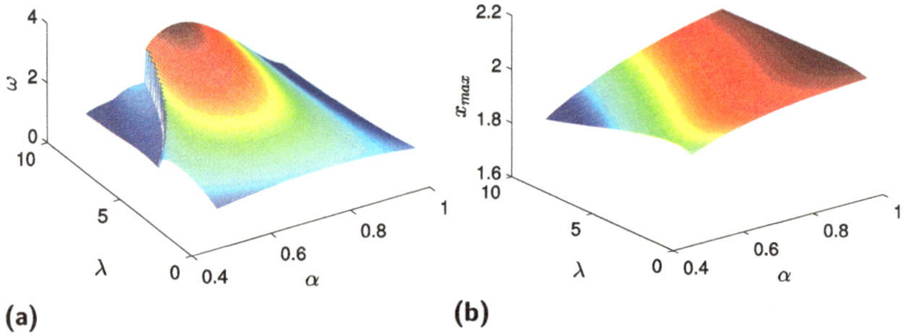

Figure 10: The limit cycle of the autonomous RF-VPO for $\lambda \in [0, 10]$ and $\alpha \in [0.5, 1]$: (a) frequency; (b) maximum amplitude.

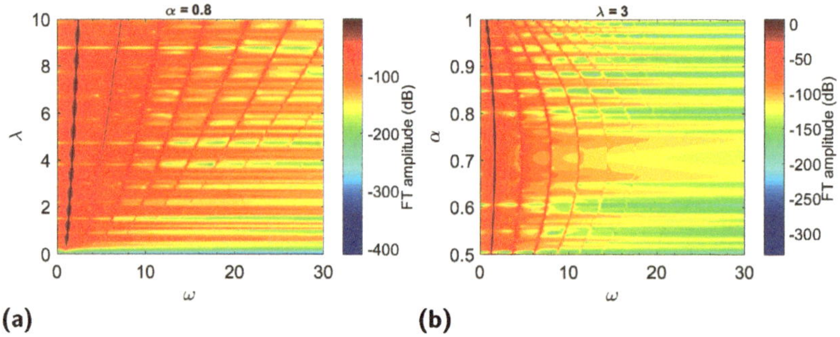

Figure 11: The amplitude Fourier spectra of x for the autonomous RF-VPO output, with $\lambda \in [0, 10]$ and $\alpha = 0.8$ and with $\alpha \in [0.5, 1]$ and $\lambda = 8$.

Figures 14 and 15 show the bifurcation diagrams and the Fourier spectra of x, respectively, for the varying parameters A, λ, ω_f, and α. Just as the classical VPO, regions of periodic, quasi-periodic, and period-locked motion are observed, revealing that the RF-VPO exhibits different types of behavior.

In synthesis, the results show that the fractional order α can act as an extra degree-of-freedom that may be useful for tuning and control, having in mind that:

- the rate of convergence to the steady-state solution of the RF-VPO varies with α, since the fractional order acts as a damping coefficient;
- both the maximum amplitude and the frequency of the RF-VPO limit cycle depend on α;
- the fractional order α can change the dynamics of the forced RF-VPO between periodic, quasi-periodic, and period-locked motions.

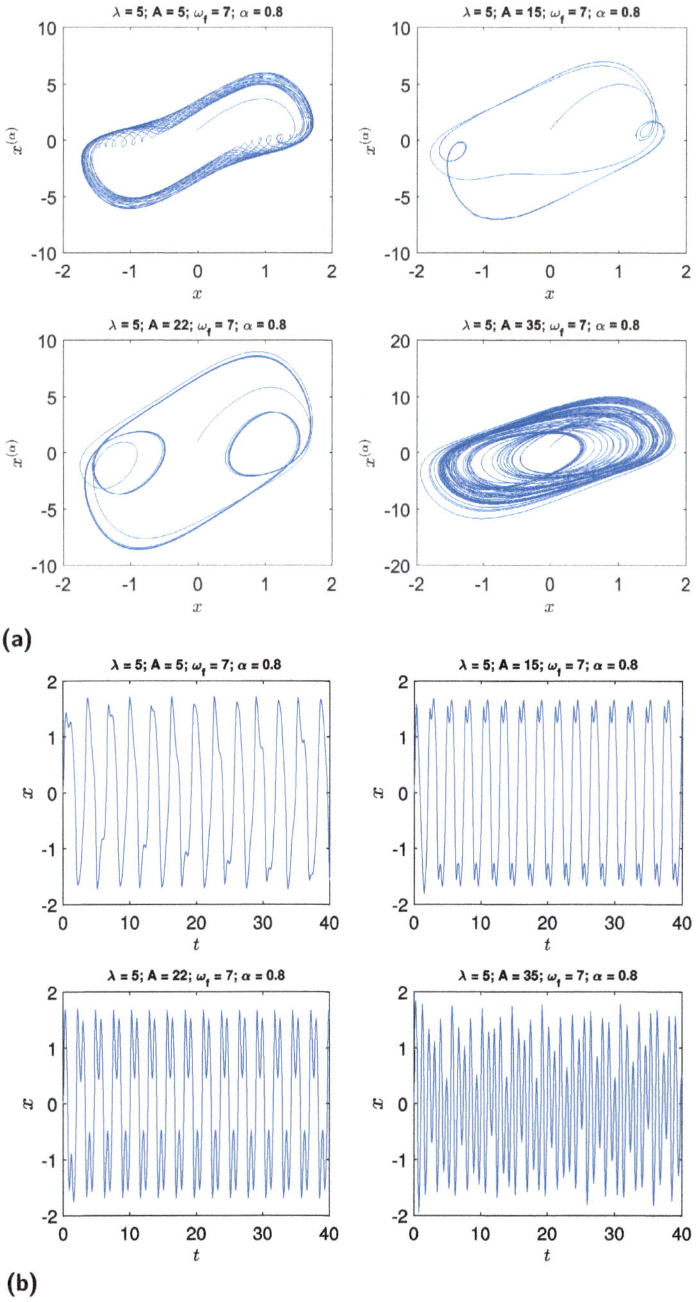

Figure 12: The forced RF-VPO dynamics for $\alpha = 0.8$, $\lambda = 5$, $\omega_f = 7$, and $A = \{5, 15, 22, 35\}$: (a) phase portraits $(x, x^{(\alpha)})$; (b) time-domain trajectories x.

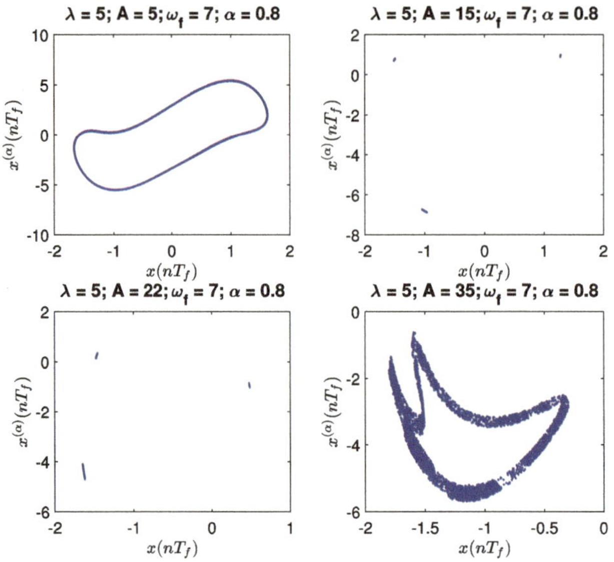

(a)

Figure 13: The Poincaré maps of the forced RF-VPO for $\alpha = 0.8$, $\lambda = 5$, $\omega_f = 7$, and $A = \{5, 15, 22, 35\}$.

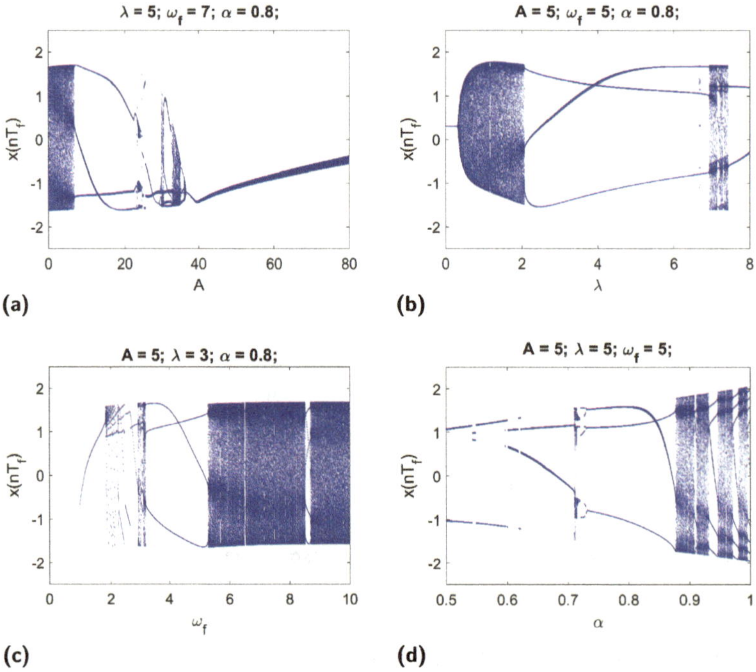

Figure 14: The bifurcation diagrams of the forced RF-VPO: (a) $\lambda = 5$, $\omega_f = 7$, $\alpha = 0.8$, $A \in [0, 80]$; (b) $A = 5$, $\omega_f = 5$, $\alpha = 0.8$, $\lambda \in [0, 8]$; (c) $A = 5$, $\lambda = 3$, $\alpha = 0.8$, $\omega_f \in [1, 10]$; (d) $A = 5$, $\lambda = 5$, $\omega_f = 5$, $\alpha \in [0.5, 1]$.

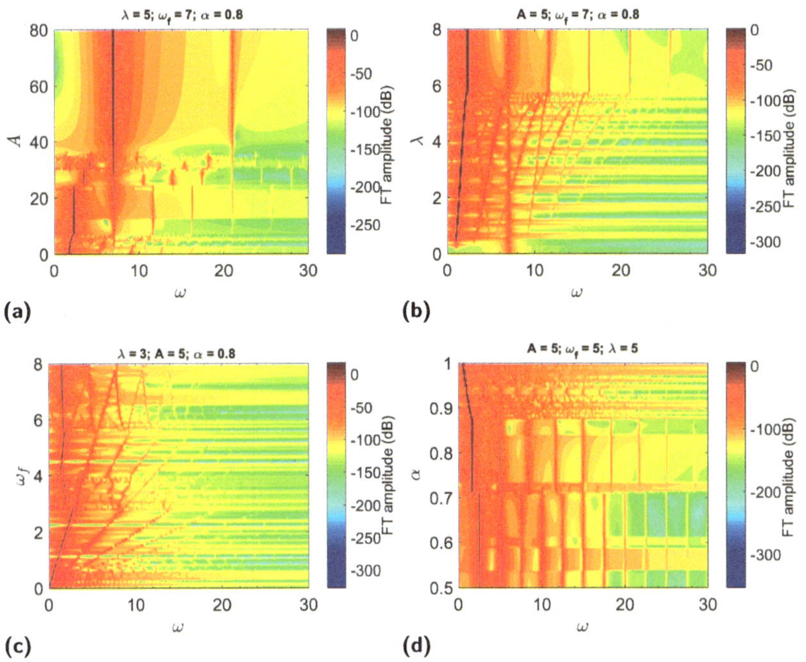

Figure 15: The Fourier spectra of x for the forced RF-VPO: (a) $\lambda = 5, \omega_f = 7, \alpha = 0.8, A \in [0, 80]$; (b) $A = 5, \omega_f = 5, \alpha = 0.8, \lambda \in [0, 8]$; (c) $A = 5, \lambda = 3, \alpha = 0.8, \omega_f \in [1, 10]$; (d) $A = 5, \lambda = 5, \omega_f = 5, \alpha \in [0.5, 1]$.

4 Complex fractional-order Van der Pol oscillator

The model (4) is generalized by introducing complex-order time derivatives. In this regard, different variants are possible [27, 28]. In this section, the complex-order fractional Van der Pol oscillator (CF-VPO) is studied, based on the model

$$x^{(1+\gamma)} + \lambda(x^2 - 1)x^{(\gamma)} + x = A\cos(\omega_f t), \quad \gamma = \alpha + j\beta, 0 < \alpha < 1, \beta \in \mathbb{R}, \tag{6}$$

where $x^{(\gamma)} = \mathbb{R}[D^\gamma x]$. This operator is implemented in the Fourier domain by means of the CRONE's recursive method [10, 14, 15, 23, 24, 30]. The number of poles and zeros is $(N, M) = (5, 4)$, and the bandwidth is $[\omega_l, \omega_h] = [10^{-2}, 10^2]$ rad s^{-1}. Other alternatives for implementing $x^{(\gamma)}$ are possible, yielding directly a real output without using the operator $\mathbb{R}[\cdot]$ [5, 27].

Figure 16 depicts the phase portraits and the time-domain trajectories of the autonomous CF-VPO for the initial conditions $(x(0), x^{(\gamma)}(0)) = (0, 1)$, $\beta = 0.8$, and different values of α and λ. This figure illustrates that, for a given β, the limit cycle of the autonomous CF-VPO is almost insensitive to the other parameters. Such behavior is also verified by the results in Figure 17, which depict the frequency ω and the amplitude of the limit cycle x_{max} versus λ, α, and β.

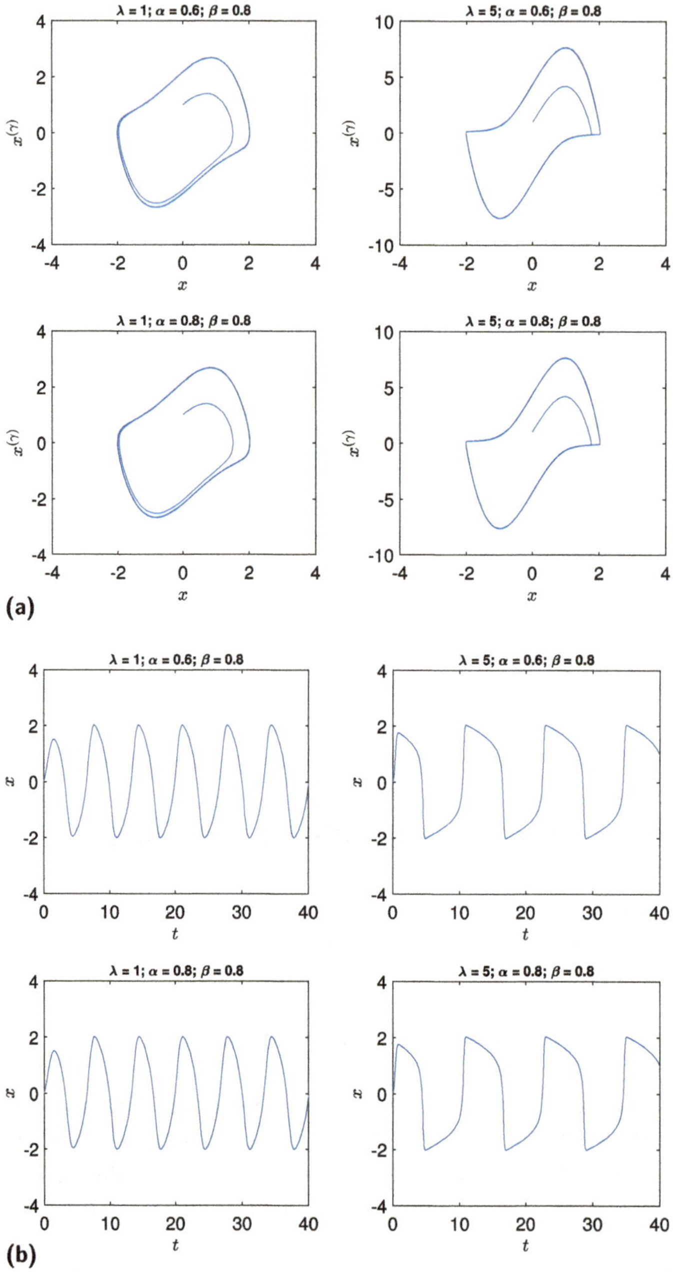

Figure 16: The autonomous CF-VPO dynamics for $\beta = 0.8$, and different values of α and λ: (a) phase portraits $(x, x^{(\gamma)})$; (b) time-domain trajectories x.

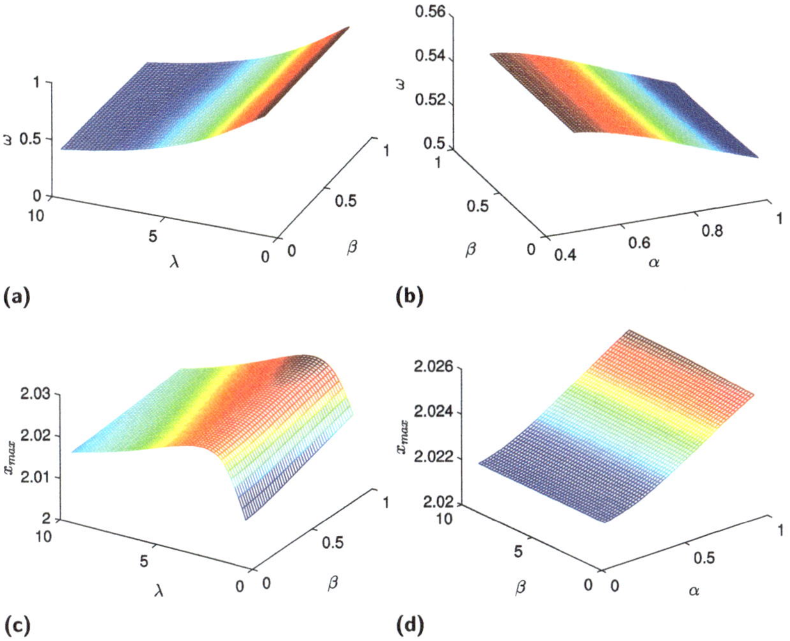

Figure 17: The limit cycle of the autonomous CF-VPO for $\lambda \in [0, 10]$, $\alpha \in [0.5, 1]$, and $\beta \in [0.1, 1]$: (a) frequency; (b) amplitude.

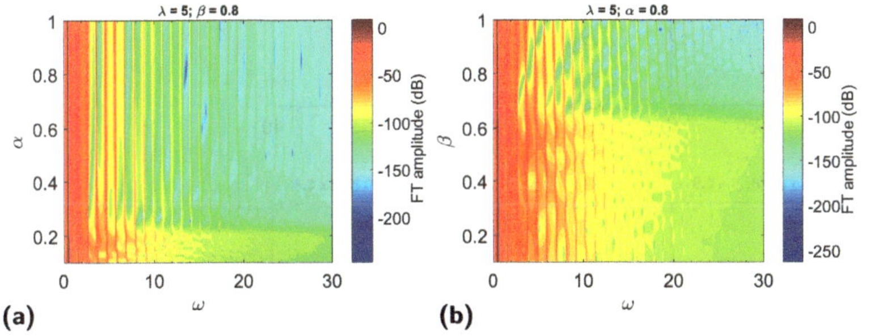

Figure 18: The amplitude Fourier spectra of x for the autonomous CF-VPO output, with $\lambda = 5$, $\alpha \in [0.1, 1]$, and $\beta \in [0.1, 1]$.

Figure 18 shows the amplitude Fourier spectra of the output x for various λ, α, and β. It can be noticed that the energy of x is not concentrated only at the fundamental and integer-odd harmonics, but is distributed also along other frequencies.

In what concerns the forced CF-VPO, Figure 19 illustrates the phase portraits and the time-domain trajectories for several distinct values of the parameters. The initial

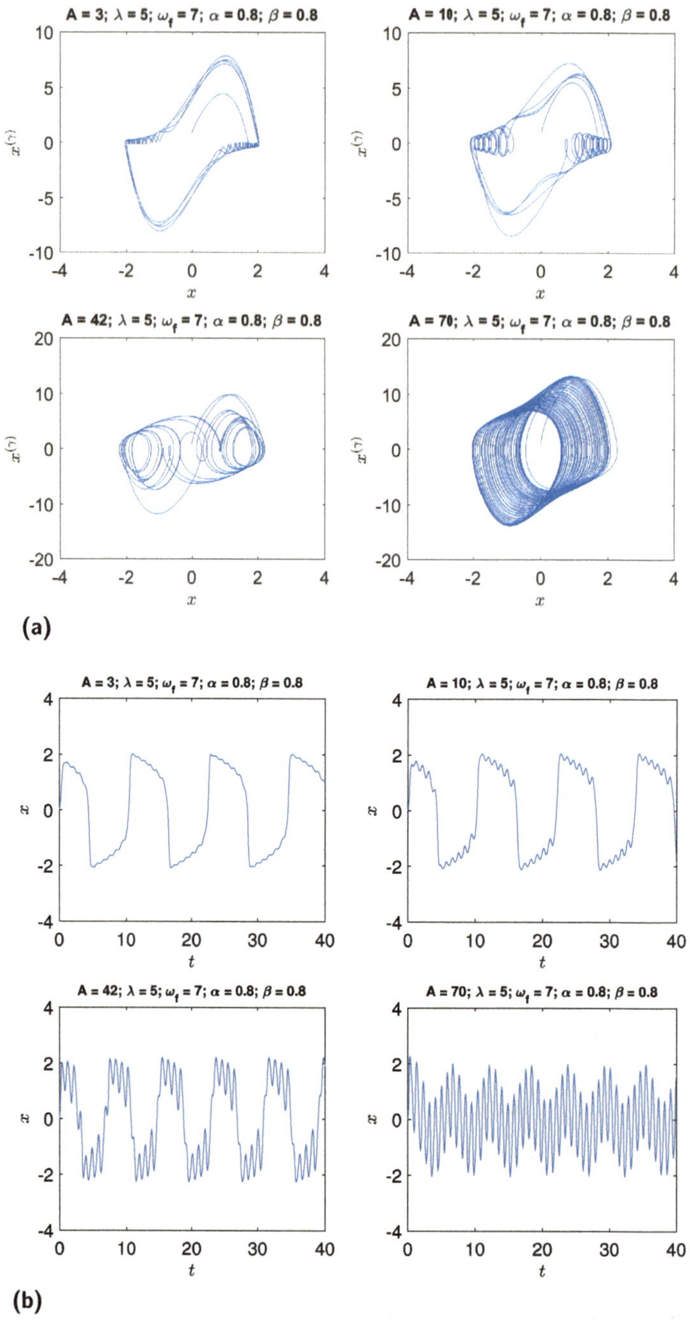

Figure 19: The forced CF-VPO dynamics for $\alpha = \beta = 0.8$, $\lambda = 5$, $\omega_f = 7$, and $A = \{3, 10, 42, 70\}$: (a) phase portraits $(x, x^{(\gamma)})$; (b) time-domain trajectories x.

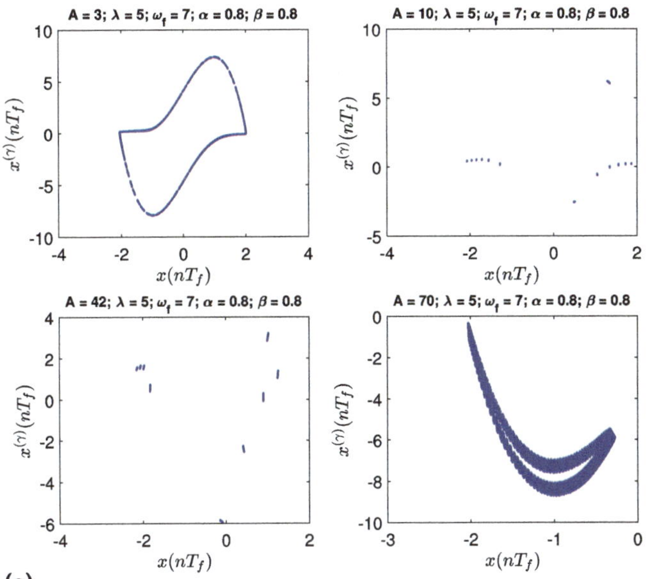

Figure 20: The Poincaré maps of the forced CF-VPO for $\alpha = \beta = 0.8, \lambda = 5, \omega_f = 7$, and $A = \{3, 10, 42, 70\}$.

conditions are $(x(0), x^{(\gamma)}(0)) = (0, 1)$, the derivative order is $\alpha = \beta = 0.8$, the damping is set to $\lambda = 5$, and the frequency and amplitudes of the exciting function are $\omega_f = 7\,\text{rad s}^{-1}$ and $A = \{3, 10, 42, 70\}$, respectively. Figure 20 represents the corresponding Poincaré maps. For $A = \{10, 42\}$, the forced RF-VPO exhibits periodic motion, whereas for $A5$ and $A = 70$, it reveals quasi-periodic and chaotic motions.

Figures 21 and 22 show the bifurcation diagrams for the parameters A, λ, ω_f, α, and β. Just as the classical VPO and the RF-VPO, regions of periodic, quasi-periodic, and period-locked motion are observed, showing that the fractional orders (α, β) act as two additional degrees-of-freedom for tuning and control the VPO.

Figures 23 and 24 depict the Fourier spectra of x for the forced CF-VPO, confirming the results obtained with the bifurcation diagrams.

In conclusion, for the CF-VPO, the results show that the fractional order β influences the system dynamics, such as:

- both the maximum amplitude and the frequency of the CF-VPO limit cycle depend on β; nevertheless, they are less sensitive to β than to α;
- the fractional order β can change the dynamics of the forced CF-VPO between periodic, quasi-periodic, and period-locked motions;
- the energy tends to be more concentrated at lower frequencies when $\beta \neq 0$.

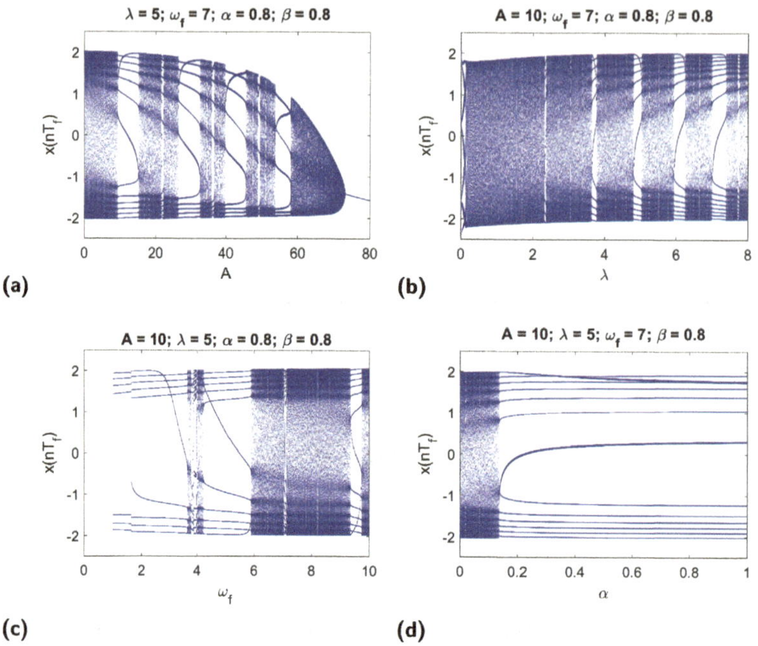

Figure 21: The bifurcation diagrams of the forced CF-VPO: (a) $\lambda = 5, \omega_f = 7, \alpha = 0.8, \beta = 0.8, A \in [0, 80]$; (b) $A = 10, \omega_f = 7, \alpha = 0.8, \beta = 0.8, \lambda \in [0, 8]$; (c) $A = 10, \lambda = 5, \alpha = 0.8, \beta = 0.8, \omega_f \in [1, 10]$; (d) $A = 10, \lambda = 5, \omega_f = 7, \beta = 0.8, \alpha \in [0, 1]$.

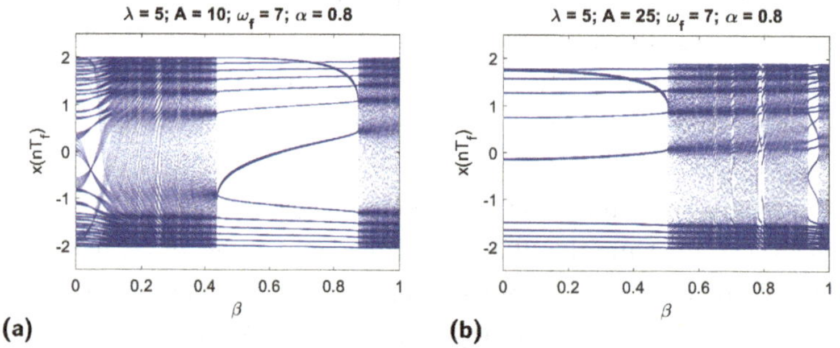

Figure 22: The bifurcation diagrams of the forced CF-VPO versus order $\beta \in [0, 1]$: (a) $\lambda = 5, A = 10, \omega_f = 7, \alpha = 0.8$; (b) $\lambda = 5, A = 25, \omega_f = 7, \alpha = 0.8$.

5 Conclusions

In this chapter several implementations of the VPO were reviewed. Such modifications consisted of the introduction of a fractional time derivatives. The unforced and forced versions of the integer- and fractional-order VPO were studied in the time and

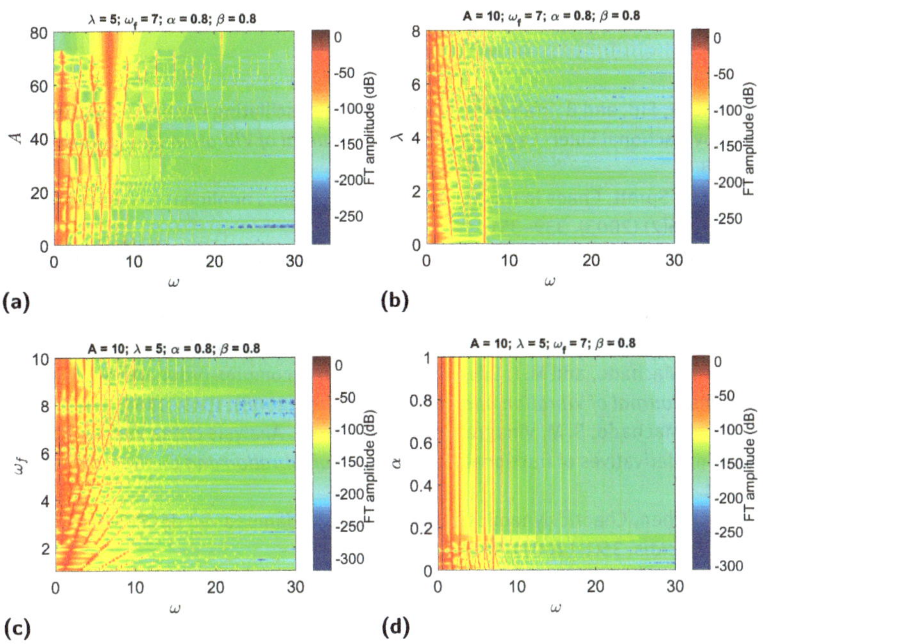

Figure 23: The Fourier spectra of x for the forced CF-VPO: (a) $\lambda = 5, \omega_f = 7, \alpha = 0.8, \beta = 0.8, A \in [0, 80]$; (b) $A = 10, \omega_f = 7, \alpha = 0.8, \beta = 0.8, \lambda \in [0, 8]$; (c) $A = 10, \lambda = 5, \alpha = 0.8, \beta = 0.8, \omega_f \in [1, 10]$; (d) $A = 10, \lambda = 5, \omega_f = 7, \beta = 0.8, \alpha \in [0, 1]$.

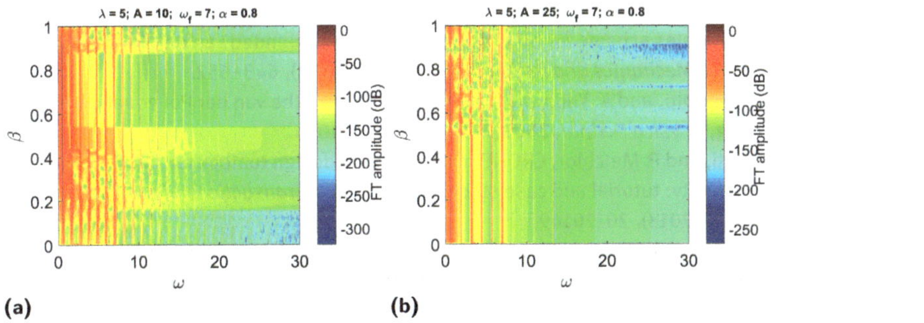

Figure 24: The Fourier spectra of x for the forced CF-VPO versus order $\beta \in [0, 1]$: (a) $\lambda = 5, A = 10, \omega_f = 7, \alpha = 0.8$; (b) $\lambda = 5, A = 25, \omega_f = 7, \alpha = 0.8$.

frequency domains by means of phase portraits, spectral analysis, and bifurcation diagrams. The results reveal that fractional-order systems can exhibit a richer dynamics than classical integer-order formulations. The fractional order acts as a modulation parameter that may be useful both for modeling and control. The complex-order system generalizes further the model and includes an additional parameter, which can be used to extend the scope in applied sciences of this seminal model.

Bibliography

[1] A. Agila, D. Baleanu, R. Eid, and B. Irfanoglu, A freely damped oscillating fractional dynamic system modeled by fractional Euler–Lagrange equations, *Journal of Vibration and Control*, **24** (2018), 1228–1238.

[2] W. M. Ahmad and J. C. Sprott, Chaos in fractional-order autonomous nonlinear systems, *Chaos, Solitons & Fractals*, **16**(2) (2003), 339–351.

[3] P. Arena, R. Caponetto, L. Fortuna, and D. Porto, *Nonlinear Noninteger Order Circuits and Systems—An Introduction*, World Scientific, Singapore, 2000.

[4] M. Attari, M. Haeri, and M. S. Tavazoei, Analysis of a fractional order Van der Pol-like oscillator via describing function method, *Nonlinear Dynamics*, **61**(1–2) (2010), 265–274.

[5] R. S. Barbosa, J. A. T. Machado, and M. F. Silva, Discretization of complex-order algorithms for control applications, *Journal of Vibration and Control*, **14**(9–10) (2008), 1349–1361.

[6] R. S. Barbosa, J. A. T. Machado, B. M. Vinagre, and A. J. Calderon, Analysis of the Van der Pol oscillator containing derivatives of fractional order, *Journal of Vibration and Control*, **13**(9–10) (2007), 1291–1301.

[7] J.-H. Chen and W.-C. Chen, Chaotic dynamics of the fractionally damped van der Pol equation, *Chaos, Solitons & Fractals*, **35**(1) (2008), 188–198.

[8] C. F. M. Coimbra, Mechanics with variable-order differential operators, *Annalen der Physik*, **12**(11–12) (2003), 692–703.

[9] Z.-M. Ge and A.-R. Zhang, Chaos in a modified van der Pol system and in its fractional order systems, *Chaos, Solitons & Fractals*, **32**(5) (2007), 1791–1822.

[10] A. Guefrachi, S. Najar, M. Amairi, and M. Aoun, Tuning of fractional complex order PID controller, *IFAC-PapersOnLine*, **50**(1) (2017), 14563–14568.

[11] T. T. Hartley, C. F. Lorenzo, and H. K. Qammer, Chaos in a fractional order Chua's system, *IEEE Transactions on Circuits and Systems I: Fundamental Theory and Applications*, **42**(8) (1995), 485–490.

[12] C. Huang, Multiple scales scheme for bifurcation in a delayed extended van der Pol oscillator, *Physica A: Statistical Mechanics and its Applications*, **490** (2018), 643–652.

[13] A. Konuralp, Ç. Konuralp, and A. Yıldırım, Numerical solution to the van der Pol equation with fractional damping, *Physica Scripta*, **2009**(T136) (2009), 014034.

[14] P. Lanusse, R. Malti, and P. Melchior, CRONE control system design toolbox for the control engineering community: tutorial and case study, *Philosophical Transactions of the Royal Society A*, **371**(1990) (2013), 20120149.

[15] P. Lanusse, J. Sabatier, and A. Oustaloup, Fractional order PID and first generation CRONE control system design, in *Fractional Order Differentiation and Robust Control Design*, pp. 63–105, Springer, 2015.

[16] A. Y. T. Leung, H. X. Yang, and Z. J. Guo, The residue harmonic balance for fractional order Van der Pol like oscillators, *Journal of Sound and Vibration*, **331**(5) (2012), 1115–1126.

[17] W. Li, M.-T. Zhang, and J.-F. Zhao, Stochastic bifurcations of generalized Duffing–van der Pol system with fractional derivative under colored noise, *Chinese Physics B*, **26**(9) (2017), 090501.

[18] G. M. Mahmoud and A. A. M. Farghaly, Chaos control of chaotic limit cycles of real and complex Van der Pol oscillators, *Chaos, Solitons & Fractals*, **21**(4) (2004), 915–924.

[19] R. Mettin, U. Parlitz, and W. Lauterborn, Bifurcation structure of the driven Van der Pol oscillator, *International Journal of Bifurcation and Chaos*, **3**(06) (1993), 1529–1555.

[20] R. E. Mickens, Fractional Van der Pol equations, *Journal of Sound and Vibration*, **259**(2) (2003), 457–460.

[21] V. Mishra, S. Das, H. Jafari, and S. H. Ong, Study of fractional order Van der Pol equation, *Journal of King Saud University-Science*, **28**(1) (2016), 55–60.

[22] B. P. Moghaddam and J. A. T. Machado, Extended algorithms for approximating variable order fractional derivatives with applications, *Journal of Scientific Computing*, **71**(3) (2017), 1351–1374.

[23] A. Oustaloup, *La commande CRONE: commande robuste d'ordre non entier*, Hermes, 1991.

[24] A. Oustaloup, F. Levron, B. Mathieu, and F. M. Nanot, Frequency-band complex noninteger differentiator: characterization and synthesis, *IEEE Transactions on Circuits and Systems I: Fundamental Theory and Applications*, **47**(1) (2000), 25–39.

[25] U. Parlitz and W. Lauterborn, Period-doubling cascades and devil's staircases of the driven Van der Pol oscillator, *Physical Review A*, **36**(3) (1987), 1428.

[26] T. Petrič, A. Gams, A. J. Ijspeert, and L. Žlajpah, On-line frequency adaptation and movement imitation for rhythmic robotic tasks, *The International Journal of Robotics Research*, **30**(14) (2011), 1775–1788.

[27] C. M. A. Pinto and J. A. T. Machado, Complex order Van der Pol oscillator, *Nonlinear Dynamics*, **65**(3) (2011), 247–254.

[28] C. M. A. Pinto and J. A. T. Machado, Complex-order forced Van der Pol oscillator, *Journal of Vibration and Control*, **18**(14) (2012), 2201–2209.

[29] S. G. Samko, Fractional integration and differentiation of variable order, *Analysis Mathematica*, **21**(3) (1995), 213–236.

[30] M. Shahiri, A. R. Noei, M. R. Karami, and R. Ghaderi, Tuning method for fractional complex order controller using standardized *k*-chart: application to pemfc control, *Asian Journal of Control*, **18**(3) (2016), 1102–1118.

[31] Y.-J. Shen, P. Wei, and S.-P. Yang, Primary resonance of fractional-order van der Pol oscillator, *Nonlinear Dynamics*, **77**(4) (2014), 1629–1642.

[32] M. S. Tavazoei, M. Haeri, M. Attari, S. Bolouki, and M. Siami, More details on analysis of fractional-order Van der Pol oscillator, *Journal of Vibration and Control*, **15**(6) (2009), 803–819.

[33] M. Tsatsos, Theoretical and numerical study of the Van der Pol equation. *Doctoral Dissertation, Aristotle University of Thessaloniki*, 2006.

[34] S. Vaidyanathan, Global chaos synchronization of the forced Van der Pol chaotic oscillators via adaptive control method, *International Journal of PharmTech Research*, **8**(6) (2015), 156–166.

[35] D. Valério and J. S. Da Costa, Variable-order fractional derivatives and their numerical approximations, *Signal Processing*, **91**(3) (2011), 470–483.

[36] D. Valério and J. S. da Costa, Variable order fractional controllers, *Asian Journal of Control*, **15**(3) (2013), 648–657.

[37] B. Van der Pol, On "relaxation-oscillations", *The London, Edinburgh, and Dublin Philosophical Magazine and Journal of Science*, **2**(11) (1926), 978–992.

[38] B. Van Der Pol, Forced oscillations in a circuit with non-linear resistance. (Reception with reactive triode), *The London, Edinburgh, and Dublin Philosophical Magazine and Journal of Science*, **3**(13) (1927), 65–80.

[39] M. Xiao, G. Jiang, W. X. Zheng, S. Yan, Y. Wan, and C. Fan, Bifurcation control of a fractional-order Van Der Pol oscillator based on the state feedback, *Asian Journal of Control*, **17**(5) (2015), 1756–1766.

[40] J.-X. Xu and J. Jiang, The global bifurcation characteristics of the forced Van der Pol oscillator, *Chaos, Solitons & Fractals*, **7**(1) (1996), 3–19.

[41] X.-J. Yang, Fractional derivatives of constant and variable orders applied to anomalous relaxation models in heat-transfer problems, *Thermal Science*, **21**(3) (2017), 1161–1171.

[42] J. Zhang and G. Xinsheng, Stability and bifurcation analysis in the delay-coupled Van der Pol oscillators, *Applied Mathematical Modelling*, **34**(9) (2010), 2291–2299.

Dumitru Baleanu and Sami I. Muslih

Fractional Lagrangian and Hamiltonian mechanics with memory

Abstract: Fractional variational principles are very important for science and engineering. Within this field of study, the fractional Lagrangian and Hamiltonian equations are challenging ones from the viewpoint of mathematics. During the last fifteen years, the field of fractional variational principles was continuously improved and developed. In this chapter, the fractional variational principles—with and without delay—will be briefly reviewed. Several illustrative examples from mechanics are presented.

Keywords: Fractional variational principles, fractional Lagrangian, fractional Hamiltonian

PACS: 45.20.Jj, 45.20.Jj, 26A333

1 Introduction

Fractional calculus is a very productive field of mathematics having many applications in various fields of science and engineering [20, 29, 37, 49, 56, 61, 67]. The calculus of variations consists of analyzing the optimization problems of quantities expressed as a functional form, which is defined as an integral over one or more independent variables [53]. The term under the integral also depends on several functions and their derivatives. Naturally, the optimization problems of this form exist in different fields of science and engineering, in which the functional, including total energy, potential, and Lagrangian, is obtained from the rules characterizing the system dynamics. However, the conventional calculus of variations is not relevant to derive the equation of motion for nonconservative phenomena. In [65, 66], some early approaches were introduced in mechanics for the expression of nonconservative system equations. In these methods, the Euler–Lagrange equations were formulated for nonconservative systems by introducing the fractional derivative terms in the functional. These days, fractional derivatives play a very important role in various sciences, including mathematics, physics, and engineering [1, 7, 18, 39, 50, 57, 62, 64]. In 2002, the Euler–Lagrange equations were presented for constrained and unconstrained frac-

Dumitru Baleanu, Department of Mathematics, Cankaya University, 06530 Balgat, Ankara, Turkey; and Institute of Space Sciences, Magurele–Bucharest, Romania, e-mail: dumitru@cankya.edu.tr
Sami I. Muslih, Department of Physics, AL-Azhar University, Gaza, Palestine, e-mail: samimuslih@gmail.com

https://doi.org/10.1515/9783110571707-002

tional variational problems [2]. When the derivatives of fractional order are replaced by integer-order ones, the results of the fractional variational principle are reduced to those obtained from the conventional calculus of variations. In [25], this approach was generalized to the classical definitions of fractional derivatives. In addition, the study [41] reviewed the equation of motion for nonconservative systems derived from the principle of variations. In [51], the described models with the Riemann–Liouville fractional derivative were discussed in Lagrangian and Hamiltonian forms. In [52], a noncommutative approach was proposed for the Lagrangian fractional mechanics. In [58], the fractional Hamiltonian formulations of motion for Lagrangians with linear velocities were investigated. In [63], taking into account generalized mechanics, the Hamiltonian formulation was derived for Lagrangians, depending on fractional derivatives of coordinates. In [10], a new form of Euler–Lagrange equations was derived for a generalized fractional variational problem. In [11], a Nöther's theorem and invariance conditions were presented for variational problems with fractional derivatives. In [9], the Hamilton's principle was generalized with fractional derivatives, so that the fractional derivative order is varied in the optimization procedure. The exact solutions of the fractional Euler–Lagrange equations, involving Caputo derivative, were studied in [31]. In [59], the Euler–Lagrange equations of motion for a pendulum were investigated in a fractional sense. In [28], the fractional discrete Lagrangians were analyzed within the Riemann–Liouville fractional derivative. Some other noticeable efforts have been carried out in [12, 13, 15, 21, 26, 27, 30, 32, 43–46, 60, 68]. Recently, a number of interesting applications have been found in the literature. In [16], the authors focused on the fractional Lagrangian and Hamiltonian of the complex Bateman–Feshbach Tikochinsky oscillator. In [17], the fractional Euler–Lagrange equation was investigated for a thin elastica model. In [19], the Euler–Lagrange equations of a harmonic oscillator on a moving platform were studied numerically in a fractional sense. In [38], the Euler–Lagrange and Hamilton equations were formulated by using the fractional derivatives with regular kernel. In [22], the fractional Lagrangian and Hamiltonian of motion were found for a bead sliding on a wire. In [14], using fractional Lagrangian, the motion of a spherical particle was examined in a rotating parabola. In [8], the Euler–Lagrange equations were investigated for fractional variational problems, where the Lagrangians contain real- and complex-order fractional derivatives. In [69], constrained fractional variational problems were considered, in which the Lagrangian depends on a variable-order Caputo fractional derivative.

The organization of this chapter is as follows: first, we formulate the fractional Euler–Lagrange equations with or without delay, based on the Riemann-Liouville and Caputo fractional derivatives. The fractional Hamiltonian equations and their applications in optimal control theory are given in Section 3. Four fractional illustrative applications are presented in Section 4. Finally, the results of this chapter are illustrated in Section 5.

2 Fractional Euler-Lagrange equations

In this section, the fractional Lagrangian problem is formulated based on the Riemann–Liouville fractional derivative. Suppose that the fractional Lagrangian takes the form

$$L := L(y(t), {}^{\mathrm{RL}}_{a}D^{\alpha_1}_{t}y(t), {}^{\mathrm{RL}}_{t}D^{\alpha_2}_{b}y(t), t), \tag{1}$$

in which ${}^{\mathrm{RL}}_{a}D^{\alpha_1}_{t}y(t)$ and ${}^{\mathrm{RL}}_{t}D^{\alpha_2}_{b}y(t)$ denote the left and right Riemann–Liouville fractional derivatives, and $\alpha_1, \alpha_2 > 0$ are the fractional derivative orders, respectively. Then, the Euler–Lagrange equation, in a fractional sense, is obtained from [2]:

$$\frac{\partial L}{\partial y(t)} + {}^{\mathrm{RL}}_{t}D^{\alpha_1}_{b}\frac{\partial L}{\partial {}^{\mathrm{RL}}_{a}D^{\alpha_1}_{t}y(t)} + {}^{\mathrm{RL}}_{a}D^{\alpha_2}_{t}\frac{\partial L}{\partial {}^{\mathrm{RL}}_{t}D^{\alpha_2}_{b}y(t)} = 0. \tag{2}$$

It should be noted that for $\alpha_1 = \alpha_2 = 1$, we have ${}^{\mathrm{RL}}_{a}D^{\alpha_1}_{t} = {}^{\mathrm{RL}}_{a}D^{\alpha_2}_{t} = \frac{d}{dt}$ and ${}^{\mathrm{RL}}_{t}D^{\alpha_1}_{b} = {}^{\mathrm{RL}}_{t}D^{\alpha_2}_{b} = -\frac{d}{dt}$, and Equation (2) is reduced to the classical Euler–Lagrange equation.

To generalize the Euler–Lagrange Equation (2) to the classical field systems, consider the following classical field action [2]:

$$S := \int L(\varphi, {}^{\mathrm{RL}}_{a}D^{\alpha_1}_{t}\varphi, {}^{\mathrm{RL}}_{t}D^{\alpha_2}_{b}\varphi, {}^{\mathrm{RL}}_{a}D^{\alpha_1}_{x}\varphi, {}^{\mathrm{RL}}_{x}D^{\alpha_2}_{b}\varphi, t)d^3x dt, \tag{3}$$

which consists of partial derivatives of fractional order. The optimization of this action, in factional sense, provides [2]:

$$\begin{aligned}\frac{\partial L}{\partial \varphi} &+ {}^{\mathrm{RL}}_{t}D^{\alpha_1}_{b}\frac{\partial L}{\partial {}^{\mathrm{RL}}_{a}D^{\alpha_1}_{t}\varphi} + {}^{\mathrm{RL}}_{a}D^{\alpha_2}_{t}\frac{\partial L}{\partial {}^{\mathrm{RL}}_{t}D^{\alpha_2}_{b}\varphi} \\ &+ {}^{\mathrm{RL}}_{a}D^{\alpha_1}_{x}\frac{\partial L}{\partial {}^{\mathrm{RL}}_{x}D^{\alpha_1}_{b}\varphi} + {}^{\mathrm{RL}}_{x}D^{\alpha_2}_{b}\frac{\partial L}{\partial {}^{\mathrm{RL}}_{a}D^{\alpha_2}_{x}\varphi} = 0.\end{aligned} \tag{4}$$

Note that the conventional Euler–Lagrange equation for the classical field is achieved by applying α_1, α_2 to 1 in Equation (4).

2.1 Fractional Euler–Lagrange equations with delay

Our opinion is that the inclusion of a delay (see, e. g., [5, 6] and the references therein) in a fractional Lagrangian can give superior results in several problems involving dynamics of complex systems. We recall that the fractional generalization of the Bloch equation [54, 55], including both fractional derivatives and time delays was developed in [34]. The existence of a time delay on the right side of the equation, balances the equation by adding a degree of system memory [34]. Several applications on the combined use of the fractional derivatives and delay can be seen in [23, 35].

2.1.1 Riemann–Liouville fractional Euler–Lagrange equations with delay

Below we present a lemma, namely [24]

Lemma 1. *Let $\alpha > 0$, $p, q \geq 1$, $r \in T = (a, b)$, and $\frac{1}{p} + \frac{1}{q} \leq 1 + \alpha$ ($p \neq 1$ and $q \neq 1$ in the case where $\frac{1}{p} + \frac{1}{q} = 1 + \alpha$).*

(a) *If $\varphi \in L_p(a, b)$ and $\psi \in L_q(a, b)$, then*

$$\int_a^r \varphi(t)\, {}^{\mathrm{RL}}_{a}I^{\alpha}_{t}\psi(t)dt = \int_a^r \psi(t)\, {}^{\mathrm{RL}}_{t}I^{\alpha}_{r}\varphi(t)dt, \tag{5}$$

and hence, if $g \in {}^{\mathrm{RL}}_{t}I^{\alpha}_{b}(L_p)$ and $f \in {}^{\mathrm{RL}}_{a}I^{\alpha}_{t}(L_q)$, then [24]

$$\int_a^r g(t)\, {}^{\mathrm{RL}}_{a}D^{\alpha}_{t}f(t)dt = \int_a^r f(t)\, {}^{\mathrm{RL}}_{t}D^{\alpha}_{r}g(t)dt. \tag{6}$$

(b) *If $\varphi \in L_p(a, b)$ and $\psi \in L_q(a, b)$, then* [24]

$$\begin{aligned}\int_r^b \varphi(t)\, {}^{\mathrm{RL}}_{a}I^{\alpha}_{t}\psi(t)dt &= \int_r^b \psi(t)\, {}^{\mathrm{RL}}_{t}I^{\alpha}_{b}\varphi(t)dt \\ &\quad + \frac{1}{\Gamma(\alpha)}\int_a^r \psi(t)\left(\int_r^b \varphi(s)(s-t)^{\alpha-1}ds\right)dt,\end{aligned} \tag{7}$$

and hence, if $g \in {}^{\mathrm{RL}}_{t}I^{\alpha}_{b}(L_p)$ and $f \in {}^{\mathrm{RL}}_{a}I^{\alpha}_{t}(L_q)$, then

$$\begin{aligned}&\int_r^b g(t)\, {}^{\mathrm{RL}}_{a}D^{\alpha}_{t}f(t)dt \\ &\quad = \int_r^b f(t)\, {}^{\mathrm{RL}}_{t}D^{\alpha}_{b}g(t)dt \\ &\quad - \frac{1}{\Gamma(\alpha)}\int_a^r {}^{\mathrm{RL}}_{a}D^{\alpha}_{t}f(t)\left(\int_r^b ({}^{\mathrm{RL}}_{s}D^{\alpha}_{b}g(s))(s-t)^{\alpha-1}ds\right)dt,\end{aligned} \tag{8}$$

which implies [24]

$$\begin{aligned}&\int_r^b g(t)\, {}^{\mathrm{RL}}_{a}D^{\alpha}_{t}f(t)dt \\ &\quad = \int_r^b f(t)\, {}^{\mathrm{RL}}_{t}D^{\alpha}_{b}g(t)dt \\ &\quad - \frac{1}{\Gamma(\alpha)}\int_a^r f(t)\, {}^{\mathrm{RL}}_{t}D^{\alpha}_{r}\left(\int_r^b ({}^{\mathrm{RL}}_{s}D^{\alpha}_{b}g(s))(s-t)^{\alpha-1}ds\right)dt.\end{aligned} \tag{9}$$

The proof of this lemma was written in [24].

Next, we analyze a modified problem, when both the fractional derivatives and delay are present in the Lagrangian. The one-dimensional problem [24] is the starting point. Minimize

$$J(y) = \int_a^b F(t, y(t), {}_a^{RL}D_t^{\alpha} y(t), y(t-\tau), y'(t-\tau))dt \tag{10}$$

such that

$$y(b) = c, \quad y(t) = \phi(t) \quad (t \in [a-\tau, a]), \tag{11}$$

where $a \neq b$ and $0 < \tau < b - a$.

By utilizing the corresponding delay notations [42]

$$y_\tau = y(t-\tau), \quad y'_\tau = y'(t-\tau), \tag{12}$$

Equation (10) becomes

$$J(y) = \int_a^b F(t, y(t), {}_a^{RL}D_t^{\alpha} y(t), y_\tau, y'_\tau)dt. \tag{13}$$

Thus, we formulate the following theorem.

Theorem 1. *Let $J(y)$ be a functional as* [24]

$$J(y) = \int_a^b F(t, y(t), {}_a^{RL}D_t^{\alpha} y(t), y_\tau, y'_\tau)dt, \tag{14}$$

defined on a set of continuous functions $y(t)$ that have continuous left Riemann–Liouville derivatives of order α in $[a, b]$ and fulfill the boundary conditions $y(t) = \phi(t)$, $t \in [a-\tau, a]$, and $y(b) = c$. The necessary condition for $J(y)$ to admit an extremum for a given function $y(t)$ is that $y(t)$ obeys the Euler–Lagrange equations

$$\begin{aligned} 0 = {} & F_y(t) + F_{y_\tau}(t+\tau) + {}_t^{RL}D_{b-\tau}^{\alpha} \frac{\partial F(t)}{\partial({}_a^{RL}D_t^{\alpha} y(t))} \\ & - \frac{dF_{y'_\tau}(t+\tau)}{dt} \\ & - \frac{1}{\Gamma(\alpha)} {}_t^{RL}D_{b-\tau}^{\alpha} \int_{b-\tau}^{b} \left[{}_t^{RL}D_b^{\alpha} \frac{\partial F(t)}{\partial({}_a^{RL}D_t^{\alpha} y(t))} \right](s)(s-t)^{\alpha-1} ds \end{aligned} \tag{15}$$

for $a \leq t \leq b - \tau$,

$$F_y(t) + {}_t^{RL}D_b^{\alpha} \left(\frac{\partial F(t)}{\partial({}_a^{RL}D_t^{\alpha} y(t))} \right) = 0 \tag{16}$$

for $b-\tau \le t \le b$, and the boundary condition

$$F_{y'_\tau}(t+\tau)\eta(t)|_a^{(b-\tau)^-} = 0. \tag{17}$$

From (15), (16), and (17), we conclude that when the delay terms are absent and when $\alpha \to 1$, the classical results are reobtained.

We briefly mention the generalization of Theorem 1 with fixed end points and several functions. To this end, let us assume that the functional $J(y_1, y_2, \dots, y_n)$ has the form [24]

$$\begin{aligned} &J(y_1, y_2, \dots, y_n) \\ &= \int_a^b F(t, y_1(t), \dots y_n(t), {}^{\mathrm{RL}}_{\ a}D_t^\alpha y_1(t), \dots, {}^{\mathrm{RL}}_{\ a}D_t^\alpha y_n(t), \\ &\qquad y_1(t-\tau), \dots, y_n(t-\tau), y'_1(t-\tau), \dots, y'_n(t-\tau))dt \end{aligned} \tag{18}$$

and that it fulfills the boundary conditions

$$y_i(b) = y_{ib}, \quad y_i(t) = \phi_i(t) \quad (i = 1, \dots, n;\ t \in [a-\tau, a]), \tag{19}$$

where $a < b$ and $0 < \tau < b-a$. Besides, we assume that F possesses continuous partial derivatives with respect to all of its parameters and that the functions ϕ_i, $i = 1, 2, \dots n$, are smooth.

Theorem 2. *A necessary condition for the curve $y_i = y_i(t)$, $i = 1, \dots, n$, fulfilling the boundary conditions* (19) *to be extremal for the functional* (18) *is that*

$$\begin{aligned} 0 &= F_{y_i}(t) + F_{(y_i)_\tau}(t+\tau) \\ &\quad + {}^{\mathrm{RL}}_{\ t}D_{b-\tau}^\alpha \frac{\partial F(t)}{\partial({}^{\mathrm{RL}}_{\ a}D_t^\alpha y_i(t))} - \frac{dF_{(y_i)'_\tau}(t+\tau)}{dt} \\ &\quad - \frac{1}{\Gamma(\alpha)} {}^{\mathrm{RL}}_{\ t}D_{b-\tau}^\alpha \int_{b-\tau}^b \left[{}^{\mathrm{RL}}_{\ t}D_b^\alpha \left(\frac{\partial F(t)}{\partial({}^{\mathrm{RL}}_{\ a}D_t^\alpha y_i(t))} \right) \right](s)(s-t)^{\alpha-1} ds \end{aligned} \tag{20}$$

for $a \le t \le b-\tau$,

$$F_{y_i}(t) + {}^{\mathrm{RL}}_{\ t}D_b^\alpha \left(\frac{\partial F(t)}{\partial({}^{\mathrm{RL}}_{\ a}D_t^\alpha y_i(t))} \right) = 0 \tag{21}$$

for $b-\tau \le t \le b$, and that the boundary conditions

$$(F_{(y_i)'_\tau})(t+\tau)\eta(t)|_a^{(b-\tau)^-} = 0 \tag{22}$$

are fulfilled for $i = 1, \dots, n$.

2.1.2 Caputo fractional Euler–Lagrange equations with delay

Below, we investigate the following problem [47]. Minimize

$$J(y)=\int_a^b F(t,y(t),{}_a^CD_t^{\alpha}y(t),{}_t^CD_b^{\beta}y(t),y_\tau,y'_\tau)dt \tag{23}$$

such that

$$y(b)=c,\quad y(t)=\phi(t)\quad (t\in[a-\tau,a]), \tag{24}$$

where $a<b$, $0<\tau<b-a$, $0<\alpha<1$, $0<\beta<1$, c is a constant, and F is a function with continuous first and second partial derivatives with respect to all of its arguments. The corresponding results are contained in the following theorem.

Theorem 3. *Let $J(y)$ be a functional of the form*

$$J(y)=\int_a^b F(t,y(t),{}_a^CD_t^{\alpha}y(t),{}_t^CD_b^{\beta}y(t),y_\tau,y'_\tau)dt \tag{25}$$

with $0<\alpha,\beta<1$, defined on a set of continuous functions $y(t)$ that have continuous left Caputo derivatives of order α and right derivatives of order β in $[a,b]$, and satisfy the conditions $y(t)=\phi(t)$ $(t\in[a-\tau,a])$ and $y(b)=c$. Moreover, let $F:[a,b]\times\mathbb{R}^5\to\mathbb{R}$ have continuous partial derivatives with respect to all of its arguments. Then, the necessary condition for $J(y)$ to possess an extremum for a given function $y(t)$ is that $y(t)$ satisfies the Euler–Lagrange equations

$$\begin{aligned}0&=\frac{\partial F}{\partial y(t)}(t)+{}_t^{RL}D_{b-\tau}^{\alpha}\Big(\frac{\partial F}{\partial\,{}_a^CD_t^{\alpha}y(t)}\Big)(t)+{}_a^{RL}D_t^{\beta}\Big(\frac{\partial F}{\partial\,{}_t^CD_b^{\beta}y(t)}\Big)(t)\\&\quad+\frac{\partial F}{\partial y(t-\tau)}(t+\tau)-\frac{d}{dt}\frac{\partial F}{\partial y'_\tau}(t+\tau)\\&\quad-\frac{1}{\Gamma(\alpha)}\,{}_t^{RL}D_{b-\tau}^{\alpha}\Big(\int_{b-\tau}^{b}\Big({}_t^{RL}D_b^{\alpha}\frac{\partial F}{\partial\,{}_a^CD_t^{\alpha}y(t)}\Big)(s)(s-t)^{\alpha-1}ds\Big)\end{aligned}\tag{26}$$

for $a\le t\le b-\tau$,

$$\begin{aligned}0&=\frac{\partial F}{\partial y(t)}(t)+{}_t^{RL}D_b^{\alpha}\Big(\frac{\partial F}{\partial\,{}_a^CD_t^{\alpha}y(t)}\Big)(t)\\&\quad+{}_{b-\tau}^{RL}D_b^{\beta}\Big(\frac{\partial F}{\partial\,{}_t^CD_b^{\beta}y(t)}\Big)(t)\\&\quad-\frac{1}{\Gamma(\beta)}\,{}_{b-\tau}^{RL}D_b^{\beta}\Big(\int_a^{b-\tau}\Big({}_a^{RL}D_t^{\beta}\frac{\partial F}{\partial\,{}_t^CD_b^{\beta}y(t)}\Big)(s)(s-t)^{\beta-1}ds\Big)\end{aligned}\tag{27}$$

for $b - \tau \leq t \leq b$, and the transversality condition

$$\frac{\partial F}{\partial y'_\tau}(t+\tau)\eta(t)\Big|_a^{b-\tau} = 0 \tag{28}$$

for any admissible function η satisfying $\eta(t) = 0$ ($t \in [a-\tau, a]$) and $\eta(b) = 0$.

3 Fractional Hamiltonian equations

Below, we are going to obtain the fractional Hamilton equation of motion. Consider the fractional Lagrangian equation of time derivatives of the coordinates given by Equation (1). Then, the generalized momenta are introduced as

$$\rho_{\alpha_1} = \frac{\partial L}{\partial {}^{\mathrm{RL}}_{a}D_t^{\alpha_1} y}, \quad \rho_{\alpha_2} = \frac{\partial L}{\partial {}^{\mathrm{RL}}_{t}D_b^{\alpha_2} y}, \tag{29}$$

and the fractional Hamiltonian function is obtained from [63]

$$\mathcal{H} = \rho_{\alpha_1}\, {}^{\mathrm{RL}}_{a}D_t^{\alpha_1} y + \rho_{\alpha_2}\, {}^{\mathrm{RL}}_{t}D_b^{\alpha_2} y - L. \tag{30}$$

Calculating the total differential of this function, we derive [63]

$$\begin{aligned} d\mathcal{H} = {} & \rho_{\alpha_1} d\, {}^{\mathrm{RL}}_{a}D_t^{\alpha_1} y + d\rho_{\alpha_1}\, {}^{\mathrm{RL}}_{a}D_t^{\alpha_1} y + \rho_{\alpha_2} d\, {}^{\mathrm{RL}}_{t}D_b^{\alpha_2} y + d\rho_{\alpha_2}\, {}^{\mathrm{RL}}_{t}D_b^{\alpha_2} y \\ & - \frac{\partial L}{\partial y} dy - \frac{\partial L}{\partial {}^{\mathrm{RL}}_{a}D_t^{\alpha_1} y} d\, {}^{\mathrm{RL}}_{a}D_t^{\alpha_1} y - \frac{\partial L}{\partial {}^{\mathrm{RL}}_{t}D_b^{\alpha_2} y} d\, {}^{\mathrm{RL}}_{t}D_b^{\alpha_2} y - \frac{\partial L}{\partial t} dt. \end{aligned} \tag{31}$$

Using the Euler–Lagrange Equation (2) concludes [63]:

$$d\mathcal{H} = d\rho_{\alpha_1}\, {}^{\mathrm{RL}}_{a}D_t^{\alpha_1} y + d\rho_{\alpha_2}\, {}^{\mathrm{RL}}_{t}D_b^{\alpha_2} y + \left({}^{\mathrm{RL}}_{a}D_t^{\alpha_2} \rho_{\alpha_2} + {}^{\mathrm{RL}}_{t}D_b^{\alpha_1} \rho_{\alpha_1}\right) dy - \frac{\partial L}{\partial t} dt. \tag{32}$$

Thus, the Hamiltonian function takes the form [63]

$$\mathcal{H} = \mathcal{H}(y, \rho_{\alpha_1}, \rho_{\alpha_2}, t), \tag{33}$$

and its total differential is computed from [63]

$$d\mathcal{H} = \frac{\partial \mathcal{H}}{\partial y} dy + \frac{\partial \mathcal{H}}{\partial \rho_{\alpha_1}} d\rho_{\alpha_1} + \frac{\partial \mathcal{H}}{\partial \rho_{\alpha_2}} d\rho_{\alpha_2} + \frac{\partial \mathcal{H}}{\partial t} dt. \tag{34}$$

By comparing Equations (32) and (34), the fractional Hamiltonian's equations are expressed as follows:

$$\frac{\partial \mathcal{H}}{\partial y} = {}^{\mathrm{RL}}_{t}D_b^{\alpha_1} \rho_{\alpha_1} + {}^{\mathrm{RL}}_{a}D_t^{\alpha_2} \rho_{\alpha_2}, \quad \frac{\partial \mathcal{H}}{\partial \rho_{\alpha_1}} = {}^{\mathrm{RL}}_{a}D_t^{\alpha_1} y, \quad \frac{\partial \mathcal{H}}{\partial \rho_{\alpha_2}} = {}^{\mathrm{RL}}_{t}D_b^{\alpha_2} y, \quad \frac{\partial \mathcal{H}}{\partial t} = -\frac{\partial L}{\partial t}. \tag{35}$$

In classical field mode, the Lagrangian contains partial fractional derivatives, namely

$$L = L(\varphi, {}^{\mathrm{RL}}_{a}D^{\alpha_1}_{t}\varphi, {}^{\mathrm{RL}}_{t}D^{\alpha_2}_{b}\varphi, {}^{\mathrm{RL}}_{a}D^{\alpha_1}_{x}\varphi, {}^{\mathrm{RL}}_{x}D^{\alpha_2}_{b}\varphi, t). \tag{36}$$

By introducing the conjugate momenta

$$\pi_{\alpha_1} = \frac{\partial L}{\partial {}^{\mathrm{RL}}_{a}D^{\alpha_1}_{t}\varphi}, \quad \pi_{\alpha_2} = \frac{\partial L}{\partial {}^{\mathrm{RL}}_{t}D^{\alpha_2}_{b}\varphi}, \tag{37}$$

the Hamiltonian is defined as

$$\mathcal{H} = \pi_{\alpha_1}\, {}^{\mathrm{RL}}_{a}D^{\alpha_1}_{t}\varphi + \pi_{\alpha_2}\, {}^{\mathrm{RL}}_{t}D^{\alpha_2}_{b}\varphi - L. \tag{38}$$

By applying total differential on both sides of the above equation, we get

$$\begin{aligned} d\mathcal{H} = {} & \pi_{\alpha_1} d\,{}^{\mathrm{RL}}_{a}D^{\alpha_1}_{t}\varphi + d\pi_{\alpha_1}\, {}^{\mathrm{RL}}_{a}D^{\alpha_1}_{t}\varphi + \pi_{\alpha_2} d\,{}^{\mathrm{RL}}_{t}D^{\alpha_2}_{b}\varphi + d\pi_{\alpha_2}\, {}^{\mathrm{RL}}_{t}D^{\alpha_2}_{b}\varphi \\ & - \frac{\partial L}{\partial \varphi} d\varphi - \frac{\partial L}{\partial {}^{\mathrm{RL}}_{a}D^{\alpha_1}_{t}\varphi} d\,{}^{\mathrm{RL}}_{a}D^{\alpha_1}_{t}\varphi - \frac{\partial L}{\partial {}^{\mathrm{RL}}_{t}D^{\alpha_2}_{b}\varphi} d\,{}^{\mathrm{RL}}_{t}D^{\alpha_2}_{b}\varphi \\ & - \frac{\partial L}{\partial {}^{\mathrm{RL}}_{a}D^{\alpha_1}_{x}\varphi} d\,{}^{\mathrm{RL}}_{a}D^{\alpha_1}_{x}\varphi - \frac{\partial L}{\partial {}^{\mathrm{RL}}_{x}D^{\alpha_2}_{b}\varphi} d\,{}^{\mathrm{RL}}_{x}D^{\alpha_2}_{b}\varphi - \frac{\partial L}{\partial t} dt. \end{aligned} \tag{39}$$

By substituting the conjugate momenta from Equation (37) into Equation (39), we get

$$\begin{aligned} d\mathcal{H} = {} & d\pi_{\alpha_1}\, {}^{\mathrm{RL}}_{a}D^{\alpha_1}_{t}\varphi + d\pi_{\alpha_2}\, {}^{\mathrm{RL}}_{t}D^{\alpha_2}_{b}\varphi - \frac{\partial L}{\partial \varphi} d\varphi \\ & - \frac{\partial L}{\partial {}^{\mathrm{RL}}_{a}D^{\alpha_1}_{x}\varphi} d\,{}^{\mathrm{RL}}_{a}D^{\alpha_1}_{x}\varphi - \frac{\partial L}{\partial {}^{\mathrm{RL}}_{x}D^{\alpha_2}_{b}\varphi} d\,{}^{\mathrm{RL}}_{x}D^{\alpha_2}_{b}\varphi - \frac{\partial L}{\partial t} dt. \end{aligned} \tag{40}$$

Within the use of Equation (4), we have

$$\begin{aligned} d\mathcal{H} = {} & d\pi_{\alpha_1}\, {}^{\mathrm{RL}}_{a}D^{\alpha_1}_{t}\varphi + d\pi_{\alpha_2}\, {}^{\mathrm{RL}}_{t}D^{\alpha_2}_{b}\varphi \\ & + \left({}^{\mathrm{RL}}_{a}D^{\alpha_2}_{t}\pi_{\alpha_2} + {}^{\mathrm{RL}}_{t}D^{\alpha_1}_{b}\pi_{\alpha_1} + {}^{\mathrm{RL}}_{a}D^{\alpha_1}_{x} \frac{\partial L}{\partial {}^{\mathrm{RL}}_{x}D^{\alpha_1}_{b}\varphi} + {}^{\mathrm{RL}}_{x}D^{\alpha_2}_{b} \frac{\partial L}{\partial {}^{\mathrm{RL}}_{a}D^{\alpha_2}_{x}\varphi} \right) d\varphi \\ & - \frac{\partial L}{\partial {}^{\mathrm{RL}}_{a}D^{\alpha_1}_{x}\varphi} d\,{}^{\mathrm{RL}}_{a}D^{\alpha_1}_{x}\varphi - \frac{\partial L}{\partial {}^{\mathrm{RL}}_{x}D^{\alpha_2}_{b}\varphi} d\,{}^{\mathrm{RL}}_{x}D^{\alpha_2}_{b}\varphi - \frac{\partial L}{\partial t} dt. \end{aligned} \tag{41}$$

As a result, the Hamiltonian function becomes

$$\mathcal{H} = \mathcal{H}(\varphi, \pi_{\alpha_1}, \pi_{\alpha_2}, {}^{\mathrm{RL}}_{a}D^{\alpha_1}_{x}\varphi, {}^{\mathrm{RL}}_{x}D^{\alpha_2}_{b}\varphi, t), \tag{42}$$

and its total differential is written as follows:

$$\begin{aligned} d\mathcal{H} = {} & \frac{\partial \mathcal{H}}{\partial \varphi} d\varphi + \frac{\partial \mathcal{H}}{\partial \pi_{\alpha_1}} d\pi_{\alpha_1} + \frac{\partial \mathcal{H}}{\partial \pi_{\alpha_2}} d\pi_{\alpha_2} \\ & + \frac{\partial \mathcal{H}}{\partial {}^{\mathrm{RL}}_{a}D^{\alpha_1}_{x}\varphi} d\,{}^{\mathrm{RL}}_{a}D^{\alpha_1}_{x}\varphi + \frac{\partial \mathcal{H}}{\partial {}^{\mathrm{RL}}_{x}D^{\alpha_2}_{b}\varphi} d\,{}^{\mathrm{RL}}_{x}D^{\alpha_2}_{b}\varphi + \frac{\partial \mathcal{H}}{\partial t} dt. \end{aligned} \tag{43}$$

By comparing Equations (41) and (43), we conclude

$$\frac{\partial \mathcal{H}}{\partial \varphi} = {}^{RL}_{a}D^{\alpha_2}_{t}\pi_{\alpha_2} + {}^{RL}_{t}D^{\alpha_1}_{b}\pi_{\alpha_1} + {}^{RL}_{a}D^{\alpha_1}_{x}\frac{\partial L}{\partial {}^{RL}_{x}D^{\alpha_1}_{b}\varphi} + {}^{RL}_{x}D^{\alpha_2}_{b}\frac{\partial L}{\partial {}^{RL}_{a}D^{\alpha_2}_{x}\varphi}, \tag{44}$$

$$\frac{\partial \mathcal{H}}{\partial t} = -\frac{\partial L}{\partial t}, \quad \frac{\partial \mathcal{H}}{\partial \pi_{\alpha_1}} = {}^{RL}_{a}D^{\alpha_1}_{t}\varphi, \quad \frac{\partial \mathcal{H}}{\partial \pi_{\alpha_2}} = {}^{RL}_{t}D^{\alpha_2}_{b}\varphi, \tag{45}$$

$$\frac{\partial \mathcal{H}}{\partial {}^{RL}_{a}D^{\alpha_1}_{x}\varphi} = -\frac{\partial L}{\partial {}^{RL}_{a}D^{\alpha_1}_{x}\varphi}, \quad \frac{\partial \mathcal{H}}{\partial {}^{RL}_{x}D^{\alpha_2}_{b}\varphi} = -\frac{\partial L}{\partial {}^{RL}_{x}D^{\alpha_2}_{b}\varphi}. \tag{46}$$

Finally, using Equations (44) and (46), we can write

$${}^{RL}_{a}D^{\alpha_2}_{t}\pi_{\alpha_2} + {}^{RL}_{t}D^{\alpha_1}_{b}\pi_{\alpha_1} = \frac{\partial \mathcal{H}}{\partial \varphi} + {}^{RL}_{a}D^{\alpha_1}_{x}\frac{\partial \mathcal{H}}{\partial {}^{RL}_{x}D^{\alpha_1}_{b}\varphi} + {}^{RL}_{x}D^{\alpha_2}_{b}\frac{\partial H}{\partial {}^{RL}_{a}D^{\alpha_2}_{x}\varphi}. \tag{47}$$

3.1 Fractional optimal control formulation

The main issue discussed below has to do with minimizing the performance index [3]

$$J(u) = \int_a^b F(x, u, t)dt \tag{48}$$

such that

$${}^{RL}_{a}D^{\alpha}_{t}x = G(x, u, t), \tag{49}$$

where x and u are the state and control vectors, respectively; G is a vector function, and the terminal conditions $x(a) = c$ and $x(b) = d$ are given.

The corresponding fractional-order counterpart formulation of this problem was proposed in [3]. More details on this topic can be seen in [3, 4].

A modified performance index is written as [3]

$$\bar{J}(u) = \int_a^b [\mathcal{H}(x, u, t) - \lambda^{T}\, {}^{RL}_{a}D^{\alpha}_{t}x]dt, \tag{50}$$

where $\mathcal{H}(x, u, \lambda, t)$ represents the Hamiltonian [3]

$$\mathcal{H}(x, u, \lambda, t) = F(x, u, t) + \lambda^{T}G(x, u, t), \tag{51}$$

and λ denotes the vector of Lagrange multipliers, also known as costate or adjoint vector variable. By utilizing (50) and (51), together with the fractional integration by parts, the necessary conditions for the fractional control problem are written as [3]:

$${}^{RL}_{t}D^{\alpha}_{b}\lambda = \frac{\partial \mathcal{H}}{\partial x}, \tag{52}$$

$$\frac{\partial \mathcal{H}}{\partial u} = 0, \tag{53}$$

$${}^{RL}_{a}D^{\alpha}_{t}x = \frac{\partial \mathcal{H}}{\partial \lambda}. \tag{54}$$

4 Illustrative examples

4.1 Fractional two electric pendulum

The model of two electric pendulum is composed of two planar pendula, with mass m and length l, suspended a distance d apart on a horizontal line, so that they swing in the same plane. The fractional Lagrangian for this model has the form [15]

$$L = 0.5m\left({}_{a}^{\mathrm{RL}}D_{t}^{\alpha_1} q_1\right)^2 + 0.5m\left({}_{a}^{\mathrm{RL}}D_{t}^{\alpha_1} q_2\right)^2 - 0.5\frac{mg}{l}\left(q_1^2 + q_2^2\right) - \frac{e^2}{d + q_2 - q_1}, \tag{55}$$

where g is the gravity constant, and e is the electron charge. Applying Equation (2), the fractional Euler–Lagrange equations are then derived as:

$${}_{t}^{\mathrm{RL}}D_{b}^{\alpha_1}\,{}_{a}^{\mathrm{RL}}D_{t}^{\alpha_1} q_1 - \frac{g}{l} q_1 - \frac{1}{m}\frac{e^2}{(d + q_2 - q_1)^2} = 0, \tag{56}$$

$${}_{t}^{\mathrm{RL}}D_{b}^{\alpha_1}\,{}_{a}^{\mathrm{RL}}D_{t}^{\alpha_1} q_2 - \frac{g}{l} q_2 + \frac{1}{m}\frac{e^2}{(d + q_2 - q_1)^2} = 0. \tag{57}$$

As $\alpha_1 \to 1$ in Equations (56) and (57), the classical Euler–Lagrange equations are achieved. Numerical simulations for different values of α_1 are depicted in Figure 1. The numerical method used relies on the Grünwald–Letnikov definition of the left and right fractional derivatives [15].

4.2 Fractional Bateman–Feshbach Tikochinsky oscillator

A time-dependent Hamiltonian to describe the dissipative systems was suggested in [33]. In addition, we recall that the time–dependent Hamiltonian describing the damped oscillation was provided by Caldirola in [36]. This section is started by the following fractional Lagrangian of the Bateman–Feshbach Tikochinsky oscillator [16]:

$$L = m\left({}_{a}^{\mathrm{RL}}D_{t}^{\alpha_1} y\right)\left({}_{a}^{\mathrm{RL}}D_{t}^{\alpha_1} q\right) + 0.5\gamma\left[q\left({}_{a}^{\mathrm{RL}}D_{t}^{\alpha_1} y\right) - y\left({}_{a}^{\mathrm{RL}}D_{t}^{\alpha_1} q\right)\right] - Kqy, \tag{58}$$

where m, K, and γ are time-independent parameters, y denotes the time-reversed counterpart, and q is the damped harmonic oscillator coordinate. According to Equation (29), the corresponding canonical momenta for Equation (58) are expressed by

$$p_{\alpha_1,y} = \frac{\partial L}{\partial\,{}_{a}^{\mathrm{RL}}D_{t}^{\alpha_1} y} = m\,{}_{a}^{\mathrm{RL}}D_{t}^{\alpha_1} q + 0.5\gamma q, \quad p_{\alpha_2,y} = \frac{\partial L}{\partial\,{}_{t}^{\mathrm{RL}}D_{b}^{\alpha_2} y} = 0, \tag{59}$$

$$p_{\alpha_1,q} = \frac{\partial L}{\partial\,{}_{a}^{\mathrm{RL}}D_{t}^{\alpha_1} q} = m\,{}_{a}^{\mathrm{RL}}D_{t}^{\alpha_1} y - 0.5\gamma y, \quad p_{\alpha_2,x} = \frac{\partial L}{\partial\,{}_{t}^{\mathrm{RL}}D_{b}^{\alpha_2} q} = 0. \tag{60}$$

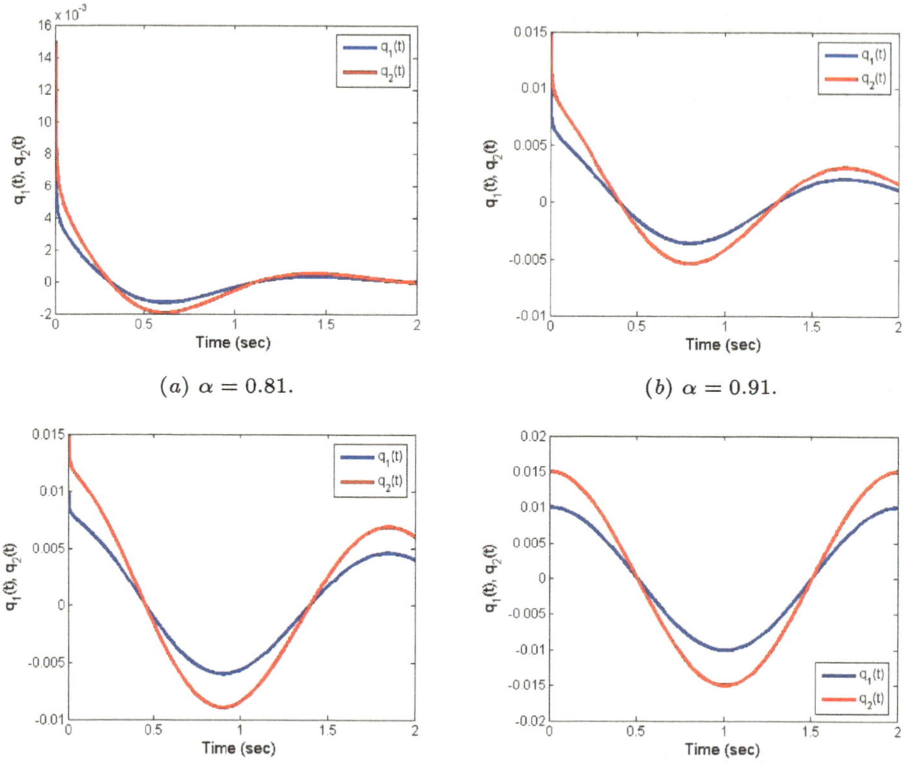

(a) $\alpha = 0.81$. (b) $\alpha = 0.91$.

(c) $\alpha = 0.96$. (d) $\alpha = 1$.

Figure 1: Fractional two electric pendulum: The graphs of $q_1(t)$ and $q_2(t)$, with $q_1(0) = 0.01$ and $q_2(0) = 0.015$, when $m = 1$, $l = 1$, $d = 1$, and $\alpha = 0.81, 0.91, 0.96, 1$.

Substituting Equations (58)–(60) into Equation (30), the fractional Hamiltonian function is derived as

$$\mathcal{H} = m\left({}^{\mathrm{RL}}_{a}D^{\alpha_1}_{t}y\right)\left({}^{\mathrm{RL}}_{a}D^{\alpha_1}_{t}q\right) + Kqy. \tag{61}$$

Finally, within the use of Equation (35), the fractional Hamiltonian equations of motion are obtained in the form

$$m\,{}^{\mathrm{RL}}_{t}D^{\alpha_1}_{b}\,{}^{\mathrm{RL}}_{a}D^{\alpha_1}_{t}y - {}^{\mathrm{RL}}_{t}D^{\alpha_1}_{b}0.5yy = Ky, \tag{62}$$

$$m\,{}^{\mathrm{RL}}_{t}D^{\alpha_1}_{b}\,{}^{\mathrm{RL}}_{a}D^{\alpha_1}_{t}q + {}^{\mathrm{RL}}_{t}D^{\alpha_1}_{b}0.5yq = Kq. \tag{63}$$

Note that, as α_1 goes to 1, Equations (62) and (63) approach the classical Hamiltonian of motion for the generalized coordinates x and y. The numerical results reported in Figure 2 clearly imply that the behaviors of the fractional Euler–Lagrange equation strongly depend on the order of the fractional derivative. In this way, the fractional dynamics are much richer than the classical formulation.

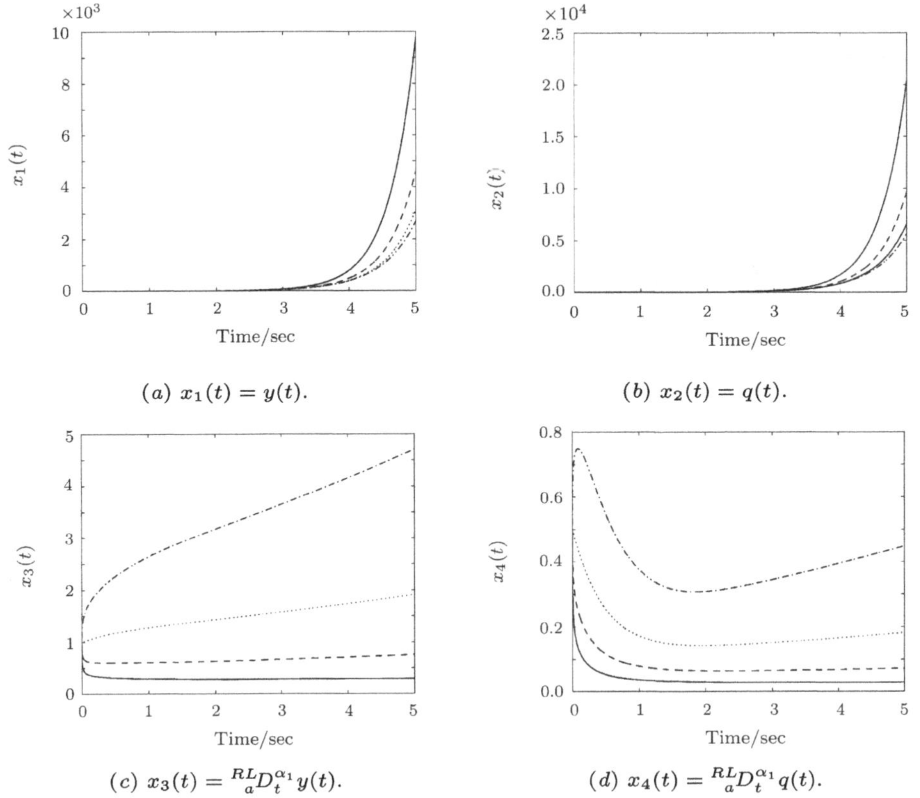

Figure 2: Fractional Bateman–Feshbach Tikochinsky oscillator: The graphs of $y(t)$, $q(t)$, ${}^{\mathrm{RL}}_{a}D^{\alpha_1}_{t}y(t)$, ${}^{\mathrm{RL}}_{a}D^{\alpha_1}_{t}q(t)$, when $m = 0.5$, $\gamma = 2$, $K = 0.1$, and different values of α ($\alpha = 0.81$ (solid), $\alpha = 0.91$ (dashed), $\alpha = 1$ (dotted), and $\alpha = 1.1$ (dashdotted)).

4.3 A fractional thin elastica model

For a thin elastica model, the fractional Lagrangian is written in the following form [17]:

$$L = 0.5(1+\varepsilon y^2)({}^{\mathrm{RL}}_{a}D^{\alpha_1}_{t}x)^2 + 0.5({}^{\mathrm{RL}}_{a}D^{\alpha_1}_{t}y)^2 - 0.5(p^2x^2+y^2), \tag{64}$$

where $x = q_1\sqrt{\frac{Jk_1}{m}}$, $y = q_2\sqrt{k_1}$, $\varepsilon = \frac{m}{Jk_1}$, $p = \sqrt{\frac{k_2/J}{k_1/m}}$, m, J, k_1, k_2 are system parameters, and q_1, q_2 are generalized coordinates denoting the rotational motion (due to the torsional motion of the elastic), and rectilinear deflection (due to the bending motion of the elastica), respectively. Applying Equation (2) to Equation (64) for both x and y, the following fractional Euler–Lagrange equations are concluded:

$$ {}^{\mathrm{RL}}_{t}D^{\alpha_1}_{b}[(1+\varepsilon y^2)\,{}^{\mathrm{RL}}_{a}D^{\alpha_1}_{t}x] - p^2x = 0, \tag{65}$$

$$ {}^{\mathrm{RL}}_{t}D^{\alpha_1}_{b}\,{}^{\mathrm{RL}}_{a}D^{\alpha_1}_{t}y - y + \varepsilon y({}^{\mathrm{RL}}_{a}D^{\alpha_1}_{t}x)^2 = 0. \tag{66}$$

Note that, as α_1 goes to 1, the above equations reduce to the classical Euler–Lagrange equations of motion.

Below, we are going to construct the Hamiltonian equations for the thin elastica model. The fractional Lagrangian, given by Equation (64), depends on two generalized coordinates; hence, four generalized momenta are introduced as follows:

$$p_{\alpha_1,x} = \frac{\partial L}{\partial {}_a^{RL}D_t^{\alpha_1} x} = (1+\varepsilon y^2)\,{}_a^{RL}D_t^{\alpha_1} x, \quad p_{\alpha_2,x} = \frac{\partial L}{\partial {}_t^{RL}D_b^{\alpha_2} x} = 0, \tag{67}$$

$$p_{\alpha_1,y} = \frac{\partial L}{\partial {}_a^{RL}D_t^{\alpha_1} y} = {}_a^{RL}D_t^{\alpha_1} y, \quad p_{\alpha_2,y} = \frac{\partial L}{\partial {}_t^{RL}D_b^{\alpha_2} y} = 0, \tag{68}$$

and the fractional Hamiltonian equation is constructed in the form

$$\mathcal{H} = \frac{0.5}{1+\varepsilon y^2} p_{\alpha_1,x}^2 + 0.5 p_{\alpha_1,y}^2 + 0.5(p^2 x^2 + y^2). \tag{69}$$

Thus, the Hamiltonian equations of motion, in a fractional sense, are derived as Equations (65) and (66), together with

$$\frac{p_{\alpha_1,x}}{1+\varepsilon y^2} = {}_a^{RL}D_t^{\alpha_1} x, \quad p_{\alpha_1,y} = {}_a^{RL}D_t^{\alpha_1} y. \tag{70}$$

The simulation results are given in Figure 3. Again, as α_1 goes to 1, the fractional Hamilton's equations approach their classical counterparts.

4.4 The fractional motion of a bead on the wire

Consider a bead of mass m, which is sliding without friction along a parabola-shaped wire $y = Ax^2$, with vertical axis in the Earth's gravitational field g. For this system, the fractional Lagrangian is written as [22]

$$L = \frac{1}{2} m[1 + 4A^2x^2]({}_a^C D_t^{\alpha_1} x)^2 - mgAx^2, \tag{71}$$

and the fractional Euler–Lagrange equation is obtained in the form

$$4mA^2x({}_a^C D_t^{\alpha_1} x)^2 - 2mgAx + m\,{}_t^C D_b^{\alpha_1}\,{}_a^C D_t^{\alpha_1} x + 4mA^2\,{}_t^C D_b^{\alpha_1}(x^2\,{}_a^C D_t^{\alpha_1} x) = 0. \tag{72}$$

To derive the fractional Hamilton's equation of motion, we introduce the following generalized momenta:

$$p_{\alpha_1} = \frac{\partial L}{\partial {}_a^C D_t^{\alpha_1} x} = m[1 + 4A^2x^2]\,{}_a^C D_t^{\alpha_1} x, \quad p_{\alpha_2} = \frac{\partial L}{\partial {}_t^C D_b^{\alpha_2} x} = 0. \tag{73}$$

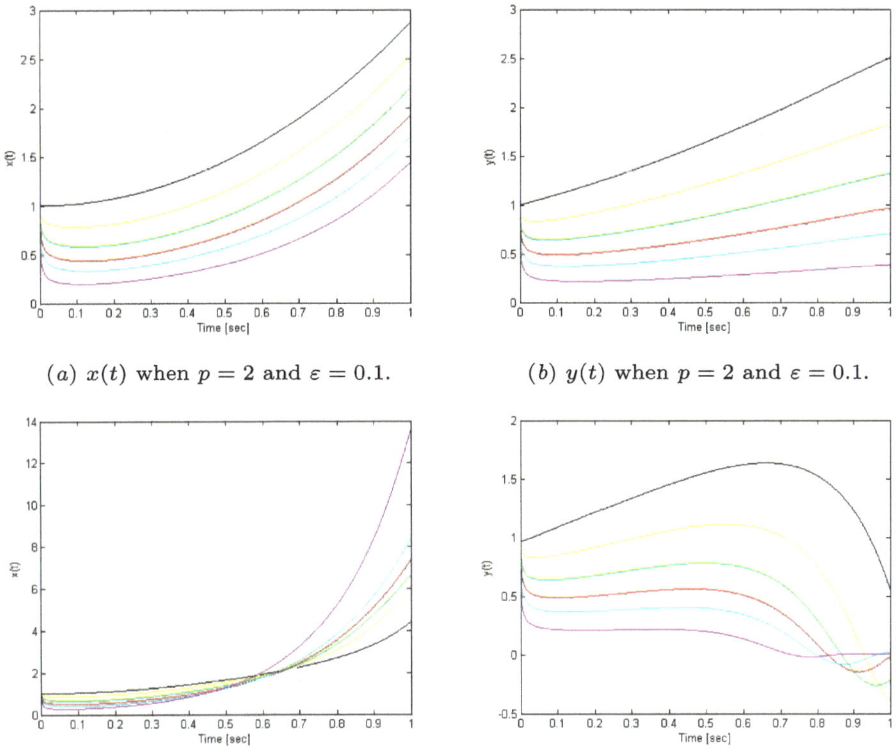

(a) $x(t)$ when $p = 2$ and $\varepsilon = 0.1$.

(b) $y(t)$ when $p = 2$ and $\varepsilon = 0.1$.

(c) $x(t)$ when $p = 3$ and $\varepsilon = 0.5$.

(d) $y(t)$ when $p = 3$ and $\varepsilon = 0.5$.

Figure 3: Fractional thin elastica model: The graphs of $x(t)$ and $y(t)$ for $\alpha_1 = 0.71$ (blue), $\alpha_1 = 0.81$ (cyan), $\alpha_1 = 0.86$ (red), $\alpha_1 = 0.91$ (green), $\alpha_1 = 0.96$ (yellow), and $\alpha_1 = 1$ (black).

Then, the fractional Hamiltonian function is obtained from Equation (30), which results in

$$H = mgAx^2 + \frac{1}{2m(1 + 4A^2x^2)} p_{\alpha_1}^2. \tag{74}$$

Finally, the fractional Hamilton's equation of motion is concluded from Equation (35), which results again in the fractional Euler–Lagrange equation (72).

In the following, we aim to solve Equation (72) numerically for some fractional orders and initial conditions. In Figures 4–6, we give the graphs of $x(t)$ and $y(t)$ different values for α, $x(0)$ and A. In these figures, we also present the solution of classical Euler–Lagrange equation, in addition to some different solutions of fractional Euler–Lagrange equation for $\alpha = 0.81, 0.86, 0.91, 0.96, 1$. These figures indicate that the numerical fractional solution tends to the classic integer solution, as α approaches 1. In addition, the fractional equation of motion exhibits different response characteristics for different values of α. Thus, considering different aspects of the fractional calculus,

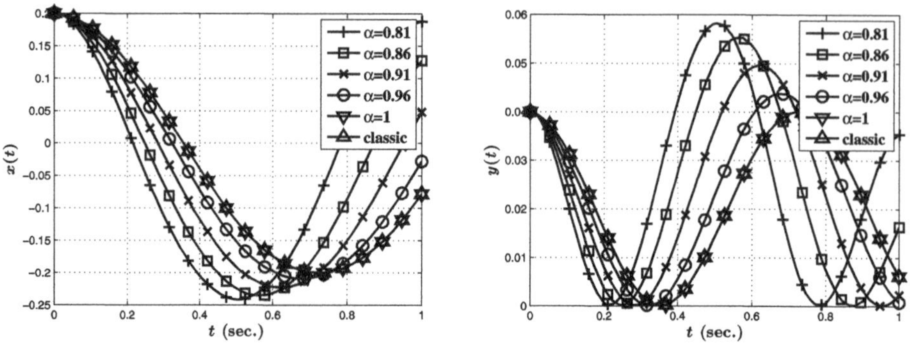

Figure 4: The motion of bead sliding on the wire: The graphs of $x(t)$ and $y(t)$ for $A = 1$, $x(0) = 0.2$, $y(0) = 0$, $m = 2$, $g = 9.81$, and $\alpha = 0.81, 0.86, 0.91, 0.96, 1$.

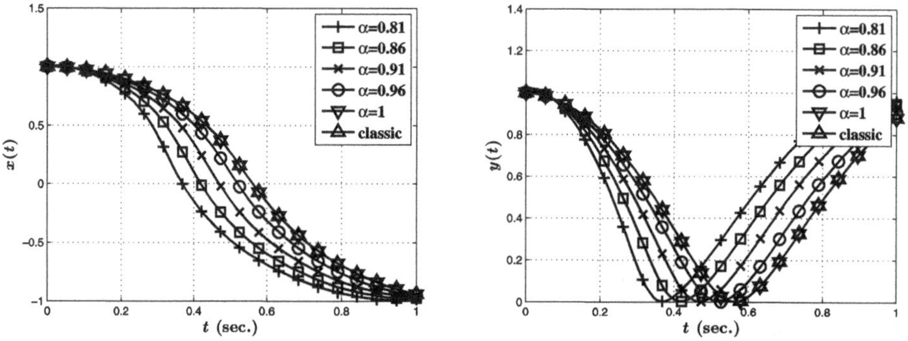

Figure 5: The motion of bead sliding on the wire: The graphs of $x(t)$ and $y(t)$ for $A = 1$, $x(0) = 1$, $y(0) = 0$, $m = 2$, $g = 9.81$, and $\alpha = 0.81, 0.86, 0.91, 0.96, 1$.

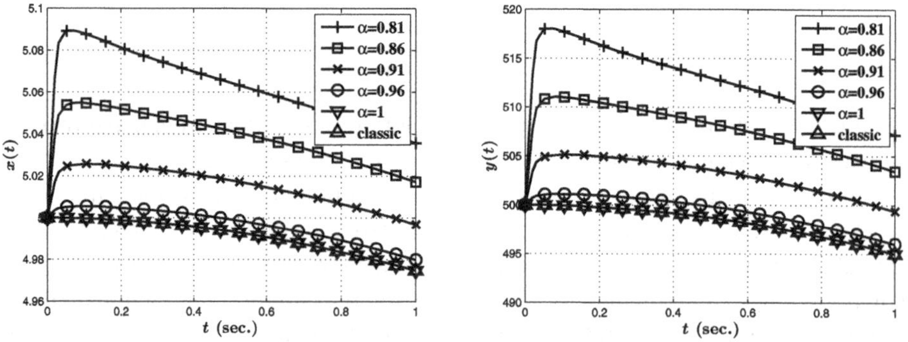

Figure 6: The motion of bead sliding on the wire: The graphs of $x(t)$ and $y(t)$ for $A = 20$, $x(0) = 5$, $y(0) = 0$, $m = 2$, $g = 9.81$, and $\alpha = 0.81, 0.86, 0.91, 0.96, 1$.

provides more flexible models to better adjust the complex behaviors of the real-world phenomena.

5 Conclusions

Fractional calculus is an old field in mathematics, but its tremendous further development has taken place during the last decades. The big confrontation of any type of fractional calculus is with experiments. Some phenomena, the standard model, for example, exhibit only local properties, but there is a big portion of nonlocal phenomena, where fractional calculus is a very good candidate to be applied. Fractional variational principles were obtained by generalizing the classical ones to the fractional case, and using the properties of the fractional integration by parts. In this way, we obtained some original fractional differential equations, containing both the left and right derivatives. For each type of fractional derivatives, we have a specific form of fractional Euler–Lagrange and Hamiltonian equations. Therefore, the fractional kernel plays an important role is solving this type of equations. In fact, within the fractional Lagrangian and Hamiltonian mechanics, we are confronted with all fundamental problems of fractional calculus, bearing on questions, such as: a) What is the physical meaning of the fractional operators? b) How can the fractional order of differentiation be observed experimentally? c) Are the fractional models consistent with the fundamental laws and fundamental symmetries of nature? So far, we have some distinct classes of fractional differential operators. An important question is: if with only a specific class of fractional operators (obtained by imposing some specific criteria which can be natural or not), can we describe the dynamics of complex nonlocal phenomena from various fields of science and engineering? The first-instance response to this question is negative, mainly because of the very complex nature of nonlocality. For each complex nonlocal system, we are able to report an optimal fractional order. As it can be seen from the illustrative examples presented above, the reported results, based on the Riemann–Liouville and Caputo derivatives, provide different behaviours than the classical one. As a result, despite many successful results reported for the fractional Lagrangian and Hamiltonian formulations in mechanics with memory, we are still at the beginning of our studies.

Bibliography

[1] O. P. Agrawal, A new Lagrangian and a new Lagrange equation of motion for fractionally damped systems, *Journal of Applied Mechanics*, **68**(2) (2001), 339–341.

[2] O. P. Agrawal, Formulation of Euler–Lagrange equations for variational problems, *Journal of Mathematical Analysis and Applications*, **272**(1) (2002), 1247–1253.

[3] O. P. Agrawal, A general formulation and solution scheme for fractional optimal control problems, *Nonlinear Dynamics*, **38** (2004), 323–337.

[4] O. P. Agrawal and D. Baleanu, A Hamiltonian formulation and direct numerical scheme for fractional optimal control problems, *Journal of Vibration and Control*, **13** (2007), 1269–1281.

[5] O. P. Agrawal and J. Gregory, O(h(2)) global pointwise algorithms for delay quadratic problems in the calculus of variations, *SIAM Journal on Numerical Analysis*, **41**(5) (2003), 1773–1784.

[6] O. P. Agrawal, J. Gregory, and K. PericakSpector, A bliss-type multiplier rule for constrained variational problems with time delay, *Journal of Mathematical Analysis and Applications*, **210**(2) (1997), 702–711.

[7] R. Almeida, A. B. Malinowska, and D. F. M. Torres, A fractional calculus of variations for multiple integrals with application to vibrating string, *Journal of Mathematics and Physics*, **51**(3) (2010), 033503.

[8] T. M. Atanacković, M. Janev, S. Pilipović, and D. Zorica, Euler–Lagrange equations for Lagrangians containing complex-order fractional derivatives, *Journal of Optimization Theory and Applications*, **174**(1) (2017), 256–275.

[9] T. M. Atanacković, S. Konjik, L. Oparnica, and S. Pilipović, Generalized Hamilton's principle with fractional derivatives, *Journal of Physics A: Mathematical and Theoretical*, **43**(25) (2010), 255203.

[10] T. M. Atanacković, S. Konjik, and S. Pilipović, Variational problems with fractional derivatives: Euler–Lagrange equations, *Journal of Physics A: Mathematical and Theoretical*, **41**(9) (2008), 095201.

[11] T. M. Atanacković, S. Konjik, S. Pilipović, and S. Simić, Variational problems with fractional derivatives: Invariance conditions and Nöther's theorem, *Nonlinear Analysis, Theory, Methods and Applications*, **71**(5–6) (2009), 1504–1517.

[12] D. Baleanu, Fractional Hamiltonian analysis of irregular systems, *Signal Processing*, **86** (2006), 2632–2636.

[13] D. Baleanu and O. P. Agrawal, Fractional Hamilton formalism within Caputo's derivative, *Czechoslovak Journal of Physics*, **56**(10–11) (2006), 1087–1092.

[14] D. Baleanu, J. Asad, M. Alipour, and T. Blaszczyk, Motion of a spherical particle in a rotating parabola using fractional Lagrangian, *Scientific Bulletin—"Politehnica" University of Bucharest. Series A, Applied Mathematics and Physics*, **79**(2) (2017), 183–192.

[15] D. Baleanu, J. H. Asad, and I. Petras, Fractional-order two-electric pendulum, *Romanian Reports in Physics*, **64**(4) (2012), 907–914.

[16] D. Baleanu, J. H. Asad, and I. Petras, Fractional Bateman–Feshbach Tikochinsky oscillator, *Communications in Theoretical Physics*, **61**(2) (2014), 221–225.

[17] D. Baleanu, J. H. Asad, and I. Petras, Numerical solution of the fractional Euler–Lagrange's equations of a thin elastica model, *Nonlinear Dynamics*, **81**(1–2) (2015), 97–102.

[18] D. Baleanu and T. Avkar, Lagrangians with linear velocities within Riemann–Liouville fractional derivatives, *Nuovo Cimento B*, **119**(1) (2004), 73–79.

[19] D. Baleanu, T. Blaszczyk, J. Asad, and M. Alipour, Numerical study for fractional Euler–Lagrange equations of a harmonic oscillator on a moving platform, *Acta Physica Polonica. A*, **130**(3) (2016), 688–691.

[20] D. Baleanu, K. Diethelm, E. Scalas, and J. J. Trujillo, *Fractional Calculus: Models and Numerical Methods*, Second, World Scientific, New York, 2017.

[21] D. Baleanu, A. K. Golmankhaneh, and A. K. Golmankhaneh, The dual action of fractional multi time Hamilton equations, *International Journal of Theoretical Physics*, **48** (2009), 2558–2569.

[22] D. Baleanu, A. Jajarmi, J. H. Asad, and T. Blaszczyk, The motion of a bead sliding on a wire in fractional sense, *Acta Physica Polonica. A*, **131**(6) (2017), 1561–1564.

[23] D. Baleanu, R. Magin, S. Bhalekar, and V. Daftardar-Gejji, Chaos in the fractional order nonlinear Bloch equation with delay, *Communications in Nonlinear Science and Numerical Simulation*, **25**(1–3) (2015), 41–49.

[24] D. Baleanu, T. Maraaba, and F. Jarad, Fractional variational principles with delay, *Journal of Physics. A, Mathematical and General*, **41** (2008), 315403.

[25] D. Baleanu and S. Muslih, Lagrangian formulation of classical fields within Riemann–Liouville fractional derivatives, *Physica Scripta*, **72**(2–3) (2005), 105.

[26] D. Baleanu and S. Muslih, Fractional Euler–Lagrange and fractional Hamilton equations for super symmetric classical model, *Fractals*, **15**(4) (2007), 379–383.

[27] D. Baleanu, S. Muslih, E. M. Rabei, A. K. Golmankhaneh, and A. K. Golmankhaneh, On fractional Hamiltonian systems possessing first-class constraints within Caputo derivatives, *Romanian Reports in Physics*, **63**(1) (2011), 3–8.

[28] D. Baleanu, S. I. Muslih, and E. M. Rabei, On fractional Euler–Lagrange and Hamilton equations and the fractional generalization of total time derivative, *Nonlinear Dynamics*, **53**(1–2) (2008), 67–74.

[29] D. Baleanu and O. Mustafa, *Asymptotic Integration and Stability for Differential Equations of Fractional Order*, World Scientific, New York, 2015.

[30] D. Baleanu, I. Petras, J. H. Asad, and M. P. Velasco, Fractional Pais–Uhlenbeck oscillator, *International Journal of Theoretical Physics*, **51**(4) (2012), 1253–1258.

[31] D. Baleanu and J. J. Trujillo, On exact solutions of a class of fractional Euler–Lagrange equations, *Nonlinear Dynamics*, **52**(4) (2008), 331–335.

[32] D. Baleanu and J. J. Trujillo, A new method of finding the fractional Euler–Lagrange and Hamilton equations within Caputo fractional derivatives, *Communications in Nonlinear Science and Numerical Simulation*, **15** (2010), 1111–1115.

[33] H. Bateman, On dissipative systems and related variational principles, *Physical Review*, **38** (1931), 815.

[34] S. Bhalekar, V. Daftardar-Gejji, and D. Baleanu, Fractional Bloch equation with delay, *Computers & Mathematics with Applications*, **61**(5) (2011), 1355–1365.

[35] S. Bhalekar, V. Daftardar-Gejji, D. Baleanu, and R. L. Magin, Generalized fractional order Bloch equation with extended delay, *International Journal of Bifurcation and Chaos in Applied Sciences and Engineering*, **22**(4) (2012), 1250071.

[36] P. Caldirola, Forze non conservative nella meccanica quantistica, *Nuovo Cimento*, **18** (1941), 393.

[37] M. Caputo, Linear models of dissipation whose Q is almost frequency independent-II, *Geophysical Journal*, **13**(5) (1967), 529–539.

[38] A. Coronel-Escamilla, J. F. Gómez-Aguilar, D. Baleanu, R. F. Escobar-Jiménez, V. H. Olivares-Peregrino, and A. Abundez-Pliego, Formulation of Euler–Lagrange and Hamilton equations involving fractional operators with regular kernel, *Advances in Difference Equations*, **2016** (2016), 283, 1–17.

[39] J. Cresson, Fractional embedding of differential operators and Lagrangian system, *Journal of Mathematical Physics*, **48** (2007), 033504.

[40] J. P. Cusumano, Low-dimensional, chaotic, nonplanar motions of the elastic, in *Ph.D. Thesis, Cornell University, NY*, 1990.

[41] D. W. Dreisigmeyer and P. M. Yoang, Nonconservative Lagrangian mechanics: a generalized functional approach, *Journal of Physics A: Mathematical and Theoretical*, **36**(30) (2003), 3297.

[42] R. D. Driver, *Ordinary and Delay Differential Equations*, Springer, Berlin, 1977.

[43] A. K. Golmankhaneh, A. K. Golmankhaneh, and D. Baleanu, Lagrangian and Hamiltonian mechanics on fractals subset of real-line, *International Journal of Theoretical Physics*, **52** (2013), 4210–4217.

[44] A. K. Golmankhaneh, A. K. Golmankhaneh, D. Baleanu, and M. C. Baleanu, Hamiltonian structure of fractional first order Lagrangian, *International Journal of Theoretical Physics*, **49** (2010), 365–375.
[45] A. K. Golmankhaneh, A. M. Yengejeh, and D. Baleanu, On the fractional Hamilton and Lagrange mechanics, *International Journal of Theoretical Physics*, **51** (2012), 2909–2916.
[46] M. A. E. Herzallah and D. Baleanu, Fractional-order Euler–Lagrange equations and formulation of Hamiltonian equations, *Nonlinear Dynamics*, **58** (2009), 385–391.
[47] F. Jarad, T. Abdeljawad, and D. Baleanu, Fractional variational control problems with delayed arguments, *Nonlinear Dynamics*, **62** (2010), 609–614.
[48] F. Jarad, T. Abdeljawad, and D. Baleanu, Fractional variational principles with delay within Caputo derivative, *Reports on Mathematical Physics*, **65** (2010), 17–28.
[49] A. A. Kilbas, H. M. Srivastava, and J. J. Trujillo, *Theory and Applications of Fractional Differential Equations*, North-Holland Mathematics Studies, vol. 204, Elsevier Science B.V., Amsterdam, 2006.
[50] M. Klimek, Fractional sequential mechanics—models with symmetric fractional derivative, *Czechoslovak Journal of Physics*, **51**(12) (2001), 1348–1354.
[51] M. Klimek, Lagrangian and Hamiltonian fractional sequential mechanics, *Czechoslovak Journal of Physics*, **52**(11) (2002), 1247–1253.
[52] M. Klimek, Lagrangian fractional mechanics—a noncommutative approach, *Czechoslovak Journal of Physics*, **55**(11) (2005), 1447–1453.
[53] C. Lanczos, *The Variational Principles of Mechanics*, Univ. Toronto Press, 1962.
[54] R. Magin, O. Abdullah, D. Baleanu, and X. H. J. Zhou, Anomalous diffusion expressed through fractional order differential operators in the Bloch–Torrey equation, *Journal of Magnetic Resonance*, **190**(2) (2008), 255–270.
[55] R. Magin, X. Feng, and D. Baleanu, Solving the fractional order Bloch equation, *Concepts in Magnetic Resonance. Part A*, **34A**(1) (2009), 16–23.
[56] R. L. Magin, *Fractional Calculus in Bioengineering*, Begell House Publishers, 2006.
[57] D. Mozyrska and F. M. D. Torres, Minimal modified energy control for fractional linear control systems with the Caputo derivative, *Carpathian Journal of Mathematics*, **26**(2) (2010), 210–221.
[58] S. I. Muslih and D. Baleanu, Hamiltonian formulation of systems with linear velocities within Riemann–Liouville fractional derivatives, *Journal of Mathematical Analysis and Applications*, **304**(2) (2005), 599–606.
[59] S. I. Muslih and D. Baleanu, Fractional Euler–Lagrange equations of motion in fractional space, *Journal of Vibration and Control*, **13**(9–10) (2007), 1209–1216.
[60] S. I. Muslih, D. Baleanu, and E. M. Rabei, Fractional Hamilton's equations of motion in fractional time, *Central European Journal of Physics*, **5**(4) (2007), 549–557.
[61] I. Podlubny, *Fractional Differential Equations*, Academic Press, San Diego, 1999.
[62] E. M. Rabei, T. S. Al-Halholy, and A. A. Taani, On Hamiltonian formulation of non-conservative systems, *Turkish Journal of Physics*, **28**(4) (2004), 213–221.
[63] E. M. Rabei, K. I. Nawafleh, R. S. Hijjawi, S. I. Muslih, and D. Baleanu, The Hamilton formalism with fractional derivatives, *Journal of Mathematical Analysis and Applications*, **327**(2) (2007), 891–897.
[64] S. S. Rekhriashvili, The Lagrangian formalism with fractional derivatives in problems of mechanics, *Technical Physics Letters*, **30**(1) (2004), 55–57.
[65] F. Riewe, Nonconservative Lagrangian and Hamiltonian mechanics, *Physical Review E*, **53**(2) (1996), 1890–1899.
[66] F. Riewe, Mechanics with fractional derivatives, *Physical Review E*, **55**(3) (1997), 3581–3592.
[67] S. G. Samko, A. A. Kilbas, and O. I. Marichev, *Fractional Integrals and Derivatives: Theory and Applications*, Gordon and Breach, Yverdon, 1993.

[68] V. E. Tarasov, Fractional variations for dynamical systems: Hamilton and Lagrange approaches, *Journal of Physics. A, Mathematical and General*, **39**(26) (2006), 8409–8425.
[69] D. Tavares, R. Almeida, and D. F. M. Torres, Constrained fractional variational problems of variable order, *IEEE/CAA Journal of Automatica Sinica*, **4**(1) (2017), 80–88.

Rudolf Gorenflo and Francesco Mainardi

Fractional relaxation-oscillation phenomena

Abstract: In this chapter we consider the basic properties of some types of differential equations of fractional order that are related to relaxation and oscillation phenomena, based on functions of the Mittag-Leffler type. We finally generalize with fractional derivatives the Basset problem known in fluid-dynamics, so exploring other fractional relaxation processes of physical relevance.

Keywords: Fractional relaxation equation, fractional oscillation equation, Mittag-Leffler function, complete monotonicity, Basset problem

MSC 2010: 26A33, 35E12, 34C26, 44A10, 76DXX

1 Introduction

In this chapter we analyze the most relevant differential equations of fractional order related to relaxation and oscillation phenomena. For this purpose, following our works [13–15, 17, 22–25, 27, 28], we choose the examples which, by means of fractional derivatives, generalize the well-known ordinary differential equations related to relaxation and oscillation phenomena.

In Section 2, we treat the simplest types, which we refer to as the *simple fractional relaxation and oscillation equations*. In Section 3 we consider the types, somewhat more cumbersome, which we refer to as the *composite fractional relaxation and oscillation equations*. Finally, in Section 4, we consider the Basset problem in fluid-dynamics, providing a fractional generalization of it, so leading to relevant relaxation phenomena of physical relevance.

For relaxations equations, governed by differential equations of fractional distributed order, see our papers [16] and [26]. For other treatments of the fractional oscillation equations, see, for example, the papers by Achar et al. [1, 2], by Duan et al. [12], and by Blaszczyk and Ciesielski [7].

Let us first recall that the classical phenomena of relaxation and oscillations in their simplest form are known to be governed by linear ordinary differential equations, of order one and two, respectively. We recall the latter, hereafter, with the cor-

Acknowledgement: The work of FM has been carried out in the framework of the activities of the National Group of Mathematical Physics (INdAM-GNFM).

Rudolf Gorenflo, Department of Mathematics, Free University Berlin, Berlin, Germany (31.07.1930–20.10.2017)
Francesco Mainardi, Department of Physics and Astronomy, University of Bologna, Via Irnerio 46, 40126 Bologna, Italy, e-mail: francesco.mainardi@bo.infn.it

https://doi.org/10.1515/9783110571707-003

responding solutions. Let us denote by $u = u(t)$ the field variable and by $q(t)$ a given continuous function with $t \geq 0$.

The *relaxation* differential equation reads:

$$u'(t) = -u(t) + q(t), \tag{1.1}$$

whose solution, under the initial condition $u(0^+) = c_0$, is

$$u(t) = c_0 \, \mathrm{e}^{-t} + \int_0^t q(t-\tau) \, \mathrm{e}^{-\tau} \, d\tau. \tag{1.2}$$

The *oscillation* differential equation reads:

$$u''(t) = -u(t) + q(t), \tag{1.3}$$

whose solution, under the initial conditions $u(0^+) = c_0$ and $u'(0^+) = c_1$, is

$$u(t) = c_0 \, \cos t + c_1 \, \sin t + \int_0^t q(t-\tau) \, \sin \tau \, d\tau. \tag{1.4}$$

From viewpoint of fractional calculus, a natural generalization of Equations (1.1) and (1.3) is obtained by replacing the ordinary derivative with a fractional one, of order α. To preserve the type of initial conditions required in the classical phenomena, we agree to replace the first and second derivative in (1.1) and (1.3) with a Caputo fractional derivative of order α with $0 < \alpha < 1$ and $1 < \alpha < 2$, respectively. We agree to refer to the corresponding equations as the *simple fractional relaxation equation* and the *simple fractional oscillation equation.*

2 Fractional relaxation and oscillation

Generally speaking, we consider the following differential equation of fractional order $\alpha > 0$ for $t \geq 0$:

$$D_*^\alpha \, u(t) = D^\alpha \left(u(t) - \sum_{k=0}^{m-1} \frac{t^k}{k!} \, u^{(k)}(0^+) \right) = -u(t) + q(t), \tag{2.1}$$

where $u = u(t)$ is the field variable, $q(t)$ is a given function continuous for $t \geq 0$, and D^α and D_*^α denote the fractional derivatives of order α in the sense of Riemann–Liouville and Caputo, respectively. Here m is a positive integer uniquely defined by $m-1 < \alpha \leq m$, which provides the number of the prescribed initial values $u^{(k)}(0^+) = c_k$, $k = 0, 1, 2, \ldots, m-1$.

In particular, we consider in detail the following cases:
(a) *fractional relaxation* $0 < \alpha \le 1$,
(b) *fractional oscillation* $1 < \alpha \le 2$.

The application of the Laplace transform yields

$$\widetilde{u}(s) = \sum_{k=0}^{m-1} c_k \frac{s^{\alpha-k-1}}{s^\alpha + 1} + \frac{1}{s^\alpha + 1} \widetilde{q}(s). \tag{2.2}$$

Then, putting for $k = 0, 1, \ldots, m-1$,

$$u_k(t) := J^k e_\alpha(t) \div \frac{s^{\alpha-k-1}}{s^\alpha + 1}, \quad e_\alpha(t) := E_\alpha(-t^\alpha) \div \frac{s^{\alpha-1}}{s^\alpha + 1}, \tag{2.3}$$

where J^k denote the k-repeated integral between 0 and t, E_α denotes the Mittag-Leffler function of order α, and using $u_0(0^+) = 1$, we find

$$u(t) = \sum_{k=0}^{m-1} c_k \, u_k(t) - \int_0^t q(t-\tau) \, u_0'(\tau) \, d\tau. \tag{2.4}$$

In particular, formula (2.4) encompasses the solutions for $\alpha = 1, 2$, since

$$\alpha = 1, \quad u_0(t) = e_1(t) = \exp(-t),$$
$$\alpha = 2, \quad u_0(t) = e_2(t) = \cos t, \quad u_1(t) = J^1 e_2(t) = \sin t.$$

When α is not integer, namely for $m - 1 < \alpha < m$, we note that $m - 1$ represents the integer part of α (usually denoted by $[\alpha]$) and m, the number of initial conditions necessary and sufficient to ensure the uniqueness of the solution $u(t)$. Thus, the m functions $u_k(t) = J^k e_\alpha(t)$, with $k = 0, 1, \ldots, m-1$ represent those particular solutions of the *homogeneous* equation, which satisfy the initial conditions $u_k^{(h)}(0^+) = \delta_{k\,h}$, $h, k = 0, 1, \ldots, m-1$. The latter, therefore, represent the *fundamental solutions* of the fractional Equation (2.1), just as in the case of $\alpha = m$. Furthermore, the function $u_\delta(t) = -u_0'(t) = -e_\alpha'(t)$ represents the *impulse-response solution.*

Now we derive the relevant properties of the basic functions $e_\alpha(t)$ directly from their Laplace representation for $0 < \alpha \le 2$,

$$e_\alpha(t) = \frac{1}{2\pi i} \int_{\mathrm{Br}} \mathrm{e}^{st} \frac{s^{\alpha-1}}{s^\alpha + 1} \, ds, \tag{2.5}$$

without detouring from the general theory of Mittag-Leffler functions in the complex plane. Here, Br denotes a Bromwich path, that is, a line $\mathrm{Re}(s) = \sigma > 0$, and $\mathrm{Im}(s)$ running from $-\infty$ to $+\infty$.

For transparency reasons, we separately discuss the cases (a) $0 < \alpha < 1$ and (b) $1 < \alpha < 2$, recalling that in the limiting cases $\alpha = 1, 2$, we know $e_\alpha(t)$ as an elementary function, namely $e_1(t) = \mathrm{e}^{-t}$ and $e_2(t) = \cos t$.

For α that is not integer, the power function s^α is uniquely defined as $s^\alpha = |s|^\alpha \, \mathrm{e}^{i \arg s}$, with $-\pi < \arg s < \pi$, that is, in the complex s-plane cut along the negative real axis.

The essential step consists of decomposing $e_\alpha(t)$ into two parts, according to $e_\alpha(t) = f_\alpha(t) + g_\alpha(t)$, as indicated below. In case (a) the function $f_\alpha(t)$ and in case (b) the function $-f_\alpha(t)$ is *completely monotone*. In both cases, $f_\alpha(t)$ tends to zero as t tends to infinity, from above in case (a), from below in case (b). The other part, $g_\alpha(t)$, is identically vanishing in case (a), but of *oscillatory* character, with exponentially decreasing amplitude, in case (b).

For the *oscillatory part*, we obtain, via the residue theorem of complex analysis,

$$g_\alpha(t) = \frac{2}{\alpha}\, \mathrm{e}^{t \cos(\pi/\alpha)} \cos\left[t \sin\left(\frac{\pi}{\alpha}\right)\right] \quad \text{if } 1 < \alpha < 2. \tag{2.6}$$

We note that this function exhibits oscillations with *circular frequency*

$$\omega(\alpha) = \sin(\pi/\alpha)$$

and with an *exponentially decaying amplitude* with the rate

$$\lambda(\alpha) = \left|\cos(\pi/\alpha)\right| = -\cos(\pi/\alpha).$$

For the *monotonic part*, we obtain

$$f_\alpha(t) := \int_0^\infty \mathrm{e}^{-rt} K_\alpha(r)\, dr, \tag{2.7}$$

with

$$K_\alpha(r) = -\frac{1}{\pi} \operatorname{Im}\left(\left. \frac{s^{\alpha-1}}{s^\alpha + 1} \right|_{s = r\,\mathrm{e}^{i\pi}} \right) = \frac{1}{\pi} \frac{r^{\alpha-1} \sin(\alpha\pi)}{r^{2\alpha} + 2r^\alpha \cos(\alpha\pi) + 1}. \tag{2.8}$$

This function $K_\alpha(r)$ vanishes identically if α is an integer; it is positive for all r if $0 < \alpha < 1$ and negative for all r if $1 < \alpha < 2$. In fact, in (2.8) the denominator is, for α not integer, always positive, being $> (r^\alpha - 1)^2 \geq 0$.

Hence, $f_\alpha(t)$ has the aforementioned monotonicity properties, decreasing towards zero in case (a) and increasing towards zero in case (b). We note that in order to satisfy the initial condition $e_\alpha(0^+) = 1$, we find

$$\int_0^\infty K_\alpha(r)\, dr = 1 \quad \text{if } 0 < \alpha \leq 1, \qquad \int_0^\infty K_\alpha(r)\, dr = 1 - 2/\alpha \quad \text{if } 1 < \alpha \leq 2.$$

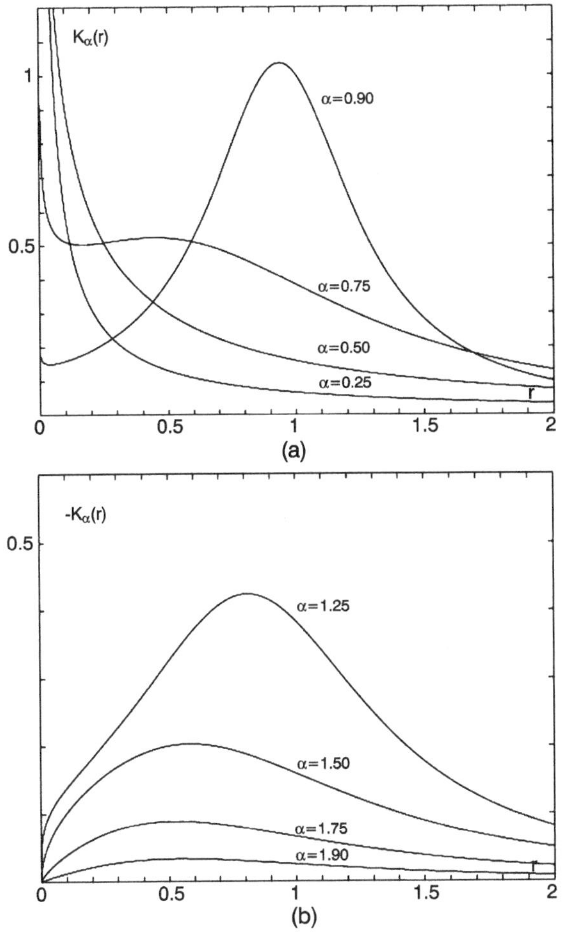

Figure 1: (a) Plots of the *basic spectral function* $K_\alpha(r)$ for $0 < \alpha < 1$: $\alpha = 0.25, \alpha = 0.50, \alpha = 0.75, \alpha = 0.90$. (b) Plots of the *basic spectral function* $-K_\alpha(r)$ for $1 < \alpha < 2$: $\alpha = 1.25, \alpha = 1.50, \alpha = 1.75, \alpha = 1.90$.

In Figures 1(a) and 1(b), we display the plots of $K_\alpha(r)$, which we denote as the *basic spectral function* for some values of α in the intervals (a) $0 < \alpha < 1$, (b) $1 < \alpha < 2$.

In addition to the basic fundamental solutions $u_0(t) = e_\alpha(t)$, we need to compute the impulse-response solutions $u_\delta(t) = -D^1\, e_\alpha(t)$ for cases (a) and (b) and, only in case (b), the second fundamental solution $u_1(t) = J^1\, e_\alpha(t)$.

For this purpose we note that in general it turns out that

$$J^k\, f_\alpha(t) = \int_0^\infty \mathrm{e}^{-rt}\, K_\alpha^k(r)\, dr \tag{2.9}$$

with

$$K_\alpha^k(r) := (-1)^k\, r^{-k}\, K_\alpha(r) = \frac{(-1)^k}{\pi} \frac{r^{\alpha-1-k}\, \sin(\alpha\pi)}{r^{2\alpha} + 2r^\alpha \cos(\alpha\pi) + 1}, \tag{2.10}$$

where $K_\alpha(r) = K_\alpha^0(r)$, and

$$J^k g_\alpha(t) = \frac{2}{\alpha}\, e^{t\cos(\pi/\alpha)} \cos\left[t \sin\left(\frac{\pi}{\alpha}\right) - k\frac{\pi}{\alpha}\right]. \tag{2.11}$$

For the impulse-response solution, we note that the effect of the differential operator D^1 is the same as that of the virtual operator J^{-1}.

In conclusion, the solutions for the *fractional relaxation* are:

(a) $0 < \alpha < 1$,

$$u(t) = c_0\, u_0(t) + \int_0^t q(t-\tau)\, u_\delta(\tau)\, d\tau, \tag{2.12a}$$

where

$$\begin{cases} u_0(t) = \int_0^\infty e^{-rt} K_\alpha^0(r)\, dr, \\ u_\delta(t) = -\int_0^\infty e^{-rt} K_\alpha^{-1}(r)\, dr, \end{cases} \tag{2.13a}$$

with

$$u_0(0^+) = 1, \quad u_\delta(0^+) = \infty,$$

and, as $t \to \infty$,

$$u_0(t) \sim \frac{t^{-\alpha}}{\Gamma(1-\alpha)}, \quad u_\delta(t) \sim \frac{t^{1-\alpha}}{\Gamma(2-\alpha)}. \tag{2.14a}$$

In conclusion, the solutions for the *fractional oscillation* are:

(b) $1 < \alpha < 2$,

$$u(t) = c_0\, u_0(t) + c_1\, u_1(t) + \int_0^t q(t-\tau)\, u_\delta(\tau)\, d\tau, \tag{2.12b}$$

$$\begin{cases} u_0(t) = \int_0^\infty e^{-rt} K_\alpha^0(r)\, dr + \frac{2}{\alpha}\, e^{t\cos(\pi/\alpha)} \cos[t \sin(\frac{\pi}{\alpha})], \\ u_1(t) = \int_0^\infty e^{-rt} K_\alpha^1(r)\, dr + \frac{2}{\alpha}\, e^{t\cos(\pi/\alpha)} \cos[t \sin(\frac{\pi}{\alpha}) - \frac{\pi}{\alpha}], \\ u_\delta(t) = -\int_0^\infty e^{-rt} K_\alpha^{-1}(r)\, dr - \frac{2}{\alpha}\, e^{t\cos(\pi/\alpha)} \cos[t \sin(\frac{\pi}{\alpha}) + \frac{\pi}{\alpha}], \end{cases} \tag{2.13b}$$

with

$$u_0(0^+) = 1, \quad u_0'(0^+) = 0, \quad u_1(0^+) = 0, \quad u_1'(0^+) = 1,$$
$$u_\delta(0^+) = 0, \quad u_\delta'(0^+) = +\infty,$$

and, as $t \to \infty$,

$$u_0(t) \sim \frac{t^{-\alpha}}{\Gamma(1-\alpha)}, \quad u_1(t) \sim \frac{t^{1-\alpha}}{\Gamma(2-\alpha)}, \quad u_\delta(t) \sim -\frac{t^{-\alpha-1}}{\Gamma(-\alpha)}. \tag{2.14b}$$

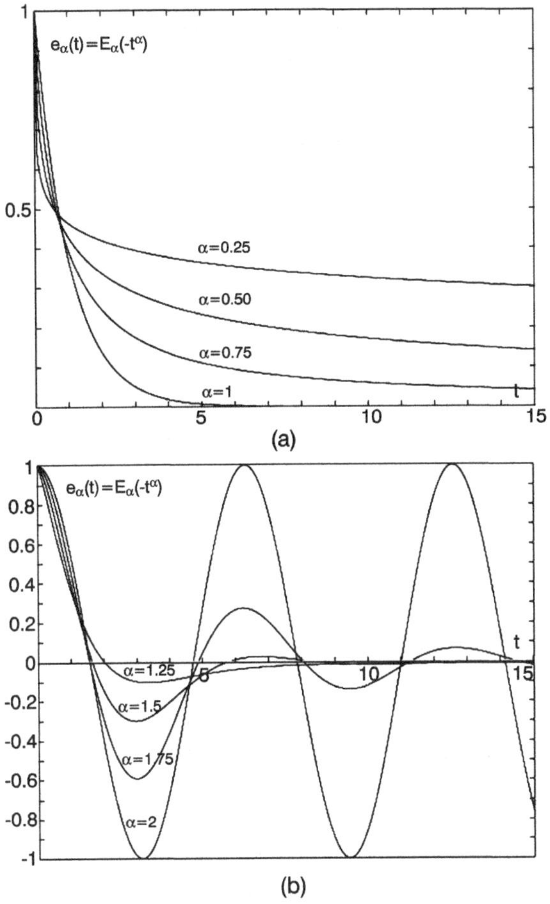

Figure 2: (a) Plots of the *basic fundamental solution* $u_0(t) = e_\alpha(t)$: $\alpha = 0.25$, $\alpha = 0.50$, $\alpha = 0.75$, $\alpha = 1$. (b) Plots of the *basic fundamental solution* $u_0(t) = e_\alpha(t)$: $\alpha = 1.25$, $\alpha = 1.50$, $\alpha = 1.75$, $\alpha = 2$.

In Figures 2(a) and 2(b), we display the plots of the basic fundamental solution for the following cases, respectively:

(a) $\alpha = 0.25, 0.50, 0.75, 1$,

(b) $\alpha = 1.25, 1.50, 1.75, 2$,

obtained from the first formula in (2.13a) and (2.13b), respectively.

We now want to point out that in both the cases (a) and (b) (in which α is not integer), that is, for *fractional relaxation* and *fractional oscillation*, all the fundamental and impulse-response solutions exhibit an *algebraic decay* as $t \to \infty$, as discussed above.

This *algebraic decay* is the most important effect of the noninteger derivative in our equations, which dramatically differs from the *exponential decay* present in the ordinary relaxation and damped-oscillation phenomena.

We would like to highlight the difference between fractional relaxation governed by the Mittag-Leffler-type function $E_\alpha(-at^\alpha)$ and stretched relaxation, governed by a stretched exponential function $\exp(-bt^\alpha)$, with $\alpha, a, b > 0$ for $t \geq 0$.

A common behavior is achieved only in a restricted range $0 \leq t \ll 1$, where we can have

$$E_\alpha(-at^\alpha) \simeq 1 - \frac{a}{\Gamma(\alpha+1)}\,t^\alpha = 1 - b\,t^\alpha \simeq \mathrm{e}^{-b\,t^\alpha}, \quad b = \frac{a}{\Gamma(\alpha+1)}.$$

In Figure 3, we compare for $\alpha = 0.25, 0.50, 0.75$, from top to bottom respectively, $E_\alpha(-t^\alpha)$ (*full* line) with its asymptotic approximations $\exp[-t^\alpha/\Gamma(1+\alpha)]$ (*dashed* line) valid for short durations, and $t^{-\alpha}/\Gamma(1-\alpha)$ (*dotted* line) valid for long durations.

We have adopted log-log plots in order to better achieve such a comparison, and the transition from the stretched exponential to the inverse power-law decay.

In Figure 4, we show some plots of the *basic fundamental solution* $u_0(t) = e_\alpha(t)$ for $\alpha = 1.25, 1.50, 1.75$, from top to down respectively.

Here, the algebraic decay of the fractional oscillation can be recognized and compared with the two contributions provided by f_α (monotonic behavior, dotted line) and $g_\alpha(t)$ (exponentially damped oscillation, dashed line).

The zeros of the solutions of the fractional oscillation

Now we find it interesting to carry out some investigations about the zeros of the basic fundamental solution $u_0(t) = e_\alpha(t)$ in the case (b) of fractional oscillations. For the second fundamental solution and the impulse-response solution, the analysis of the zeros can be easily carried out analogously.

Recalling the first Equation in (2.13b), the required zeros of $e_\alpha(t)$ are the solutions of the equation

$$e_\alpha(t) = f_\alpha(t) + \frac{2}{\alpha}\,\mathrm{e}^{t\cos(\pi/\alpha)}\cos\left[t\sin\left(\frac{\pi}{\alpha}\right)\right] = 0. \tag{2.15}$$

We first note that the function $e_\alpha(t)$ exhibits an *odd* number of zeros, in that $e_\alpha(0) = 1$, and for sufficiently large t, $e_\alpha(t)$ turns out to be permanently negative, as shown in (2.14b) by the sign of $\Gamma(1-\alpha)$.

The smallest zero lies in the first positivity interval of $\cos[t\sin(\pi/\alpha)]$. Hence, in the interval $0 < t < \pi/[2\sin(\pi/\alpha)]$. All other zeros can only lie in the succeeding positivity intervals of $\cos[t\sin(\pi/\alpha)]$. In each of these two zeros are present as long as

$$\frac{2}{\alpha}\,\mathrm{e}^{t\cos(\pi/\alpha)} \geq |f_\alpha(t)|. \tag{2.16}$$

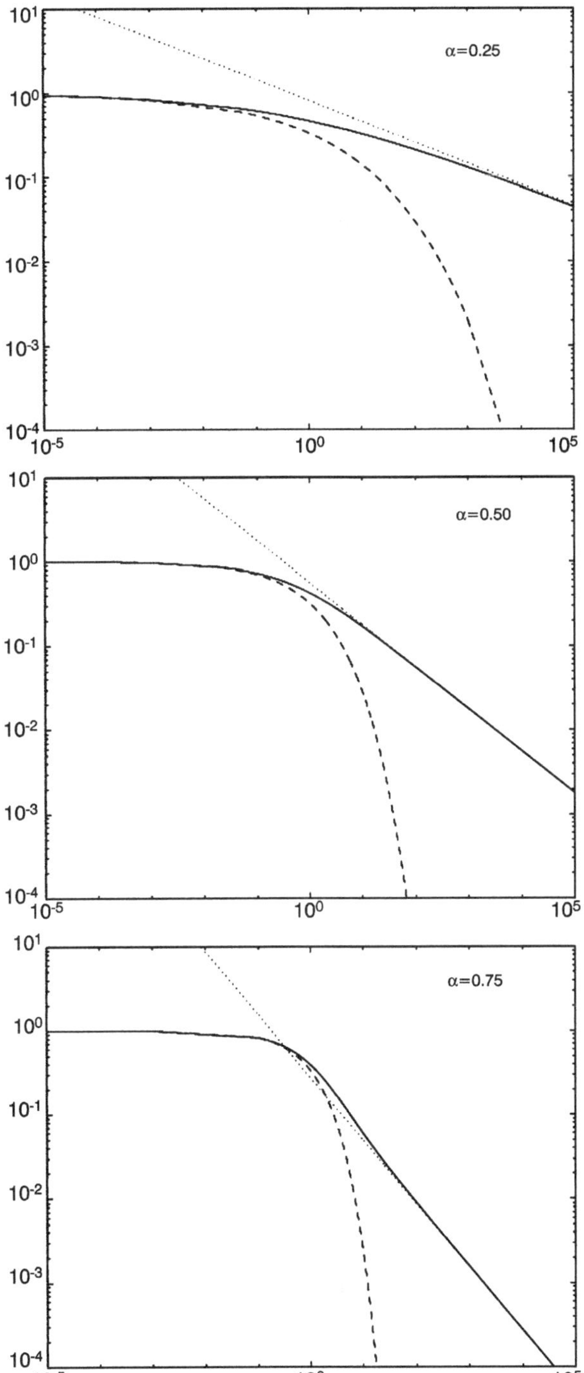

Figure 3: Log-log plot of $E_\alpha(-t^\alpha)$ for $\alpha = 0.25, 0.50, 0.75$, and for $10^{-6} \le t \le 10^6$.

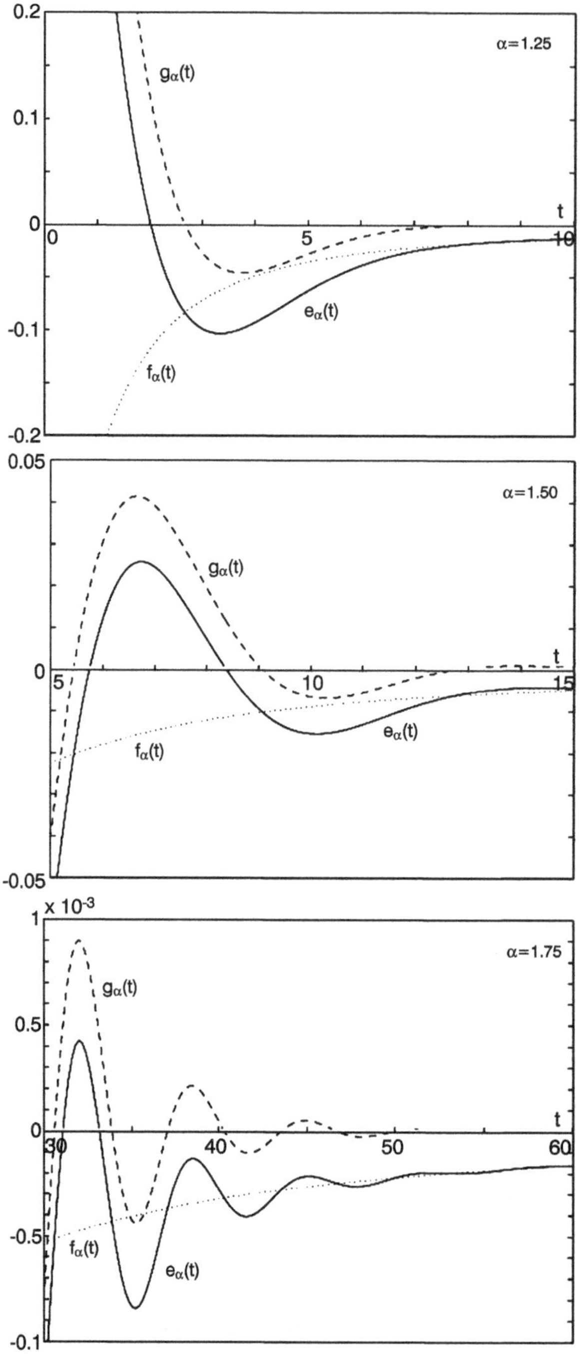

Figure 4: Decay of the *basic fundamental solution* $u_0(t) = e_\alpha(t)$ for $\alpha = 1.25, 1.50, 1.75$, from top to down, respectively. *Full* line = $e_\alpha(t)$, *dashed* line = $g_\alpha(t)$, *dotted* line = $f_\alpha(t)$.

When t is sufficiently large, the zeros are expected to be found approximately from the equation

$$\frac{2}{\alpha}\,\mathrm{e}^{t\cos(\pi/\alpha)} \approx \frac{t^{-\alpha}}{|\Gamma(1-\alpha)|}, \tag{2.17}$$

obtained from (2.15) by ignoring the oscillation factor of $g_\alpha(t)$, and taking the first term in the asymptotic expansion of $f_\alpha(t)$. As we have shown in our 1996 report [14], such approximate equation turns out to be useful when $\alpha \to 1^+$ and $\alpha \to 2^-$.

For $\alpha \to 1^+$, only one zero is present, which is expected to be very far from the origin, in view of the large period of the function $\cos[t\sin(\pi/\alpha)]$. In fact, since there is no zero for $\alpha = 1$, and by increasing α, more and more zeros arise, we are sure that only one zero exists for α to be sufficiently close to 1. Putting $\alpha = 1+\epsilon$, the asymptotic position T_* of this zero can be found from the relation (2.17), in the limit $\epsilon \to 0^+$. Assuming in this limit the first-order approximation, we get

$$T_* \sim \log\left(\frac{2}{\epsilon}\right), \tag{2.18}$$

which shows that T_* tends to infinity slower than $1/\epsilon$ as $\epsilon \to 0$.

For $\alpha \to 2^-$, there is an increasing number of zeros up to infinity, since $e_2(t) = \cos t$ has infinitely many zeros [in $t_n^* = (n+1/2)\pi$, $n = 0, 1, \dots$]. Putting now $\alpha = 2-\delta$, the asymptotic position T_* for the largest zero can be found again from (2.17) in the limit $\delta \to 0^+$. Assuming in this limit the first-order approximation, we get

$$T_* \sim \frac{12}{\pi\,\delta}\log\left(\frac{1}{\delta}\right). \tag{2.19}$$

For details, see [14].

Now, for $\delta \to 0^+$, the length of the positivity intervals of $g_\alpha(t)$ tends to π and, as long as $t \le T_*$, there are two zeros in each positivity interval. Hence, in the limit $\delta \to 0^+$, there is, in average, one zero per interval of length π. So, we expect that $N_* \sim T_*/\pi$.

Remark. For the above considerations we got inspiration from an interesting 1905 paper by Wiman [39], who—after having treated the Mittag-Leffler function in the complex plane—considered the position of the zeros of the function on the negative real axis (without providing any detail). Our expressions of T_* are in disagreement with those by Wiman for numerical factors; however, the results of our numerical studies carried out in our 1996 report [14] confirm and illustrate the validity of our analysis.

Here, we find it interesting to analyze the phenomenon of the transition of the (odd) number of zeros as $1.4 \le \alpha \le 1.8$. For this purpose, in Table 1, we report the intervals of amplitude $\Delta\alpha = 0.01$, where these transitions occur, and the location T_* (evaluated within a relative error of 0.1 %) of the largest zeros found at the two extreme

Table 1: The transition of the (odd) zeros.

N_*	α	T_*
$1 \div 3$	$1.40 \div 1.41$	$1.730 \div 5.726$
$3 \div 5$	$1.56 \div 1.57$	$8.366 \div 13.48$
$5 \div 7$	$1.64 \div 1.65$	$14.61 \div 20.00$
$7 \div 9$	$1.69 \div 1.70$	$20.80 \div 26.33$
$9 \div 11$	$1.72 \div 1.73$	$27.03 \div 32.83$
$11 \div 13$	$1.75 \div 1.76$	$33.11 \div 38.81$
$13 \div 15$	$1.78 \div 1.79$	$39.49 \div 45.51$
$15 \div 17$	$1.79 \div 1.80$	$45.51 \div 51.46$

N_* = number of zeros, α = fractional order, T_* location of the largest zero.

values of the above intervals. We recognize that the transition from 1 to 3 zeros occurs as $1.40 \le \alpha \le 1.41$; that transition from 3 to 5 zeros occurs as $1.56 \le \alpha \le 1.57$, and so on. The last transition is from 15 to 17 zeros, and it occurs as $1.79 \le \alpha \le 1.80$.

3 Composite fractional relaxation-oscillation

In this section, we shall consider the following fractional differential equations for $t \ge 0$, equipped with the necessary initial conditions,

$$\frac{du}{dt} + a\, D_*^\alpha\, u(t) + u(t) = q(t), \quad u(0^+) = c_0, \quad 0 < \alpha < 1, \tag{3.1}$$

$$\frac{d^2v}{dt^2} + a\, D_*^\alpha\, v(t) + v(t) = q(t), \quad v(0^+) = c_0, \quad v'(0^+) = c_1, \quad 0 < \alpha < 2, \tag{3.2}$$

where a is a positive constant. The unknown functions $u(t)$ and $v(t)$ (the field variables) are required to be sufficiently well-behaved to be treated with their derivatives $u'(t)$, $v'(t)$, $v''(t)$ by the technique of Laplace transform. The given function $q(t)$ is supposed to be continuous. In the above equations the fractional derivative of order α is assumed to be provided by the operator D_*^α and the *Caputo derivative*, in agreement with our choice followed in the previous section. Note that in (3.2) we must distinguish the cases: (a) $0 < \alpha < 1$, (b) $1 < \alpha < 2$, and $\alpha = 1$.

The Equations (3.1) and (3.2) will be referred to as the *composite fractional relaxation equation* and the *composite fractional oscillation equation*, respectively, to be distinguished from the corresponding *simple* fractional equations treated in Section 2.

The fractional differential equation in (3.1) with $\alpha = 1/2$ corresponds to the *Basset problem*, a classical problem in fluid dynamics concerning the unsteady motion of a particle accelerating in a viscous fluid under the action of the gravity; that will be dealt along with its fractional generalization in Section 4.

The fractional differential equation in (3.2) with $0 < \alpha < 2$ models an oscillation process with fractional damping term. It was formerly treated by Caputo [10], who provided a preliminary analysis by the Laplace transform. The special cases $\alpha = 1/2$ and $\alpha = 3/2$, but with the standard definition D^α for the fractional derivative, have been discussed by Bagley [3]. Recently, Beyer and Kempfle [6] discussed (3.2) for $-\infty < t < +\infty$ to investigate the uniqueness and *causality* of the solutions. As they let t running in all of $\mathbb{R}$, they used the Fourier transform and characterized the fractional derivative D^α by its properties in frequency space, thereby requiring that for noninteger α, the principal branch of $(i\omega)^\alpha$ should be taken. Under the global condition that the solution is square-summable, they showed that the system described by (3.2) is *causal if* $a > 0$.

Also here we shall apply the Laplace transform method to solve the fractional differential equations and get some insight into their *fundamental* and *impulse-response solutions*. However, in contrast with the previous section, we now find it more convenient to apply directly the well-known formula for the Laplace transform of fractional and integer derivatives, than reduce the equations with the prescribed initial conditions, such as equivalent (fractional) integral equations to be treated by the Laplace transform.

3.1 The composite fractional relaxation equation

Let us apply the Laplace transform to the fractional relaxation equation (3.1). We are led to the transformed algebraic equation

$$\widetilde{u}(s) = c_0 \frac{1 + a\, s^{\alpha-1}}{w_1(s)} + \frac{\widetilde{q}(s)}{w_1(s)}, \quad 0 < \alpha < 1, \tag{3.3}$$

where

$$w_1(s) := s + a\, s^\alpha + 1, \tag{3.4}$$

and $a > 0$. Putting

$$u_0(t) \div \widetilde{u}_0(s) := \frac{1 + a\, s^{\alpha-1}}{w_1(s)}, \quad u_\delta(t) \div \widetilde{u}_\delta(s) := \frac{1}{w_1(s)}, \tag{3.5}$$

and recognizing that

$$u_0(0^+) = \lim_{s\to\infty} s\, \widetilde{u}_0(s) = 1, \quad \widetilde{u}_\delta(s) = -\left[s\, \widetilde{u}_0(s) - 1\right], \tag{3.6}$$

we can conclude that

$$u(t) = c_0\, u_0(t) + \int_0^t q(t-\tau)\, u_\delta(\tau)\, d\tau, \quad u_\delta(t) = -\,u_0'(t). \tag{3.7}$$

We thus recognize that $u_0(t)$ and $u_\delta(t)$ are, respectively, the *fundamental solution* and *impulse-response solution* for Equation (3.1).

Let us first consider the problem to get $u_0(t)$ as the inverse Laplace transform of $\widetilde{u}_0(s)$. We easily see that the function $w_1(s)$ has no zero in the main sheet of the Riemann surface, including its boundaries on the cut (simply show that $\mathrm{Im}\{w_1(s)\}$ does not vanish if s is not a real positive number); so that the inversion of the Laplace transform $\widetilde{u}_0(s)$ can be carried out by deforming the original Bromwich path into the Hankel path $\mathrm{Ha}(\epsilon)$, introduced in the previous section, that is, into the loop constituted by a small circle $|s| = \epsilon$, with $\epsilon \to 0$ and by the two borders of the cut negative real axis. As a consequence, we write

$$u_0(t) = \frac{1}{2\pi i} \int\limits_{\mathrm{Ha}(\epsilon)} \mathrm{e}^{st} \frac{1 + as^{\alpha-1}}{s + a\, s^{\alpha} + 1}\, ds. \tag{3.8}$$

It is now an exercise in complex analysis to show that the contribution from the Hankel path $\mathrm{Ha}(\epsilon)$, as $\epsilon \to 0$, is provided by

$$u_0(t) = \int\limits_0^\infty \mathrm{e}^{-rt}\, H^{(1)}_{\alpha,0}(r; a)\, dr \tag{3.9}$$

with

$$\begin{aligned} H^{(1)}_{\alpha,0}(r; a) &= -\frac{1}{\pi} \,\mathrm{Im}\left\{ \frac{1 + as^{\alpha-1}}{w_1(s)} \bigg|_{s = r\,\mathrm{e}^{i\pi}} \right\} \\ &= \frac{1}{\pi} \frac{a\, r^{\alpha-1} \sin(\alpha\pi)}{(1-r)^2 + a^2 r^{2\alpha} + 2(1-r)\, a\, r^{\alpha} \cos(\alpha\pi)}. \end{aligned} \tag{3.10}$$

For $a > 0$ and $0 < \alpha < 1$, the function $H^{(1)}_{\alpha,0}(r; a)$ is positive for all $r > 0$, since it has the sign of the numerator. In fact, in (3.10) the denominator is strictly positive, being equal to $|w_1(s)|^2$ as $s = r\,\mathrm{e}^{\pm i\pi}$. Hence, the *fundamental solution* $u_0(t)$ has the peculiar property to be *completely monotone*, and $H^{(1)}_{\alpha,0}(r; a)$ is its *spectral function.*

Now the determination of $u_\delta(t) = -u_0'(t)$ is straightforward. We see that also the *impulse-response solution* $u_\delta(t)$ is *completely monotone*, since it can be represented by

$$u_\delta(t) = \int\limits_0^\infty \mathrm{e}^{-rt}\, H^{(1)}_{\alpha,-1}(r; a)\, dr \tag{3.11}$$

with the *spectral function*

$$H^{(1)}_{\alpha,-1}(r; a) = r\, H^{(1)}_{\alpha,0}(r; a) = \frac{1}{\pi} \frac{a\, r^{\alpha} \sin(\alpha\pi)}{(1-r)^2 + a^2 r^{2\alpha} + 2(1-r)\, a\, r^{\alpha} \cos(\alpha\pi)}. \tag{3.12}$$

We recognize that both the solutions $u_0(t)$ and $u_\delta(t)$ turn out to be strictly decreasing from 1 towards 0, as t runs from 0 to ∞. Their behavior as $t \to 0^+$ and $t \to \infty$ can be inspected by means of a proper asymptotic analysis.

The behavior of the solutions, as $t \to 0^+$, can be determined from the behavior of their Laplace transforms as $\mathrm{Re}\{s\} \to +\infty$, as well known from the theory of the Laplace transform; see, for example, [11]. We obtain as $\mathrm{Re}\{s\} \to +\infty$,

$$\widetilde{u}_0(s) = s^{-1} - s^{-2} + O(s^{-3+\alpha}), \quad \widetilde{u}_\delta(s) = s^{-1} - a\, s^{-(2-\alpha)} + O(s^{-2}), \tag{3.13}$$

so that, as $t \to 0^+$,

$$u_0(t) = 1 - t + O(t^{2-\alpha}), \quad u_\delta(t) = 1 - a\,\frac{t^{1-\alpha}}{\Gamma(2-\alpha)} + O(t). \tag{3.14}$$

The spectral representations (3.9) and (3.11) are suitable to obtain the asymptotic behavior of $u_0(t)$ and $u_\delta(t)$ as $t \to +\infty$ by using the Watson lemma. In fact, expanding the spectral functions for small r and taking the dominant term in the corresponding asymptotic series, we obtain

$$u_0(t) \sim a\,\frac{t^{-\alpha}}{\Gamma(1-\alpha)}, \quad u_\delta(t) \sim -a\,\frac{t^{-\alpha-1}}{\Gamma(-\alpha)}, \quad \text{as } t \to \infty. \tag{3.15}$$

We note that the limiting case $\alpha = 1$ can be easily treated, extending the validity of Equations (3.3)–(3.7) to $\alpha = 1$, as it is legitimate. In this case we obtain

$$u_0(t) = \mathrm{e}^{-t/(1+a)}, \quad u_\delta(t) = \frac{1}{1+a}\,\mathrm{e}^{-t/(1+a)}, \quad \alpha = 1. \tag{3.16}$$

Of course, in the case $a \equiv 0$, we recover the standard solutions $u_0(t) = u_\delta(t) = \mathrm{e}^{-t}$.

We conclude this sub-section with some considerations regarding the solutions when the order α is just a rational number. If we take $\alpha = p/q$, where $p, q \in \mathbb{N}$ are assumed (for convenience) to be relatively prime, a factorization in (3.4) is possible, by using the procedure indicated by Miller and Ross [32]. In these cases the solutions can be expressed in terms of a linear combination of q Mittag-Leffler functions of fractional order $1/q$, which, in turn, can be expressed in terms of incomplete gamma functions; see for example, the treatise by Gorenflo et al. [13].

Here, we shall illustrate the factorization in the simplest case $\alpha = 1/2$, and provide the solutions $u_0(t)$ and $u_\delta(t)$ in terms of the functions $e_\alpha(t;\lambda)$ (with $\alpha = 1/2$), introduced in the previous section. In this case, in view of the application to the *Basset problem* (see [23]), Equation (3.1) deserves particular attention. For $\alpha = 1/2$, we can write

$$w_1(s) = s + a\,s^{1/2} + 1 = (s^{1/2} - \lambda_+)\,(s^{1/2} - \lambda_-), \quad \lambda_\pm = -a/2 \pm (a^2/4 - 1)^{1/2}. \tag{3.17}$$

Here, $\lambda_\pm$ denote the two roots (real or conjugate complex) of the second-degree polynomial with positive coefficients $z^2 + az + 1$, which, in particular, satisfy the following binary relations

$$\lambda_+ \cdot \lambda_- = 1, \quad \lambda_+ + \lambda_- = -a, \quad \lambda_+ - \lambda_- = 2(a^2/4 - 1)^{1/2} = (a^2 - 4)^{1/2}. \tag{3.18}$$

We recognize that we must treat separately the following two cases:

$$\text{i)}\quad 0 < a < 2, \quad \text{or} \quad a > 2, \quad \text{and} \quad \text{ii)} \quad a = 2,$$

which correspond to two distinct roots ($\lambda_+ \neq \lambda_-$), or two coincident roots ($\lambda_+ \equiv \lambda_- = -1$), respectively. For this purpose, using the notation introduced in [23], we write

$$\widetilde{M}(s) := \frac{1 + a\, s^{-1/2}}{s + a\, s^{1/2} + 1} = \begin{cases} \text{i)} & \frac{A_-}{s^{1/2}\,(s^{1/2}-\lambda_+)} + \frac{A_+}{s^{1/2}(s^{1/2}-\lambda_-)}, \\ \text{ii)} & \frac{1}{(s^{1/2}+1)^2} + \frac{2}{s^{1/2}(s^{1/2}+1)^2}, \end{cases} \tag{3.19}$$

and

$$\widetilde{N}(s) := \frac{1}{s + a\, s^{1/2} + 1} = \begin{cases} \text{i)} & \frac{A_+}{s^{1/2}\,(s^{1/2}-\lambda_+)} + \frac{A_-}{s^{1/2}(s^{1/2}-\lambda_-)}, \\ \text{ii)} & \frac{1}{(s^{1/2}+1)^2}, \end{cases} \tag{3.20}$$

where

$$A_\pm = \pm \frac{\lambda_\pm}{\lambda_+ - \lambda_-}. \tag{3.21}$$

Using (3.18), we note that

$$A_+ + A_- = 1, \quad A_+\lambda_- + A_-\lambda_+ = 0, \quad A_+\lambda_+ + A_-\lambda_- = -a. \tag{3.22}$$

Recalling known Laplace transforms with the Mittag-Leffler function of order 1/2, we obtain

$$u_0(t) = M(t) := \begin{cases} \text{i)} & A_-\, E_{1/2}(\lambda_+ \sqrt{t}) + A_+\, E_{1/2}\,(\lambda_- \sqrt{t}), \\ \text{ii)} & (1 - 2t)\, E_{1/2}\,(-\sqrt{t}) + 2\,\sqrt{t/\pi}, \end{cases} \tag{3.23}$$

and

$$u_\delta(t) = N(t) := \begin{cases} \text{i)} & A_+\, E_{1/2}(\lambda_+ \sqrt{t}) + A_-\, E_{1/2}(\lambda_- \sqrt{t}), \\ \text{ii)} & (1 + 2t)\, E_{1/2}(-\sqrt{t}) - 2\,\sqrt{t/\pi}. \end{cases} \tag{3.24}$$

We thus recognize in (3.23)–(3.24) the presence of the functions $e_{1/2}(t; -\lambda_\pm) = E_{1/2}(\lambda_\pm \sqrt{t})$ and $e_{1/2}(t) = e_{1/2}(t; 1) = E_{1/2}(-\sqrt{t})$.

In particular, the solution of the *Basset problem* can be easily obtained from (3.7), with $q(t) = q_0$, by using (3.23)–(3.24) and noting that $\int_0^t N(\tau)\, d\tau = 1 - M(t)$. Denoting this solution by $u_B(t)$, we get

$$u_B(t) = q_0 - (q_0 - c_0)\, M(t). \tag{3.25}$$

When $a \equiv 0$, that is, in the absence of a term containing the fractional derivative (due to the Basset force), we recover the classical Stokes solution, which we denote by $u_S(t)$,

$$u_S(t) = q_0 - (q_0 - c_0)\, \mathrm{e}^{-t}.$$

In the particular case $q_0 = c_0$, we get the steady-state solution $u_B(t) = u_S(t) \equiv q_0$.

For vanishing initial condition $c_0 = 0$, we have the creep-like solutions

$$u_B(t) = q_0\,[1 - M(t)], \quad u_S(t) = q_0\,[1 - \mathrm{e}^{-t}],$$

which we compare in the normalized plots of Figure 5 of [23]. In this case, it is instructive to compare the behaviors of the two solutions as $t \to 0^+$ and $t \to \infty$. Recalling the general asymptotic expressions of $u_0(t) = M(t)$ in (3.14) and (3.15) with $\alpha = 1/2$, we recognize that

$$u_B(t) = q_0\,[t + O(t^{3/2})], \quad u_S(t) = q_0\,[t + O(t^2)], \quad \text{as } t \to 0^+,$$

and

$$u_B(t) \sim q_0\,[1 - a/\sqrt{\pi t}\,], \quad u_S(t) \sim q_0\,[1 - EST\,], \quad \text{as } t \to \infty,$$

where *EST* denotes *exponentially small terms*. In particular, we note that the normalized plot of $u_B(t)/q_0$ remains under that of $u_S(t)/q_0$ as t runs from 0 to ∞.

The reader is invited to convince himself of the fact that follows. In the general case $0 < \alpha < 1$, the solution $u(t)$ has the particular property of being equal to 1 for all $t \geq 0$ if $q(t)$ has this property and $u(0^+) = 1$, whereas $q(t) = 1$ for all $t \geq 0$, and $u(0^+) = 0$ implies that $u(t)$ is a creep function tending to 1 as $t \to \infty$.

3.2 The composite fractional oscillation equation

Let us now apply the Laplace transform to the fractional oscillation Equation (3.2). We are led to the transformed algebraic equations

$$\text{(a)}\quad \widetilde{v}(s) = c_0\frac{s + a\,s^{\alpha-1}}{w_2(s)} + c_1\frac{1}{w_2(s)} + \frac{\widetilde{q}(s)}{w_2(s)}, \quad 0 < \alpha < 1, \tag{3.26a}$$

or

$$\text{(b)}\quad \widetilde{v}(s) = c_0\frac{s + a\,s^{\alpha-1}}{w_2(s)} + c_1\frac{1 + a\,s^{\alpha-2}}{w_2(s)} + \frac{\widetilde{q}(s)}{w_2(s)}, \quad 1 < \alpha < 2, \tag{3.26b}$$

where

$$w_2(s) := s^2 + a\,s^\alpha + 1, \tag{3.27}$$

and $a > 0$. Putting

$$\widetilde{v}_0(s) := \frac{s + a\,s^{\alpha-1}}{w_2(s)}, \quad 0 < \alpha < 2, \tag{3.28}$$

we recognize that

$$v_0(0^+) = \lim_{s\to\infty} s\,\widetilde{v}_0(s) = 1, \quad \frac{1}{w_2(s)} = -\,[s\,\widetilde{v}_0(s) - 1] \div -v_0'(t), \tag{3.29}$$

and

$$\frac{1 + a\,s^{\alpha-2}}{w_2(s)} = \frac{\widetilde{v}_0(s)}{s} \div \int_0^t v_0(\tau)\,d\tau. \tag{3.30}$$

Thus, we can conclude that

$$\text{(a)} \quad v(t) = c_0\,v_0(t) - c_1\,v_0'(t) - \int_0^t q(t-\tau)\,v_0'(\tau)\,d\tau, \quad 0 < \alpha < 1, \tag{3.31a}$$

or

$$\text{(b)} \quad v(t) = c_0\,v_0(t) + c_1 \int_0^t v_0(\tau)\,d\tau - \int_0^t q(t-\tau)\,v_0'(\tau)\,d\tau, \quad 1 < \alpha < 2. \tag{3.31b}$$

In both of the above equations, the term $-v_0'(t)$ represents the *impulse-response solution* $v_\delta(t)$ for the *composite fractional oscillation Equation* (3.2), namely, the particular solution of the inhomogeneous equation with $c_0 = c_1 = 0$ and with $q(t) = \delta(t)$. For the *fundamental solutions* of (3.2), we recognize from Equation (3.31a)–(3.31b) that we have two distinct couples of solutions according to the case (a) and (b), which read:

$$\text{(a)} \quad \{v_0(t),\, v_{1\,\mathrm{a}}(t) = -v_0'(t)\}, \quad \text{(b)} \quad \left\{v_0(t),\, v_{1\,\mathrm{b}}(t) = \int_0^t v_0(\tau)\,d\tau\right\}. \tag{3.32}$$

We first consider the particular case $\alpha = 1$, for which the fundamental and impulse response solutions are known in terms of elementary functions. This limiting case can also be treated by extending the validity of Equations (3.31a) and (3.31b) to $\alpha = 1$, as it is legitimate. From

$$v_0(s) = \frac{s + a}{s^2 + a\,s + 1} = \frac{s + a/2}{(s + a/2)^2 + (1 - a^2/4)} - \frac{a/2}{(s + a/2)^2 + (1 - a^2/4)}, \tag{3.33}$$

we obtain the *basic fundamental solution*

$$v_0(t) = \begin{cases} \mathrm{e}^{-at/2}\,[\cos(\omega t) + \frac{a}{2\omega}\,\sin(\omega t)] & \text{if } 0 < a < 2, \\ \mathrm{e}^{-t}\,(1 - t) & \text{if } a = 2, \\ \mathrm{e}^{-at/2}\,[\cosh(\chi t) + \frac{a}{2\chi}\,\sinh(\chi t)] & \text{if } a > 2, \end{cases} \tag{3.34}$$

where

$$\omega = \sqrt{1 - a^2/4}, \quad \chi = \sqrt{a^2/4 - 1}. \tag{3.35}$$

By a differentiation of (3.34), we easily obtain the *second fundamental solution* $v_{1\,a}(t)$ and the *impulse-response solution* $v_\delta(t)$, since $v_{1\,\mathrm{a}}(t) = v_\delta(t) = -v_0'(t)$. We point out that all the solutions exhibit an *exponential decay* as $t \to \infty$.

Let us now consider the problem to get $v_0(t)$ as the inverse Laplace transform of $\widetilde{v}_0(s)$, as given by (3.26a), (3.26b)–(3.27),

$$v_0(t) = \frac{1}{2\pi i} \int_{\mathrm{Br}} \mathrm{e}^{st} \, \frac{s + a\, s^{\alpha-1}}{w_2(s)} \, ds, \tag{3.36}$$

where Br denotes the usual Bromwich path. Using a result by Beyer and Kempfle [6], we know that the function $w_2(s)$ (for $a > 0$ and $0 < \alpha < 2$, $\alpha \neq 1$) has exactly *two simple, conjugate complex zeros on the principal branch in the open left half-plane*, cut along the negative real axis, say $s_+ = \rho\, \mathrm{e}^{+i\gamma}$ and $s_- = \rho\, \mathrm{e}^{-i\gamma}$, with $\rho > 0$ and $\pi/2 < \gamma < \pi$. This enables us to repeat the considerations carried out for the simple fractional oscillation equation to decompose the basic fundamental solution $v_0(t)$ into two parts, according to $v_0(t) = f_\alpha(t; a) + g_\alpha(t; a)$. In fact, the evaluation of the Bromwich integral (3.36) can be achieved by adding the contribution $f_\alpha(t; a)$ from the Hankel path $\mathrm{Ha}(\epsilon)$ as $\epsilon \to 0$ to the residual contribution $g_\alpha(t; a)$, from the two poles $s_\pm$.

As an exercise in complex analysis, we obtain

$$f_\alpha(t; a) = \int_0^\infty \mathrm{e}^{-rt} \, H^{(2)}_{\alpha,0}(r; a)\, dr \tag{3.37}$$

with *spectral function*

$$\begin{aligned} H^{(2)}_{\alpha,0}(r; a) &= -\frac{1}{\pi} \,\mathrm{Im} \left\{ \left. \frac{s + a s^{\alpha-1}}{w_2(s)} \right|_{s = r\,\mathrm{e}^{i\pi}} \right\} \\ &= \frac{1}{\pi} \, \frac{a\, r^{\alpha-1} \, \sin(\alpha\pi)}{(r^2+1)^2 + a^2 r^{2\alpha} + 2\,(r^2+1)\, a\, r^\alpha \, \cos(\alpha\pi)} . \end{aligned} \tag{3.38}$$

Since in (3.38) the denominator is strictly positive, being equal to $|w_2(s)|^2$, as $s = r\,\mathrm{e}^{\pm i\pi}$, the *spectral function* $H^{(2)}_{\alpha,0}(r; a)$ turns out to be positive for all $r > 0$ for $0 < \alpha < 1$ and negative for all $r > 0$ for $1 < \alpha < 2$. Hence, in case (a), the function $f_\alpha(t)$ and, in case (b), the function $-f_\alpha(t)$ is *completely monotone*. In both cases, $f_\alpha(t)$ tends to zero as $t \to \infty$, from above in case (a) and from below in case (b), according to the asymptotic behavior

$$f_\alpha(t; a) \sim a \, \frac{t^{-\alpha}}{\Gamma(1-\alpha)} \quad \text{as } t \to \infty, \quad 0 < \alpha < 1, \quad 1 < \alpha < 2, \tag{3.39}$$

as derived by applying the Watson lemma in (3.37) and considering (3.38).

The other part, $g_\alpha(t; a)$, is obtained as

$$\begin{cases} g_\alpha(t; a) = \mathrm{e}^{s_+ t} \, \mathrm{Res} \left[\frac{s + a\, s^{\alpha-1}}{w_2(s)} \right]_{s_+} + \quad \text{conjugate complex} \\ \qquad\qquad = 2\, \mathrm{Re} \{ \frac{s_+ + a\, s_+^{\alpha-1}}{2 s_+ + a\, \alpha\, s_+^{\alpha-1}} \, \mathrm{e}^{s_+ t} \}. \end{cases} \tag{3.40}$$

Thus, this term exhibits an *oscillatory* character with exponentially decreasing amplitude like $\exp(-\rho\, t \,|\cos\gamma|)$.

Then, we recognize that the basic fundamental solution $v_0(t)$ exhibits a *finite* number of zeros and that for sufficiently large t it turns out to be permanently positive if $0 < \alpha < 1$, and permanently negative if $1 < \alpha < 2$, with an *algebraic decay* provided by (3.39).

For the second fundamental solutions $v_{1\,\mathrm{a}}(t)$, $v_{1\,\mathrm{b}}(t)$ and for the impulse-response solution $v_\delta(t)$, the corresponding analysis is straightforward in view of their connection with $v_0(t)$, pointed out in (3.31a), (3.31b)–(3.32). The *algebraic decay* of all the solutions, as $t \to \infty$, for $0 < \alpha < 1$ and $1 < \alpha < 2$, is henceforth resumed in the relations

$$v_0(t) \sim a\,\frac{t^{-\alpha}}{\Gamma(1-\alpha)}, \quad v_{1\,\mathrm{a}}(t) = v_\delta(t) \sim -a\,\frac{t^{-\alpha-1}}{\Gamma(-\alpha)}, \quad v_{1\,\mathrm{b}}(t) \sim a\,\frac{t^{1-\alpha}}{\Gamma(2-\alpha)}. \tag{3.41}$$

In conclusion, except in the particular case $\alpha = 1$, all the present solutions of the composite fractional oscillation equation exhibit similar characteristics with the corresponding solutions of the simple fractional oscillation equation, namely a *finite number of damped oscillations*, followed by a *monotonic algebraic decay* as $t \to \infty$.

4 Application: the Basset problem via fractional calculus

The dynamics of a sphere immersed in an incompressible viscous fluid represents a classical problem, which has many applications in flows of geophysical and engineering interest. Usually, the low Reynolds number limit (slow motion approximation) is assumed, so that the Navier–Stokes equations describing the fluid motion may be linearized.

The particular, but relevant, situation of a sphere subjected to gravity was first considered independently by Boussinesq [9] in 1885 and by Basset [4] in 1888, who introduced a special hydrodynamic force, related to the history of the relative acceleration of the sphere, which is nowadays referred to as *Basset force*. The relevance of these studies was in that, up to then, only steady motions or small oscillations of bodies in a viscous liquid had been considered, starting from Stokes' celebrated memoir on pendulums [37], in 1851. The subject matter was considered with more details in 1907 by Picciati [35] and Boggio [8], in some notes presented by the great Italian scientist, Levi-Civita. The whole was summarized by Basset himself in a later paper [5], and, in more recent times, by Hughes and Gilliand [18].

Nowadays the dynamics of impurities in unsteady flows is quite relevant, as shown by several publications, whose aim is to provide more general expressions for the hydrodynamic forces, including the Basset force, in order to fit experimental data and numerical simulations; see, for example, [19–21, 29–31, 33, 34, 36].

In the following, we shall recall the general equation of motion for a spherical particle in a viscous fluid, pointing out the different force contributions due to effects of inertia, viscous drag, and buoyance. In particular, the so-called Basset force will be interpreted in terms of a fractional derivative of order 1/2 of the particle velocity, relative to the fluid. Based on our works [27, 28], we shall introduce the *generalized Basset force*, which is expressed in terms of a fractional derivative of any order α, ranging in the interval $0 < \alpha < 1$. This generalization, suggested by a mathematical speculation, is expected to provide a phenomenological insight for the experimental data.

We shall consider the simplified problem, originally investigated by Basset, where the fluid is quiescent and the particle moves under the action of gravity, starting at $t = 0$ with a certain vertical velocity. For the sake of generality, we prefer to consider the problem with the generalized Basset force and will provide the solution for the particle velocity in terms of Mittag-Leffler-type functions. The most evident effect of this generalization will be to modify the long-term behavior of the solution, changing its algebraic decay from $t^{-1/2}$ to $t^{-\alpha}$. This effect can be of some interest for a better fit of experimental data.

4.1 The equation of motion for the Basset problem

Let us consider a small rigid sphere of radius r_0, mass m_p, density ρ_p, initially centered in $\mathbf{X}(t)$ and moving with velocity $\mathbf{V}(t)$ in a homogeneous fluid, of density ρ_f and kinematic viscosity ν, characterized by a flow field $\mathbf{u}(\mathbf{x}, t)$. In general, the equation of motion is required to take into account effects due to inertia, viscous drag, and buoyance, so it can be written as

$$m_p \frac{d\mathbf{V}}{dt} = \mathbf{F}_i + \mathbf{F}_d + \mathbf{F}_g, \tag{4.1}$$

where the forces on the R.H.S. correspond, in turn, to the above effects. According to Maxey and Riley [29], these forces read (adopting our notation):

$$\mathbf{F}_i = m_f \left.\frac{D\mathbf{u}}{Dt}\right|_{\mathbf{X}(t)} - \frac{1}{2} m_f \left(\frac{d\mathbf{V}}{dt} - \left.\frac{D\mathbf{u}}{Dt}\right|_{\mathbf{X}(t)} \right), \tag{4.2}$$

$$\mathbf{F}_d = -\frac{1}{\mu} \left\{ [\mathbf{V}(t) - \mathbf{u}(\mathbf{X}(t), t)] + \sqrt{\frac{\tau_0}{\pi}} \int_{-\infty}^{t} \frac{d[\mathbf{V}(\tau) - \mathbf{u}(\mathbf{X}(\tau), \tau)]/d\tau}{\sqrt{t-\tau}} \, d\tau \right\} \tag{4.3}$$

$$\mathbf{F}_g = (m_p - m_f)\,\mathbf{g}, \tag{4.4}$$

where $m_f = (4/3)\pi r_0^3 \rho_f$ denotes the mass of the fluid displaced by the spherical particle, and

$$\tau_0 := \frac{r_0^2}{\nu}, \tag{4.5}$$

$$\frac{1}{\mu} := 6\pi r_0 \nu \rho_f = \frac{9}{2} m_f \tau_0^{-1}. \tag{4.6}$$

The time constant τ_0 represents a sort of time scale induced by viscosity, whereas the constant μ is usually referred to as the *mobility coefficient*.

In (4.2), we note two different time derivatives, D/Dt, d/dt, which represent the time derivatives following a fluid element and the moving sphere, respectively. So

$$\left.\frac{D\mathbf{u}}{Dt}\right|_{\mathbf{X}(t)} = \left[\frac{\partial \mathbf{u}}{\partial t} + (\mathbf{u}\cdot\nabla)\,\mathbf{u}(\mathbf{x},t)\right], \quad \frac{d}{dt}\mathbf{u}[\mathbf{X}(t),t] = \left[\frac{\partial \mathbf{u}}{\partial t} + (\mathbf{V}\cdot\nabla)\,\mathbf{u}(\mathbf{x},t)\right],$$

where the brackets are computed at $\mathbf{x} = \mathbf{X}(t)$.

The terms on the R.H.S. of (4.2) correspond, in turn, to the effects of pressure gradient of the undisturbed flow and of added mass, whereas those of (4.3) represent, respectively, the well-known viscous Stokes drag, that we shall denote by $\mathbf{F}_S$, and to the augmented viscous Basset drag, denoted by $\mathbf{F}_B$. Using the characteristic time τ_0, the Stokes and Basset forces read, respectively:

$$F_S = -\frac{9}{2} m_f \tau_0^{-1}\,[\mathbf{V}(t) - \mathbf{u}(\mathbf{X}(t),t)], \tag{4.7}$$

$$F_B = -\frac{9}{2} m_f \tau_0^{-1/2} \left\{ \frac{1}{\sqrt{\pi}} \int_{-\infty}^{t} \frac{d[\mathbf{V}(\tau) - \mathbf{u}(\mathbf{X}(\tau),\tau)]/d\tau}{\sqrt{t-\tau}}\, d\tau \right\}. \tag{4.8}$$

We thus recognize that the time constant τ_0 provides the natural time scale for the diffusive processes, related to the fluid viscosity, and that the integral expression in brackets at the right-hand side of (4.8) only represents the *Caputo fractional derivative* of order 1/2, with starting point $-\infty$, of the particle velocity relative to the fluid.

Presumably, the first scientist who has pointed out the relationship between the Basset force and fractional calculus has been Tatom [38] in 1988. However, Tatom has limited himself to this achievement, and has not treated any related problem by the methods of fractional calculus.

We now introduce the *generalized Basset force* by the definition

$$F_B^\alpha = -\frac{9}{2} m_f \tau_0^{\alpha-1} \frac{d^\alpha}{dt^\alpha}[\mathbf{V}(t) - \mathbf{u}(\mathbf{X}(t),t)], \quad 0 < \alpha < 1, \tag{4.9}$$

where the fractional derivative of order α is in Caputo's sense.

Introducing the so-called *effective mass*

$$m_e := m_p + \frac{1}{2} m_f \tag{4.10}$$

and allowing for the *generalized Basset force* in (4.3), we can rewrite the equation of motion (4.1)–(4.4) in a more compact and significant form

$$m_e \frac{d\mathbf{V}}{dt} = \frac{3}{2} m_f \frac{D\mathbf{u}}{Dt} - \frac{9}{2} m_f \left[\frac{1}{\tau_0} + \frac{1}{\tau_0^{1-\alpha}} \frac{d^\alpha}{dt^\alpha}\right] (\mathbf{V} - \mathbf{u}) + (m_p - m_f)\,\mathbf{g}, \tag{4.11}$$

which we refer to as the *generalized equation of motion*. Of course, if in (4.11) we put $\alpha = 1/2$, then we recover the *basic equation of motion* with the original Basset force.

4.2 The generalized Basset problem

Let us now assume that the fluid is quiescent, namely $\mathbf{u}(\mathbf{x}, t) = 0$, $\forall\, \mathbf{x}, t$, and that the particle starts to move under the action of gravity, from a given instant $t_0 = 0$, with a certain velocity $V(0^+) = V_0$, in the vertical direction. This was the problem considered by Basset [4] and first solved by Boggio [8]—in a cumbersome way—in terms of Gauss and Fresnel integrals.

Introducing the nondimensional quantities (related to the densities ρ_f and ρ_p of the fluid and particle),

$$\chi := \frac{\rho_p}{\rho_f}, \quad \beta := \frac{9\rho_f}{2\rho_p + \rho_f} = \frac{9}{1 + 2\chi}, \tag{4.12}$$

we find it convenient to define a new characteristic time

$$\sigma_e := \mu\, m_e = \tau_0/\beta\,; \tag{4.13}$$

see (4.5), (4.10), (4.12), and a characteristic velocity (related to the gravity)

$$V_S = (2/9)\,(\chi - 1)\, g\, \tau_0. \tag{4.14}$$

Then, we can eliminate the mass factors and the gravity acceleration in (4.11), and obtain the *equation of motion* in the form

$$\frac{dV}{dt} = -\frac{1}{\sigma_e}\left[1 + {\tau_0}^{\alpha}\, \frac{d^\alpha}{dt^\alpha}\right] \mathbf{V} + \frac{1}{\sigma_e}\, V_S. \tag{4.15}$$

If the Basset term were absent, we would obtain the classical Stokes solution

$$V(t) = V_S + (V_0 - V_S)\, \mathrm{e}^{-t/\sigma_e}, \tag{4.16}$$

where σ_e represents the characteristic time of the motion, and V_S, the final value assumed by the velocity. Later, we shall show that in the presence of the Basset term, the same final value is still attained by the solution $V(t)$, but with an algebraic rate, which is much slower than the exponential one found in (4.16).

To investigate the effect of the (generalized) Basset term, we compare the exact solution of (4.15) with the Stokes solution (4.16); for this aim, we found it convenient to scale times and velocities in (4.15) with $\{\sigma_e, V_S\}$, that is, to refer to the nondimensional quantities $t' = t/\sigma_e$, $V' = V/V_S$, $V_0' = V_0/V_S$. The resulting equation of motion reads (suppressing the apices):

$$\left[\frac{d}{dt} + a\,\frac{d^\alpha}{dt^\alpha} + 1\right] V(t) = 1, \quad V(0^+) = V_0, \quad a = \beta^\alpha > 0, \quad 0 < \alpha < 1. \tag{4.17}$$

This is the *composite fractional relaxation equation* treated in Section 3.1 by using the Laplace transform method. Recalling that in an obvious notation, we have

$$V(t) \div \widetilde{V}(s), \quad \frac{d^\alpha}{dt^\alpha} V(t) \div s^\alpha\, \widetilde{V}(s) - s^{\alpha-1}\, V_0, \quad 0 < \alpha \le 1, \tag{4.18}$$

the transformed solution of (4.17) reads:

$$\widetilde{V}(s) = \widetilde{M}(s)\, V_0 + \frac{1}{s}\, \widetilde{N}(s), \tag{4.19}$$

where

$$\widetilde{M}(s) = \frac{1 + a\, s^{\alpha-1}}{s + a\, s^{\alpha} + 1}, \quad \widetilde{N}(s) = \frac{1}{s + a\, s^{\alpha} + 1}. \tag{4.20}$$

Noting that

$$\frac{1}{s}\, \widetilde{N}(s) = \frac{1}{s} - \widetilde{M}(s) \div \int_0^t N(\tau)\, d\tau = 1 - M(t) \iff N(t) = -M'(t), \tag{4.21}$$

the actual solution of (4.17) turns out to be

$$V(t) = 1 + (V_0 - 1)\, M(t), \tag{4.22}$$

which is "similar" to the Stokes solution (4.16) if we consider the substitution of e^{-t} with the function $M(t)$.

In [27, 28] Mainardi, Pironi, and Tampieri have used a factorization method to invert $\widetilde{N}(s)$, and henceforth $\widetilde{M}(s)$, using a procedure indicated by Miller and Ross [32], which is valid when α is a rational number, say $\alpha = p/q$, where $p, q \in \mathbb{N}$, $p < q$. In this way, the actual solution can be finally expressed as a linear combination of certain incomplete gamma functions. This algebraic method is of course convenient for the ordinary Basset problem ($\alpha = 1/2$), but becomes cumbersome for $q > 2$.

Here, following the analysis in [15], we prefer to adopt the general method of inversion, based on the complex Bromwich formula. In this way, we are free from the restriction of α being a rational number and, furthermore, we are able to provide an integral representation of the solution convenient for numerical computation, which allows us to recognize the monotonicity properties of the solution, without need of plotting.

We now resume the relevant results from [15] using the present notation. The integral representation for $M(t)$ turns out to be

$$M(t) = \int_0^\infty e^{-rt} K(r)\, dr, \tag{4.23}$$

where

$$K(r) = \frac{1}{\pi}\, \frac{a\, r^{\alpha-1} \sin(\alpha\pi)}{(1-r)^2 + a^2 r^{2\alpha} + 2(1-r)\, a\, r^{\alpha} \cos(\alpha\pi)} > 0. \tag{4.24}$$

Thus, $M(t)$ is a *completely monotone* function (with spectrum $K(r)$), which is decreasing from 1 towards 0 as t runs from 0 to ∞. The behavior of $M(t)$ as $t \to 0^+$ and $t \to \infty$ can be inspected by means of a proper asymptotic analysis, as indicated below.

The behavior as $t \to 0^+$ can be determined from the behavior of the Laplace transform $\widetilde{M}(s) = s^{-1} - s^{-2} + O(s^{-3+\alpha})$ as $\mathrm{Re}\{s\} \to +\infty$. We obtain

$$M(t) = 1 - t + O(t^{2-\alpha}) \quad \text{as } t \to 0^+. \tag{4.25}$$

The spectral representation (4.23)–(4.24) is suitable to obtain the asymptotic behavior of $M(t)$ as $t \to +\infty$ by using the Watson lemma. In fact, expanding the spectrum $K(r)$ for small r and taking the dominant term in the corresponding asymptotic series, we obtain

$$M(t) \sim a\,\frac{t^{-\alpha}}{\Gamma(1-\alpha)} = a\,\frac{\sin(\alpha\pi)}{\pi}\int_0^\infty \mathrm{e}^{-rt}\, r^{\alpha-1}\,dr \quad \text{as } t \to \infty. \tag{4.26}$$

Furthermore, we recognize that $1 > M(t) > \mathrm{e}^{-t} > 0$, $0 < t < \infty$, namely, the decreasing plot of $M(t)$ remains above that of the exponential, as t runs from 0 to ∞. Although the two functions tend monotonically to 0, the difference between the two plots increases with t: at the initial point $t = 0$, both curves assume the unitary value and decrease with the same initial rate, but as $t \to \infty$, they exhibit very different decays, algebraic (slow) against exponential (fast).

For the ordinary Basset problem, it is convenient to report the result obtained by the factorization method [27, 28]. In this case, we must note that $a = \sqrt{\beta}$ (see (4.17)) ranges from 0 to 3, since from (4.12) we recognize that β runs from 0 ($\chi = \infty$, infinitely heavy particle) to 9 ($\chi = 0$, infinitely light particle).

The actual solution is obtained by expanding $\widetilde{M}(s)$ into partial fractions and then inverting. Considering the two roots $\lambda_\pm$ of the polynomial $P(z) \equiv z^2 + az + 1$, with $z = s^{1/2}$, we must treat separately the following two cases:

$$\text{i)}\quad 0 < a < 2, \quad \text{or} \quad 2 < a < 3, \quad \text{and} \quad \text{ii)} \quad a = 2,$$

which correspond to two distinct roots ($\lambda_+ \neq \lambda_-$) or two coincident roots ($\lambda_+ \equiv \lambda_- = -1$), respectively. We obtain

i) $a \neq 2 \Longleftrightarrow \beta \neq 4,\ \chi \neq 5/8$,

$$\widetilde{M}(s) = \frac{1 + a\,s^{-1/2}}{s + a\,s^{1/2} + 1} = \frac{A_-}{s^{1/2}\,(s^{1/2} - \lambda_+)} + \frac{A_+}{s^{1/2}\,(s^{1/2} - \lambda_-)} \tag{4.27}$$

with

$$\lambda_\pm = \frac{-a \pm (a^2 - 4)^{1/2}}{2} = \frac{1}{\lambda_\mp}, \quad A_\pm = \pm\frac{\lambda_\pm}{\lambda_+ - \lambda_-}; \tag{4.28}$$

ii) $a = 2 \Longleftrightarrow \beta = 4,\ \chi = 5/8$,

$$\widetilde{M}(s) = \frac{1 + 2s^{-1/2}}{s + 2s^{1/2} + 1} = \frac{1}{(s^{1/2} + 1)^2} + \frac{2}{s^{1/2}\,(s^{1/2} + 1)^2}. \tag{4.29}$$

The Laplace inversion of (4.27)–(4.29) can be expressed in terms of Mittag-Leffler functions of order 1/2, $E_{1/2}(\lambda\sqrt{t}) = \exp(\lambda^2 t)\ \mathrm{erfc}(-\lambda\sqrt{t})$, as shown in the Appendix of [15]. We obtain

$$M(t) = \begin{cases} \text{i)} & A_-\, E_{1/2}\,(\lambda_+\,\sqrt{t}) + A_+\, E_{1/2}\,(\lambda_-\,\sqrt{t}), \\ \text{ii)} & (1-2t)\, E_{1/2}\,(-\sqrt{t}) + 2\,\sqrt{t/\pi}. \end{cases} \tag{4.30}$$

We recall that the analytical solution to the classical Basset problem was formerly provided by Boggio [8] in 1907, with a different (cumbersome) method. One can show that our solution (4.30), derived by the tools of the Laplace transform and fractional calculus, coincides with Boggio's solution. Also, Boggio arrived at the analysis of the two roots $\lambda_\pm$, but his expression of the solution—in the case of two conjugate complex roots ($\chi > 5/8$), given as a sum of Fresnel integrals—could induce one to forecast unphysical oscillations in the absence of numerical tables or plots. This disturbed Basset who, when he summarized the state of art about his problem in a later paper of 1910 [5], thought there was some physical deficiency in his own theory. With our integral representation of the solution, see (4.23)–(4.24), we can prove the monotone character of the solution, even if the arguments of the exponential and error functions are complex.

To have some insight about the effects of the two parameters, α and a, on the (generalized) Basset problem, we exhibit some (normalized) plots for the particle velocity $V(t)$, corresponding to the solution of (4.17), assuming for simplicity a vanishing initial velocity ($V_0 = 0$).

We consider three cases for α, namely, $\alpha = 1/2$ (the ordinary Basset problem) and $\alpha = 1/4, 3/4$ (the generalized Basset problem), corresponding to Figures 5, 6, and 7, respectively. For each α, we consider four values of a corresponding to $\chi := \rho_p/\rho_f = 0.5, 2, 10, 100$. For each couple $\{\alpha, \chi\}$, we compare the Basset solution with its asymptotic expression (in dotted line) for large times, and the Stokes solution

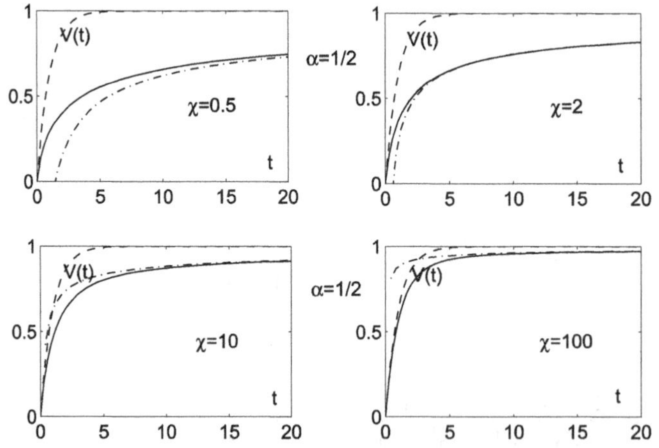

Figure 5: The normalized velocity $V(t)$ for $\alpha = 1/2$ and $\chi = 0.5, 2, 10, 100$. Basset: continuous line; Basset asymptotic: dotted-dashed line; Stokes: dashed line.

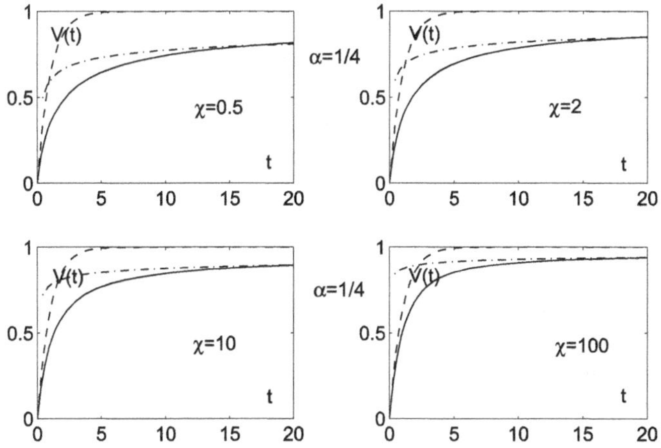

Figure 6: The normalized velocity $V(t)$ for $\alpha = 1/4$ and $\chi = 0.5, 2, 10, 100$. Basset: continuous line; Basset asymptotic: dotted-dashed line; Stokes: dashed line.

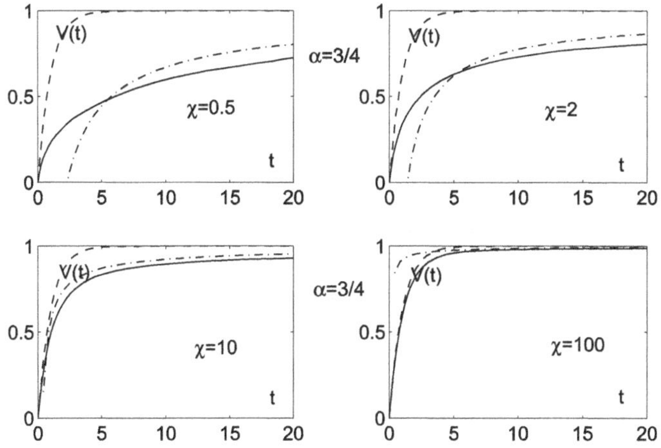

Figure 7: The normalized velocity $V(t)$ for $\alpha = 3/4$ and $\chi = 0.5, 2, 10, 100$. Basset: continuous line; Basset asymptotic: dotted-dashed line; Stokes: dashed line.

($a = 0$). From these figures, we can recognize the retarding effect of the (generalized) Basset force, which is more relevant for lighter particles in reaching the final value of the velocity. This effect is, of course, due to the algebraic decay of the function $M(t)$—see (4.26)—which is much slower than the exponential decay of the Stokes solution.

Bibliography

[1] B. N. N. Achar, J. W. Hanneken, and T. Clarke, Damping characteristics of a fractional oscillator, *Physica A*, **339** (2004), 311–319.

[2] B. N. N. Achar, J. W. Hanneken, T. Enck, and T. Clarke, Dynamics of the fractional oscillator, *Physica A*, **297** (2001), 361–367.

[3] R. L. Bagley, On the fractional order initial value problem and its engineering applications, in K. Nishimoto (ed.) *Fractional Calculus and Its Applications*, pp. 12–20, College of Engineering, Nihon University, Tokyo, 1990.

[4] A. B. Basset, *A Treatise on Hydrodynamics, Vol. 2, Deighton Bell*, Cambridge, 1888, Chap. 22.

[5] A. B. Basset, On the descent of a sphere in a viscous liquid, *Quarterly Journal of Mathematics*, **41** (1910), 369–381.

[6] H. Beyer and S. Kempfle, Definition of physically consistent damping laws with fractional derivatives, *Zeitschrift für Angewandte Mathematik und Mechanik*, **75** (1995), 623–635.

[7] T. Blaszczyk and M. Ciesielski, Fractional oscillation equation: analytical solution and algorithm for its approximate computation, *Journal of Vibration and Control*, **22**(8) (2016), 2045–2052. [First published on line, 20 Jan 2015.]

[8] T. Boggio, Integrazione dell'equazione funzionale che regge la caduta di una sfera in un liquido viscoso, *Rend. R. Acc. Naz. Lincei (ser. 5)*, **16** (1907), 613–620, 730–737. [2-nd sem.].

[9] J. Boussinesq, Sur la résistance qu'oppose un liquid indéfini en repos, san pesanteur, au mouvement varié d'une sphère solide qu'il mouille sur toute sa surface, quand les vitesses restent bien continues et assez faibles pour que leurs carrés et produits soient négligeables, *Comptes Rendus Mathematique. Academie Des Sciences. Paris*, **100** (1885), 935–937.

[10] M. Caputo, *Elasticità e Dissipazione*, Znichelli, Bologna, 1969 [in Italian].

[11] G. Doetsch, *Introduction to the Theory and Applications of the Laplace Transformation*, Springer Verlag, Berlin, 1974.

[12] J.-S. Duan, Z. Wang, and S.-Z. Fu, The zeros of the solution of the fractional oscillation equation, *Fractional Calculus & Applied Analysis. An International Journal for Theory and Applications*, **17**(1) (2014), 10–22, DOI:10.2478/s13540-014-0152-x.

[13] R. Gorenflo, A. A. Kilbas, F. Mainardi, and S. V. Rogosin, *Mittag-Leffler Functions, Related Topics and Applications*, Springer, Heidelberg, 2014. [2nd edition in preparation.]

[14] R. Gorenflo and F. Mainardi, Fractional oscillations and Mittag-Leffler functions, Preprint A–14/96, Fachbereich Mathematik und Informatik, Freie Universität, Berlin (1996), 22 pages. [E-print http://www.math.fu-berlin.de/publ/index.html].

[15] R. Gorenflo and F. Mainardi, Fractional calculus, integral and differential equations of fractional order, in *Fractals and Fractional Calculus in Continuum Mechanics*, A. Carpinteri and F. Mainardi (eds.), pp. 223–276, Springer Verlag, Wien, 1997. [E-print http://arxiv.org/abs/0805.3823.]

[16] R. Gorenflo and F. Mainardi, Fractional relaxation of distributed order, in *Complexus Mundi: Emergent Patterns in Nature*, M. Novak (ed.), pp. 33–42, World Scientific, Singapore, 2006.

[17] R. Gorenflo and R. Rutman, On ultraslow and intermediate processes, in *Transform Methods and Special Functions, Sofia 1994 Science Culture Technology, Singapore*, P. Rusev, I. Dimovski, and V. Kiryakova (eds.), pp. 61–81, 1995. [Proc. Int. Workshop, Sofia, Bulgaria, 12–17 August 1994.]

[18] R. R. Hughes and E. R. Gilliand, The mechanics of drops, *Chemical Engineering Progress*, **48** (1952), 497–504.

[19] C. J. Lawrence and R. Mei, Long-time behaviour of the drag on a body in impulsive motion, *Journal of Fluid Mechanics*, **283** (1995), 307–327.

[20] P. M. Lovalenti and J. F. Brady, The hydrodynamic force on a rigid particle undergoing arbitrary time-dependent motion at small Reynolds number, *Journal of Fluid Mechanics*, **256** (1993), 561–605.

[21] P. M. Lovalenti and J. F. Brady, The temporal behaviour of the hydrodynamic force on a body in response to an abrupt change in velocity at small but finite Reynolds number, *Journal of Fluid Mechanics*, **293** (1995), 35–46.

[22] F. Mainardi, Fractional relaxation-oscillation and fractional diffusion-wave phenomena, *Chaos, Solitons & Fractals*, **7** (1996), 1461–1477.
[23] F. Mainardi, Fractional calculus, some basic problems in continuum and statistical mechanics, in *Fractals and Fractional Calculus in Continuum Mechanics*, A. Carpinteri and F. Mainardi (eds.), pp. 291-348, Springer Verlag, Wien, 1997. [E-print http://arxiv.org/abs/1201.0863.]
[24] F. Mainardi and R. Gorenflo, The Mittag-Leffler function in the Riemann–Liouville fractional calculus, in *Boundary Value Problems, Special Functions and Fractional Calculus*, A. A. Kilbas (ed.), pp. 215–225, Byelorussian State University, Minsk, 1996. [Proc. Int. Conf., 90-th Birth Anniversary of Academician F. D. Gakhov, Minsk, Byelorussia, 16–20 February 1996.]
[25] F. Mainardi and R. Gorenflo, On Mittag-Leffler-type functions in fractional evolution processes, *Journal of Computational and Applied Mathematics*, **118**(1–2) (2000), 283–299.
[26] F. Mainardi, A. Mura, R. Gorenflo, and M. Stojanovic, The two forms of fractional relaxation of distributed order, *Journal of Vibration and Control*, **13** (2007), 1249–1268. [E-print http://arxiv.org/abs/cond-mat/0701131.]
[27] F. Mainardi, P. Pironi, and F. Tampieri, A numerical approach to the generalized Basset problem for a sphere accelerating in a viscous fluid, in *Proceedings CFD 95*, P. A. Thibault and D. M. Bergeron (eds.) vol. II, pp. 105–112, 1995. [3-rd Annual Conference of the Computational Fluid Dynamics Society of Canada, Banff, Alberta, Canada, 25–27 June 1995.]
[28] F. Mainardi, P. Pironi, and F. Tampieri, On a generalization of the Basset problem via fractional calculus, in *Proceedings CANCAM 95*, B. Tabarrok and S. Dost (eds.) vol. II, pp. 836–837, 1995. [15-th Canadian Congress of Applied Mechanics, Victoria, British Columbia, Canada, 28 May–2 June 1995.]
[29] M. R. Maxey and J. J. Riley, Equation of motion for a small rigid sphere in a nonuniform flow, *Physics of Fluids*, **26** (1983), 883–889.
[30] S. McKee and A. Stokes, Product integration methods for the nonlinear Basset equation, *SIAM Journal on Numerical Analysis*, **20** (1983), 143–160.
[31] R. Mei, History forces on a sphere due to a step change in the free-stream velocity, *International Journal of Multiphase Flow*, **19** (1993), 509–525.
[32] K. S. Miller and B. Ross, *An Introduction to the Fractional Calculus and Fractional Differential Equations*, Wily, New York, 1993.
[33] F. Odar, Verification of the proposed equation for calculation of forces on a sphere accelerating in a viscous fluid, *Journal of Fluid Mechanics*, **25** (1966), 591–592.
[34] F. Odar, and W. S. Hamilton, Forces on a sphere accelerating in a viscous fluid, *Journal of Fluid Mechanics*, **18** (1964), 302–314.
[35] G. Picciati, Sul moto di una sfera in un liquido viscoso, *Rend. R. Acc. Naz. Lincei (ser. 5)*, **16** (1907), 943–951. [1-st sem.].
[36] M. V. Reeks and S. McKee, The dispersive effects of Basset history forces on particle motion in a turbulent flow, *Physics of Fluids*, **27** (1984), 1573–1582.
[37] G. G. Stokes, On the effect of the internal friction of fluids on the motion of pendulums, *Cambridge Phil. Trans. (Ser. VIII)*, **9** (1851), 8–141, reprinted in *Mathematical and Physical Papers*, vol. III, pp. 1–141, Cambridge Univ. Press, 1922.
[38] F. B. Tatom, The Basset term as a semiderivative, *Applied Science Research*, **45** (1988), 283–285.
[39] A. Wiman, Über die Nullstellen der Funktionen $E_\alpha(x)$, *Acta Mathematica*, **29** (1905), 217–234.

Vasily E. Tarasov
Fractional calculus and long-range interactions

Abstract: In this chapter, connection between equations with fractional integro-differential operators and lattice models with long-range interactions is described. The fractional derivatives and integers on physical lattices and exact discretizations of derivatives of integer and noninteger orders are considered. The dynamics of system of particles with long-range interactions is related with the fractional continuous medium equations by transform operation. Exact discrete (lattice) analogs of fractional derivatives and integrals of integer and noninteger orders are described.

Keywords: Long-range interaction, lattice, chain, fractional dynamics, fractional calculus, fractional derivative, lattice operator, exact difference, fractional difference

PACS: 45.10.Hj, 04.60.Nc, 45.05.+x, 45.90.+t

1 Introduction

Long-range interactions have been studied in various fields of physics. One of the first who began studies of long-range interactions in discrete (lattice) physical models was Freeman J. Dyson [16–18]. Then, classical and quantum models of spins, with long-range interactions were investigated [25, 34, 54–57, 70]. Recently, the nonlocal properties and nonlinear effects in discrete and continuous systems with long-range interparticle interactions were discussed in [1–9, 11, 23, 24, 26–28, 31, 32, 42, 52, 53, 64, 68, 118]. The following effects and phenomena in systems with long-range interactions can be noted.

(1) Kinks in the Frenkel–Kontorova model with long-range interparticle interactions [6–8];
(2) Solitons in lattice with the long-range interaction of Lennard–Jones-type [32].
(3) Breathers in discrete systems with algebraically decaying long-range interactions [23, 28], and in systems that are described by the Klein–Gordon [4, 6, 7, 23], and discrete nonlinear Schrodinger equations [26, 27, 52, 53, 68].
(4) Synchronization and spatial correlations in coupled map lattices with long-range interactions are discussed in [118].
(5) Nonequilibrium phase transitions [3].
(7) Statistical mechanics and solvable models with long-range interactions are considered in [3, 5, 9, 11, 31, 42].

Vasily E. Tarasov, Skobeltsyn Institute of Nuclear Physics, Lomonosov Moscow State University, Moscow 119991, Russia, e-mail: tarasov@theory.sinp.msu.ru

https://doi.org/10.1515/9783110571707-004

All these works, devoted to long-range interactions, did not use the concepts and methods of the fractional calculus [35, 36, 69]. It should be emphasized that the equations with derivatives and integrals of noninteger orders is a powerful tool for describing nonlocal properties and long-range interactions. The derivatives of fractional orders have important nonstandard properties [62, 76, 95, 96, 108]. For example, the derivatives of noninteger order violate the standard Leibniz rule, and this violation is a characteristic property for all types of fractional derivatives [76, 95]. Fractional-order differentiation is characterized by a violation of the usual chain rule [96]. These nonstandard properties of fractional derivatives make it possible to describe nonlocal processes [76, 108]. Note that the dynamics, which is described by the equations with fractional space derivatives, can be characterized by the solutions that have power-like tails. Similar features were observed in the lattice models with power-like long-range interactions [1, 2, 23, 28, 64].

The direct connection between equations with fractional derivatives and lattice models with long-range interactions has been proved in [71, 72, 74, 75, 79, 87, 97, 109, 110] and [88, 98, 99, 105]. In these works, we consider chains and lattices with long-range interactions, and continuous limits of these discrete systems. It has been shown how the continuous limit for the systems of oscillators with long-range interaction can be described by the corresponding fractional equation. A connection between the dynamics of lattice system of particles, with long-range interactions, and the fractional continuum equations, has been proved by using the transform operation.

The relationship between long-range interactions and fractional dynamics has been applied to describe different systems in the continuum mechanics and the elasticity theory [77, 80–83, 89–91, 100, 101] (see also [78, 92, 93, 102, 106] and [12, 13, 33]); in the nonlinear dynamics and theory of deterministic chaos [37, 38, 109, 110, 116, 117, 119] (see also [40, 44, 111]); in statistical physics [84, 85, 94]; in economics [112–115]; in quantum mechanics and quantum field theory [86, 103, 104, 107].

In recent papers [79, 87, 97] and [88, 98, 99, 105], we proposed the lattice fractional partial derivatives and integrals of order α. These integro-differentiations are directly connected with the usual partial derivatives for all integer orders. The continuum limit of the lattice derivatives with $\alpha = n \in \mathbb{N}$ gives the usual (local) partial derivatives of integer orders n. From a physical point of view, the suggested lattice integro-differentiation describes long-range particle interactions in physical (crystal) lattices. Mathematically, the proposed lattice operators describe an exact discretization of the standard continuum integro-differentiation of integer and noninteger orders. The suggested lattice integro-differentiation have the same algebraic properties that are performed for standard partial derivatives. For example, the standard Leibniz rule, which is a characteristic property of the integer-order derivatives [76, 95], is satisfied by lattice operators of first order. The action of lattice operators on entire functions gives rise to standard expressions as an action of corresponding derivatives. The suggested lattice integro-differentiation allows us to get an exact discretization of integro-differential equations. In particular, this means that the lattice analogs of solutions of differential equations are solutions of the corresponding equations with lattice operators. In addition to the lattice integro-differentiation, which is an exact discretization of continuum integro-differentiation, we consider different asymptotic lattice fractional operators. These asymptotic lattice operators manifest the algebraic properties of continuum operators only when the primitive lattice vectors tend to zero.

2 Model of physical lattice with long-range interaction

Physical lattices are characterized by space periodicity. If lattices are unbounded, we can define noncoplanar vectors $\mathbf{a}_i$, $i = 1, 2, 3$. These vectors describe the fact that displacement of the lattice by any of these vectors brings it back to itself. All lattice sites are defined by the vector $\mathbf{n} = (n_1, n_2, n_3)$, where n_i are integer numbers. The vector $\mathbf{n}$ can be considered as a "number vector" of lattice particle. If the coordinate origin is chosen at one of the sites, the position vector of an arbitrary lattice site has the form $\mathbf{r}(\mathbf{n}) = \sum_{i=1}^{3} n_i \mathbf{a}_i$. For simplification, we consider a lattice with mutually perpendicular vectors $\mathbf{a}_i$, $i = 1, 2, 3$, such that the axes of the Cartesian coordinate system coincide with these vectors. In this case, we can write $\mathbf{a}_i = a_i\, \mathbf{e}_i$, where $a_i = |\mathbf{a}_i| > 0$ and $\mathbf{e}_i = \mathbf{a}_i / |\mathbf{a}_i|$ are the vectors of the basis of the Cartesian coordinate system.

Let us assume that equilibrium positions of particles coincide with the lattice sites. If particle is displaced with respect to the equilibrium, the coordinate of the corresponding particle differs from coordinates of lattice site. In this case, the displacement, with respect to equilibrium of particle with vector $\mathbf{n}$, is described by the vector field $\mathbf{u}(\mathbf{n}, t) = \sum_{k=1}^{3} u_k(\mathbf{n}, t)\, \mathbf{e}_k$.

For the physical lattices, equations of motion for particle of the site $\mathbf{n}$ and mass M have the form

$$M \frac{d^2 u_i(\mathbf{n}, t)}{dt^2} = -\sum_{\alpha} \sum_{k=1}^{3} \sum_{\mathbf{m}} K_{\alpha}^{ik}(\mathbf{n}, \mathbf{m})\, u_k(\mathbf{m}, t) + F_i(\mathbf{n}, t). \tag{1}$$

For physical lattices $i, k \in \{1, 2, 3\}$ are the coordinate indices. The coefficients $K_{\alpha}^{ik}(\mathbf{n}, \mathbf{m})$ describe an interaction of $\mathbf{n}$-particle with the $\mathbf{m}$-particles in the lattice. We can consider $K_{\alpha}^{ik}(\mathbf{n}, \mathbf{m})$ as a two-order elastic stiffness tensor kernel, which characterizes nonlocality of long-range interactions of the α-type [74, 75]. The interaction kernel $K_{\alpha}^{ik}(\mathbf{n}, \mathbf{m})$ can be interpreted as effective stiffness coefficients for virtual discrete mass-spring system that corresponds to the suggested lattice model. The interaction of lattice particles is described by $K_{\alpha}^{ik}(\mathbf{n}, \mathbf{m})$, with $\mathbf{n} \neq \mathbf{m}$, that is, when there is at least one n_j ($j = 1, 2, 3$) of the components of the vector $\mathbf{n}$ that is different from m_j. The terms with $K_{\alpha}^{ik}(0)$ can be interpreted as measure of self-interaction of lattice particles. The sum $\sum_{\mathbf{m}}$ means the summations from $-\infty$ to $+\infty$ over n_i, $i = 1, 2, 3$. The sum $\sum_{\alpha}$ means a sum over the different values of α. The parameter α is a positive real number that characterizes a decrease rate of the long-range interaction. This parameter can also be considered as a degree of the power law of spatial dispersion in the lattice [77, 80]. For an unbounded homogeneous lattice, due to its homogeneity, the kernels $K_\alpha^{ik}(\mathbf{n},\mathbf{m})$ have the form

$$K_\alpha^{ik}(\mathbf{n},\mathbf{m}) = K_\alpha^{ik}(\mathbf{n}-\mathbf{m}).$$

Equation (1) allows describing a wider class of long-range interactions. To formulate models of physical lattices we can define a lattice analog of partial derivative of noninteger order α with respect to n_i in the direction $\mathbf{e}_i = \mathbf{a}_i/|\mathbf{a}_i|$. The kernels $K_\alpha^{ik}(\mathbf{n}-\mathbf{m})$ can be constructed as sum and product of these lattice derivatives and integrals of integer and noninteger orders.

3 Lattice integro-differential operators

Let us consider an unbounded lattice $\mathbb{Z}^N$ that is characterized by N non-coplanar vectors $\mathbf{a}_j$, $j = 1,\dots,N$, called primitive lattice vectors. For simplification, we assume that $\mathbf{a}_j$, $j = 1,\dots,N$, are mutually perpendicular vectors such that the axes of the Cartesian coordinate system coincide with the vector $\mathbf{a}_j$. Then $\mathbf{a}_j = a_j\,\mathbf{e}_j$, where $a_j = |\mathbf{a}_j|$ and the vectors $\mathbf{e}_j$, $j = 1,\dots,N$ form a basis of the Cartesian coordinate system for $\mathbb{R}^N$. The position vector of an arbitrary lattice site is $\mathbf{r}(\mathbf{n}) = \sum_{j=1}^N n_j\,\mathbf{a}_j$, where $n_j \in \mathbb{Z}$. Then lattice sites are numbered by $\mathbf{n} = \sum_{j=1}^N n_j\,\mathbf{e}_j$. For simplification, we will consider scalar functions $f(\mathbf{n})$ on $\mathbb{Z}^N$. In many cases, we can assume that $f(\mathbf{n})$ belongs to the Hilbert space l_2 of square-summable sequences to apply the Fourier transform. We will consider integro-differential operators that act on the functions $f(\mathbf{n})$.

Let us define lattice operator of integro-differentiation. Lattice integro-differential operators $\mathbb{D}_L^\pm\left[{\alpha \atop j}\right]$ of order $\alpha \in \mathbb{R}$, in the direction $\mathbf{e}_j = \mathbf{a}_j/|\mathbf{a}_j|$, are defined by the equation

$$\left(\mathbb{D}_L^\pm\begin{bmatrix}\alpha\\ j\end{bmatrix} f\right)(\mathbf{n}) = \frac{1}{a_j^\alpha}\sum_{m_j=-\infty}^{+\infty} K_\alpha^\pm(n_j-m_j)f(\mathbf{m}) \quad (j=1,\dots,N), \tag{2}$$

where $\mathbf{n},\mathbf{m} \in \mathbb{Z}^N$, and $\mathbf{n} = \mathbf{m} + (n_j - m_j)\mathbf{e}_j$. The kernels $K_\alpha^\pm(n)$ are real-valued functions of integer variable $n \in \mathbb{Z}$ such that the kernel $K_\alpha^+(n)$ is an even function, and $K_\alpha^-(n)$ is an odd function:

$$K_\alpha^\pm(-n) = \pm K_\alpha^\pm(n) \quad (n \in \mathbb{Z}). \tag{3}$$

The parameter α is called the order of the lattice operator (2). The kernels $K^\pm(n)$ of lattice operators (2) are defined by the equations:

$$K_\alpha^+(n) = \frac{\pi^\alpha}{\alpha+1}\,{}_1F_2\left(\frac{\alpha+1}{2};\frac{1}{2},\frac{\alpha+3}{2};-\frac{\pi^2 n^2}{4}\right), \tag{4}$$

where $\alpha > -1$, and

$$K_\alpha^-(n) = -\frac{\pi^{\alpha+1} n}{\alpha+2} \, {}_1F_2\left(\frac{\alpha+2}{2}; \frac{3}{2}, \frac{\alpha+4}{2}; -\frac{\pi^2 n^2}{4}\right), \tag{5}$$

where $\alpha > -2$. In Equations (4) and (5), we use the generalized hypergeometric function that is defined (see Section 1.6 of [35]) by the equation

$${}_1F_2(a; b, c; z) := \sum_{k=0}^{\infty} \frac{(a)_k}{(b)_k \, (c)_k} \frac{z^k}{k!} = \sum_{k=0}^{\infty} \frac{\Gamma(a+k)\,\Gamma(b)\,\Gamma(c)}{\Gamma(a)\,\Gamma(b+k)\,\Gamma(c+k)} \frac{z^k}{k!}, \tag{6}$$

where $a_l \in \mathbb{C}$ $(l = 1, \ldots, p)$, $b_j \in \mathbb{C}$, $b_j \neq 0, -1, -2, \ldots$ $(j = 1, \ldots, q)$, and $(a)_k$ is the Pochhammer symbol (rising factorial), defined by $(a)_0 = 1$.

It should be noted that expression (4) of the kernel $K_\alpha^+(n)$, which has been expressed through the Lommel function, was first proposed in [71, 72, 74], (see Equation (41) in [71], page 14900 and 14908 of [72], and Equation (8.44) in [74, 75]). Note that the Lommel function can be represented by the generalized hypergeometric function (6) (for example, see Section 6.2.9. on page 217 of [43] and page 428 of [19]). Then, it can be represented by the generalized hypergeometric function in [79, 87, 97], where the expression (4) of the kernel $K_\alpha^+(n)$ also has also been proposed.

The kernels $K_\alpha^\pm(n)$ of the lattice operators (2) are defined [87] such that the Fourier series transforms

$$\hat{K}_\alpha^\pm(k) = \mathrm{F}_{a,\Delta}\{K_\alpha^\pm(n)\} := \sum_{n=-\infty}^{+\infty} e^{-ikn} K_\alpha^\pm(n) \tag{7}$$

are represented in the form

$$\hat{K}_\alpha^+(k) = |k|^\alpha, \tag{8}$$

$$\hat{K}_\alpha^-(k) = i \, \mathrm{sgn}(k) \, |k|^\alpha. \tag{9}$$

These conditions are used in order of the lattice operators (2), with $\alpha = m \in \mathbb{N}$ to be lattice analogs of standard derivatives of integer orders $m \in \mathbb{N}$.

It is important to note that Equations (4) and (5) are valid, not only for the positive α, but also for some negative values of α. This fact allows us to interpret the lattice integro-differential operators (2) with negative α as lattice analogs of integration of order α. Note also that the terms

$$K_\alpha^+(0) = \frac{\pi^\alpha}{\alpha+1}, \quad K_\alpha^-(0) = 0 \tag{10}$$

characterizes a self-interaction of lattice particles. The interaction of different particles is described by $K_\alpha^\pm(n-m)$, with $n - m \neq 0$.

In papers [71, 72, 74, 79, 87], the lattice-continuum transform operation $\mathrm{T}_{L\to C}$ has been suggested in the form

$$\mathrm{T}_{L\to C} := \mathrm{F}^{-1} \circ \mathrm{Lim} \circ \mathrm{F}_\Delta, \tag{11}$$

where F_Δ is the Fourier series transform; Lim denotes the passage to the limits $a_j \to 0$ ($j = 1, \ldots, N$), and F^{-1} is the inverse Fourier integral transform. Let us give a more detailed definition of $T_{L\to C}$. The lattice-continuum transform operator $T_{L\to C}$, which maps the lattice functions and operators into the continuum functions and operators, is defined by the equation

$$T_{L\to C}(\,) := \frac{1}{(2\pi)^N} \int_{-\infty}^{+\infty} dk_1 \cdots \int_{-\infty}^{+\infty} dk_N \, e^{i\sum_{j=1}^N k_j x_j} \circ \lim_{a_1,\ldots a_N \to 0} \circ \sum_{n_1,\ldots,n_N=-\infty}^{+\infty} e^{-i(\mathbf{k},\mathbf{r}(\mathbf{n}))}(\,), \tag{12}$$

where $\mathbf{r}(\mathbf{n}) = \sum_{j=1}^N n_j \, \mathbf{a}_j$ is the position vector of a lattice site, and $n_j \in \mathbb{Z}$.

For the case $N = 1$, operator (12) is represented by the equation

$$T_{L\to C}(\,) := \frac{1}{2\pi} \int_{-\infty}^{+\infty} dk \, e^{ikx} \lim_{a\to 0} \sum_{n=-\infty}^{+\infty} e^{-ik\,a\,n}(\,). \tag{13}$$

The lattice-continuum transform operation $T_{L\to C}$ maps [87] the lattice fractional derivatives $\mathbb{D}_L^{\pm}\left[{\alpha \atop j}\right]$ of order α into the continuum fractional operators

$$T_{L\to C}\left(\left(\mathbb{D}_L^{\pm}\begin{bmatrix}\alpha\\ j\end{bmatrix} f\right)(\mathbf{n})\right) = \mathbb{D}_C^{\pm}\begin{bmatrix}\alpha\\ j\end{bmatrix} f(\mathbf{r}), \tag{14}$$

where $\mathbb{D}_C^{\pm}\left[{\alpha \atop j}\right]$ are the fractional integro-differentiation of the Riesz type [87]. The fractional operators $\mathbb{D}_C^{\pm}\left[{\alpha \atop j}\right]$ can be defined by expressions

$$\mathbb{D}_C^{\pm}\begin{bmatrix}\alpha\\ j\end{bmatrix} f(\mathbf{r}) = F^{-1}(\hat{K}_\alpha^{\pm}(k_j)), \tag{15}$$

so that

$$\mathbb{D}_C^{+}\begin{bmatrix}\alpha\\ j\end{bmatrix} f(\mathbf{r}) = F^{-1}(|k_j|^\alpha), \tag{16}$$

$$\mathbb{D}_C^{-}\begin{bmatrix}\alpha\\ j\end{bmatrix} f(\mathbf{r}) = F^{-1}(i \,\mathrm{sgn}(k_j)\,|k_j|^\alpha). \tag{17}$$

For details, see [79, 87, 97, 98] and [84, 89, 90, 94].

The lattice fractional derivatives $\mathbb{D}_L^{\pm}\left[{\alpha \atop j}\right]$ of integer positive orders $\alpha \in \mathbb{N}$ are connected [87] with the usual partial derivatives of integer orders by the equations

$$T_{L\to C}\left(\left(\mathbb{D}_L^{+}\begin{bmatrix}2m\\ j\end{bmatrix} f\right)(\mathbf{n})\right) = (-1)^m \frac{\partial^{2m} f(\mathbf{r})}{\partial x_j^{2m}}, \tag{18}$$

$$\mathrm{T}_{L\to C}\left(\left(\mathbb{D}_L^-\left[\begin{matrix}2m+1\\ j\end{matrix}\right] f\right)(\mathbf{n})\right) = (-1)^m \frac{\partial^{2m+1} f(\mathbf{r})}{\partial x_j^{2m+1}}, \tag{19}$$

where $m \in \mathbb{N}$.

The lattice operators (2) of orders $\alpha \in \mathbb{R}$ are used in the different fractional nonlocal generalization of local continuum models and corresponding microstructural (lattice) models (for example, see [79, 84, 86, 89, 90, 94]).

4 Lattices fractional integro-differentiation

Let us give a definition of the lattice fractional integro-differentiation.

Lattice integro-differential operators ${}^T\mathbb{D}_L\left[\begin{smallmatrix}\alpha\\ j\end{smallmatrix}\right]$ of order $\alpha > -1$, in the direction $\mathbf{e}_j = \mathbf{a}_j/|\mathbf{a}_j|$, is an operator that is defined by the equation

$$\left({}^T\mathbb{D}_L\left[\begin{matrix}\alpha\\ j\end{matrix}\right] f\right)(\mathbf{n}) = \frac{1}{a_j^\alpha}\sum_{m_j=-\infty}^{+\infty} K_\alpha(n_j - m_j) f(\mathbf{m}), \tag{20}$$

where $j = 1, \dots, N$, $\mathbf{n}, \mathbf{m} \in \mathbb{Z}^N$, and

$$\mathbf{n} = \mathbf{m} + (n_j - m_j)\mathbf{e}_j.$$

The kernel $K_\alpha(n)$ is a real-valued function of integer variable $n \in \mathbb{Z}$, which is defined by the equation

$$K_\alpha(n) = \sum_{k=0}^{\infty} \frac{(-1)^k \pi^{2k+1/2} n^{2k}}{2^{2k} k!\, \Gamma(k+1/2)} \left(\frac{\cos(\frac{\pi\alpha}{2})}{\alpha+2k+1} - \frac{\pi n \sin(\frac{\pi\alpha}{2})}{(\alpha+2k+2)(2k+1)}\right), \tag{21}$$

where $\alpha > -1$. If $\alpha > 0$, then the lattice operator (20) will be called the lattice fractional derivative. For $-1 < \alpha < 0$, the lattice operator (20) is called the lattice fractional integral.

The lattice fractional integro-differentiation (20) can be represented as the lattice operator in the form

$${}^T\mathbb{D}_L\left[\begin{matrix}\alpha\\ j\end{matrix}\right] := \cos\left(\frac{\pi\alpha}{2}\right)\mathbb{D}_L^+\left[\begin{matrix}\alpha\\ j\end{matrix}\right] + \sin\left(\frac{\pi\alpha}{2}\right)\mathbb{D}_L^-\left[\begin{matrix}\alpha\\ j\end{matrix}\right] \quad (\alpha > -1), \tag{22}$$

where $\mathbb{D}_L^\pm\left[\begin{smallmatrix}\alpha\\ j\end{smallmatrix}\right]$ are defined by Equations (2), with kernels (4) and (5). The kernel (21) can be represented in the form

$$K_\alpha(n) = \cos\left(\frac{\pi\alpha}{2}\right)K_\alpha^+(n) + \sin\left(\frac{\pi\alpha}{2}\right)K_\alpha^-(n) \quad (\alpha > -1), \tag{23}$$

where $K_\alpha^\pm(n)$ are defined by Equations (4) and (5).

It should be noted that expression (20), with the kernels (21) in the form of lattice operators and as exact fractional differences, was proposed in [88, 97–99, 105] and [102–104, 107].

The Fourier series transform F_Δ of the lattice fractional integro-differentiation (20) has the form

$$\mathrm{F}_\Delta\left(\left({}^T\mathbb{D}_L\begin{bmatrix}\alpha\\ j\end{bmatrix} f\right)(\mathbf{n})\right)(\mathbf{k}) = e^{i\pi\alpha\,\mathrm{sgn}(k_j)/2}|k_j|^\alpha(\mathrm{F}f)(\mathbf{k}). \quad (24)$$

The lattice fractional derivatives ${}^T\mathbb{D}_L\left[\begin{smallmatrix}\alpha\\ j\end{smallmatrix}\right]$ for integer positive values of $\alpha = m \in \mathbb{N}$ gives [97] the lattice integro-differential operators $\mathbb{D}_L^{\pm}\left[\begin{smallmatrix}m\\ j\end{smallmatrix}\right]$ in the form

$${}^T\mathbb{D}_L\begin{bmatrix}2m\\ j\end{bmatrix} = (-1)^m\mathbb{D}_L^{+}\begin{bmatrix}2m\\ j\end{bmatrix}, \quad (25)$$

$${}^T\mathbb{D}_L\begin{bmatrix}2m+1\\ j\end{bmatrix} = (-1)^m\mathbb{D}_L^{-}\begin{bmatrix}2m+1\\ j\end{bmatrix}, \quad (26)$$

where $m \in \mathbb{N}$.

Let us consider a continuum fractional integro-differentiation of order α, of the Riesz type, which is defined by the equation

$${}^{RT}\mathbb{D}_C\begin{bmatrix}\alpha\\ j\end{bmatrix} f(\mathbf{r}) = \mathrm{T}_{L\to C}\left(\left({}^T\mathbb{D}_L\begin{bmatrix}\alpha\\ j\end{bmatrix} f\right)(\mathbf{n})\right) \quad (\alpha > -1). \quad (27)$$

The fractional operators ${}^{RT}\mathbb{D}_C\left[\begin{smallmatrix}\alpha\\ j\end{smallmatrix}\right]$ can be defined by the following expressions:

$${}^{RT}\mathbb{D}_C\begin{bmatrix}\alpha\\ j\end{bmatrix} f(\mathbf{r}) = \mathrm{F}^{-1}(e^{i\pi\alpha\,\mathrm{sgn}(k_j)/2}|k_j|^\alpha(\mathrm{F}f)(\mathbf{k})). \quad (28)$$

For details, see [79, 87, 97, 98].

The lattice fractional derivatives of orders α, for integer positive values $\alpha = m \in \mathbb{N}$, give [97] usual partial derivatives of integer orders m:

$${}^{RT}\mathbb{D}_C\begin{bmatrix}m\\ j\end{bmatrix} f(\mathbf{r}) = \mathrm{T}_{L\to C}\left(\left({}^T\mathbb{D}_L\begin{bmatrix}m\\ j\end{bmatrix} f\right)(\mathbf{n})\right) = \frac{\partial^m f(\mathbf{r})}{\partial x_j^m}, \quad (29)$$

where $m \in \mathbb{N}$.

As a result, the continuum limits of lattice integro-differentiations of integer orders α give the partial derivatives of these orders for all $\alpha \in \mathbb{N}$. Relation (29) implements the principle of correspondence with nonfractional case. This simplifies construction of lattice models with long-range interactions for nonlocal theories of continua, media, and fields.

5 Lattice operators as exact discretization of differential operators

An exact discretization of differential equations is discussed in several papers, for example, see [45, 47, 49, 65, 66] and [46, 48, 50, 51]. The proposed exact discretization is considered as a scheme for which the difference (lattice) equation has the same general solution as the associated differential equation. The condition of exact discretization for differential equations is usually formulated in the following form: an exact discretization is a map of differential equation into a difference (lattice) equation for which the solution $f[n]$ of the difference (lattice) equation, and the solution $f(x)$ of associated differential equation, are the same; that is, if and only if the discrete function $f[n]$ is exactly equal to the function $f(x)$ for $x = a\,n$; that is, $f[n] = f(a\,n)$ $(n \in \mathbb{Z})$ for arbitrary values of $a > 0$.

In paper [98], the following definition of exact discretization for derivatives of integer and non-integer orders has been suggested:

Exact discretization of integro-differentiation. Let $A(\mathbb{R}^N)$ be a function space, and let $\mathbb{D}_C\left[\begin{smallmatrix}\alpha\\ j\end{smallmatrix}\right]$ be a differential operator on the space $A(\mathbb{R}^N)$ such that

$$\mathbb{D}_C\begin{bmatrix}\alpha\\ j\end{bmatrix} f(\mathbf{r}) = g(\mathbf{r}) \quad (\mathbf{r} \in \mathbb{R}^N) \tag{30}$$

for all $f(\mathbf{r}) \in A(\mathbb{R}^N)$, where $g(\mathbf{r}) \in A(\mathbb{R}^N)$. The lattice operator ${}^T\mathbb{D}_L\left[\begin{smallmatrix}\alpha\\ j\end{smallmatrix}\right]$ will be called an exact discretization of $\mathbb{D}_C\left[\begin{smallmatrix}\alpha\\ j\end{smallmatrix}\right]$ if the equation

$$\mathbb{D}_L\begin{bmatrix}\alpha\\ j\end{bmatrix} f(\mathbf{n}) = g(\mathbf{n}) \quad (\mathbf{n} \in \mathbb{Z}^N) \tag{31}$$

holds for all $f(\mathbf{n}) \in A(\mathbb{Z}^N)$, where $g(\mathbf{n}) \in A(\mathbb{Z}^N)$.

Therefore, we can state that condition (31) means that the equality

$$\left(\mathbb{D}_C\begin{bmatrix}\alpha\\ j\end{bmatrix} f(\mathbf{r})\right)_{\mathbf{r}=\mathbf{n}} = {}^T\mathbb{D}_L\begin{bmatrix}\alpha\\ j\end{bmatrix} f(\mathbf{n}) \tag{32}$$

holds for all $\mathbf{n} \in \mathbb{Z}^N$.

In article [98], a principle of algebraic correspondence for an exact discretization of theory of differential and integral operators has been proposed. This correspondence principle for discrete theories states: *The correspondence between the theories of difference and differential (integro-differential) equations lies not so much in the limiting condition when the steps tend to zero as in the fact that mathematical operations on these two theories should obey in many cases the same laws.*

For the lattice fractional calculus, this principle can be formulated [98] in the form that follows.

Principle of exact correspondence between lattice and continuum theories. *The correspondence between the theories of difference (lattice) and differential (integro-differential) equations lies not so much in the limiting condition when the primitive lattice vectors $a_j \to 0$ as in the fact that mathematical operations on these two theories should obey in many cases the same laws.*

This principle is similar to the Dirac's principle of correspondence between the quantum and classical theories (see page 649 of [15] and Chapter 8 of [73]).

To specify our consideration, we formulate [98] the principle of algebraic correspondence for integro-differential operators: *The lattice integro-differential operators, which are exact discretization of continuum integro-differential operators of integer or non-integer orders, should satisfy the same algebraic characteristic relations as the continuum integro-differential operators.*

The proposed lattice operators ${}^T\mathbb{D}_L\left[\begin{smallmatrix} m \\ j \end{smallmatrix}\right]$ of integer orders $m \in \mathbb{N}$ can be considered as exact discretization of the derivatives $\partial^m/\partial x_j^m$, since these differences preserve the following characteristic properties of the lattice derivative of first order.

(1) The Leibniz rule:

$$ {}^T\mathbb{D}_L\begin{bmatrix} 1 \\ j \end{bmatrix}(f[\mathbf{n}]\,g[\mathbf{n}]) = g[\mathbf{n}]\,{}^T\mathbb{D}_L\begin{bmatrix} 1 \\ j \end{bmatrix} f[\mathbf{n}] + f[\mathbf{n}]\,{}^T\mathbb{D}_L\begin{bmatrix} 1 \\ j \end{bmatrix} g[\mathbf{n}]. \tag{33} $$

(2) The differentiation of power-law functions:

$$ {}^T\mathbb{D}_L\begin{bmatrix} 1 \\ j \end{bmatrix} n_j^k = k\, n_j^{k-1} \quad (k \geq 1). \tag{34} $$

(3) The semigroup property:

$$ {}^T\mathbb{D}_L\begin{bmatrix} 1 \\ j \end{bmatrix} {}^T\mathbb{D}_L\begin{bmatrix} 1 \\ j \end{bmatrix} = {}^T\mathbb{D}_L\begin{bmatrix} 2 \\ j \end{bmatrix}. \tag{35} $$

To derive an exact discretization of continuum integro-differentiation $\mathbb{D}_C\left[\begin{smallmatrix} \alpha \\ j \end{smallmatrix}\right]$, the necessary conditions of exact and asymptotic (approximate) discretizations of the operators $\mathbb{D}_C\left[\begin{smallmatrix} \alpha \\ j \end{smallmatrix}\right]$ has been proposed [98].

The lattice operator (difference) ${}^T\mathbb{D}_L\left[\begin{smallmatrix} \alpha_j \\ j \end{smallmatrix}\right]$ is called satisfying the necessary condition of exact discretization of the continuum operator $\mathbb{D}_C\left[\begin{smallmatrix} \alpha_j \\ j \end{smallmatrix}\right]$ of order α_j if the condition

$$ \frac{1}{a_j}\mathcal{F}^{-1}\,\mathcal{F}_{a,\Delta}\left({}^T\mathbb{D}_L\begin{bmatrix} \alpha_j \\ j \end{bmatrix}\right) = \mathbb{D}_C\begin{bmatrix} \alpha_j \\ j \end{bmatrix} \tag{36} $$

holds for arbitrary values of $a_j > 0$. If condition (36) is not satisfied, but the condition

$$ \lim_{a_j \to 0+} \frac{1}{a_j}\mathcal{F}^{-1}\,\mathcal{F}_{a,\Delta}\left({}^T\mathbb{D}_L\begin{bmatrix} \alpha_j \\ j \end{bmatrix}\right) = \mathbb{D}_C\begin{bmatrix} \alpha_j \\ j \end{bmatrix} \tag{37} $$

holds, then the lattice operator ${}^{T}\mathbb{D}_L\left[\begin{smallmatrix}\alpha_j\\ j\end{smallmatrix}\right]$ is called the asymptotic (approximate) discretization of the continuum operator ${}^{T}\mathbb{D}_L\left[\begin{smallmatrix}\alpha_j\\ j\end{smallmatrix}\right]$. Here, F^{-1} is the inverse Fourier integral transform, and $\mathrm{F}_{a,\Delta}$ is the Fourier series transform.

It is obvious that the lattice operator, which is an exact discretization, also satisfies the condition (37) of the asymptotic discretization.

It should be emphasized that Equation (36) cannot be considered as a sufficient condition proving that the principle of algebraic correspondence was carried out. It is a necessary condition only, but it can be used as a constructive requirement, which allows us to obtain [98] explicit expressions for the discrete operators that claim to be the exact discretization of the differential operators.

6 Lattice derivatives of integer order

Let us consider the one-dimensional lattice $\mathbb{Z}^1$ ($N = 1$) and the lattice operators of integer orders [88, 98, 103]. For this case, we will use the following notation:

$$ {}^{\mathrm{T}}\mathbf{\Delta}^{s} = {}^{T}\mathbb{D}_L\begin{bmatrix} s\\ 1\end{bmatrix} \quad (s \in \mathbb{N}). \tag{38}$$

The kernels of these lattice operators can be represented in a simpler form [88, 98, 103].

Let us give exact expressions of lattice derivatives, which are exact discretization of the integer-order derivatives [88, 98].

The lattice derivatives of integer orders are defined by the equations

$$ {}^{\mathrm{T}}\mathbf{\Delta}^{2s} f[n] := \sum_{\substack{m=-\infty\\ m\neq 0}}^{+\infty} K_{2s}(m) f[n-m] + K_{2s}(0) f[n] \quad (s \in \mathbb{N}), \tag{39}$$

$$ {}^{\mathrm{T}}\mathbf{\Delta}^{2s-1} f[n] := \sum_{\substack{m=-\infty\\ m\neq 0}}^{+\infty} K_{2s-1}(m) f[n-m] \quad (s \in \mathbb{N}), \tag{40}$$

where the kernels $K_{2s}(m)$ and $K_{2s-1}(m)$ are defined by the following equations:

$$ K_{2s}(m) = \sum_{k=0}^{s-1} \frac{(-1)^{m+k+s}\,(2s)!\,\pi^{2s-2k-2}}{(2s-2k-1)!} \frac{1}{m^{2k+2}}, \tag{41}$$

$$ K_{2s-1}(m) = \sum_{k=0}^{s-1} \frac{(-1)^{m+k+s+1}\,(2s-1)!\,\pi^{2s-2k-2}}{(2s-2k-1)!} \frac{1}{m^{2k+1}} \tag{42}$$

where $m \in \mathbb{Z}$, $m \neq 0$, and

$$ K_{2s}(0) = \frac{(-1)^s\,\pi^{2s}}{2s+1}, \quad K_{2s-1}(0) = 0. \tag{43}$$

Note that $K_{2s}(0)$ describes a self-interaction of lattice particles. The interaction of different particles is described by $K_s(n-m)$, with $n-m \neq 0$.

Let us give examples of the suggested lattice derivatives of integer orders. The first-order lattice operators is

$$ {}^{\mathrm{T}}\mathbf{\Delta}^1 f[n] := \sum_{\substack{m=-\infty \\ m\neq 0}}^{+\infty} \frac{(-1)^m}{m} f[n-m]. \tag{44}$$

The second-order lattice operators is

$$ {}^{\mathrm{T}}\mathbf{\Delta}^2 f[n] := -\sum_{\substack{m=-\infty \\ m\neq 0}}^{+\infty} \frac{2\,(-1)^m}{m^2} f[n-m] - \frac{\pi^2}{3} f[n]. \tag{45}$$

The $(n+1)$th-order lattice operator is

$$ {}^{\mathrm{T}}\mathbf{\Delta}^{n+1} f[n] := {}^{\mathrm{T}}\mathbf{\Delta}^{1\,\mathrm{T}}\mathbf{\Delta}^{n} f[n]. \tag{46}$$

Using the even and odd property of the kernels (41) and (42), the lattice derivatives (39), (40) of even and odd orders can be represented by the following expressions:

$$ {}^{\mathrm{T}}\mathbf{\Delta}^{2s} f[n] := \sum_{m=1}^{+\infty} K_{2s}(m)\,(f[n-m] + f[n+m]) + K_{2s}(0) f[n], \tag{47}$$

$$ {}^{\mathrm{T}}\mathbf{\Delta}^{2s-1} f[n] := \sum_{m=1}^{+\infty} K_{2s-1}(m)\,(f[n-m] - f[n+m]), \tag{48}$$

where $s \in \mathbb{N}$. These expressions simplify the direct calculation of the lattice derivatives for some cases.

Let us define lattice operator for negative integer order $n = -1$. The lattice antiderivative is defined by the equation

$$ {}^{\mathrm{T}}\mathbf{\Delta}^{-1} f[n] := \sum_{\substack{m=-\infty \\ m\neq 0}}^{+\infty} \pi^{-1}\,\mathrm{Si}(\pi m) f[n-m], \tag{49}$$

where $\mathrm{Si}(z)$ is the sine integral. Using $\mathrm{Si}(-z) = -\mathrm{Si}(z)$, Equation (49) can be represented in the form

$$ {}^{\mathrm{T}}\mathbf{\Delta}^{-1} f[n] := \sum_{m=1}^{+\infty} K_{-1}(m)\,(f[n-m] - f[n+m]). \tag{50}$$

Using the suggested lattice derivatives and antiderivative, we can obtain exact lattice analogues of the differential equations without approximation, which is based on deleting the terms $O(a_j^m)$. It should be noted that this discretization allows obtaining lattice (difference) equations, whose solutions are equal to the solutions of corresponding differential equations [98, 106, 107, 113, 114].

7 Properties of lattice derivatives of integer order

The suggested lattice derivatives can be used for discrete function $f[n]$ defined for $n \in \mathbb{Z}$, if the corresponding expression are represented by convergent series.

Using the concept of generalized convergence, in the sense of Cesaro and Poisson–Abel, we can generalize the lattice derivatives of integer order. Let us give a definition of the generalized lattice derivatives of first order.

The Poisson–Abel-type lattice derivative is defined by the equation

$$ {}^{\mathrm{T}}\mathbf{\Delta}^1 f[n] := \lim_{x \to 1-0} \sum_{m=1}^{\infty} \frac{(-x)^m}{m} (f[n-m] - f[n+m]), \tag{51} $$

if this limit exists. The Cesaro-type lattice derivative is defined if the following limit exists:

$$ {}^{\mathrm{T}}\mathbf{\Delta}^1 f[n] := \lim_{N \to \infty} \frac{1}{N} \sum_{k=1}^{N} \sum_{m=1}^{k} \frac{(-1)^m}{m} (f[n-m] - f[n+m]), \tag{52} $$

and the sum of the series is equal to its Cesaro's sum. If the series is Cesaro- summable, then this series is Poisson–Abel-summable, with the same sum (see Sections 11–12 of [20] and [21, 22, 30]). Likewise, we can define the generalized lattice derivatives of all integer and fractional orders by Cesaro and Poisson–Abel summation.

Using the Poisson–Abel-type lattice derivative, we can get the lattice derivative of power-law functions [97, 98]. The action of lattice derivative of the first order on the discrete function $f[n] = n^k$ has the form

$$ {}^{\mathrm{T}}\mathbf{\Delta}^1 n^k = k\, n^{k-1} \quad (k \in \mathbb{N}, n \in \mathbb{Z}). \tag{53} $$

Equation (53) allows us to consider the lattice derivatives for discrete analogs of analytic functions and polynomials.

We can define the following space of discrete functions. We will say that the discrete function $f[n]$ $(n \in \mathbb{Z})$ belongs to the space $\mathrm{A}(\mathbb{Z})$ if there exists a real-valued function $f(x)$, with $x \in \mathbb{R}$, that can be represented in the form

$$ f(x) = \sum_{k=0}^{\infty} f_k\, x^k \quad (|x| < \infty), \tag{54} $$

and $f[n]$ coincides with $f(x)$ at all points $x = n \in \mathbb{Z}$, so that $f[n] = f(n)$ for all $n \in \mathbb{Z}$. The definition means that $f[n] \in \mathrm{A}(\mathbb{Z})$ can be represented in the form

$$ f[n] = \sum_{k=0}^{\infty} f_k\, n^k \quad (|n| < \infty). \tag{55} $$

The function $f[n] \in \mathrm{A}(\mathbb{Z})$ can be considered as an entire function $f(z)$, with $z \in \mathbb{C}$ at $z = n \in \mathbb{Z}$, that is, $f[n] = f(z)$ for all $z = n \in \mathbb{Z}$. It is well known that any series (55) such that the Cauchy–Hadamar condition

$$\lim_{k\to\infty} \sqrt[k]{|f_k|} = 0 \tag{56}$$

is satisfied represents a discrete function from $A(\mathbb{Z})$.

Let us give an important property, which states that the lattice derivative of first order can be considered as an exact discrete analog of the derivative of first order: if a discrete function $f[n]$ with $n \in \mathbb{Z}$ belongs to the space $A(\mathbb{Z})$, then the lattice derivative of first order of this function is equal to the first derivative of $f(x)$ at $x = n$, that is,

$$^{\mathrm{T}}\boldsymbol{\Delta}^1 f[n] = \left(\frac{df(x)}{dx}\right)_{x=n}, \tag{57}$$

where $^{\mathrm{T}}\boldsymbol{\Delta}^1$ is the lattice derivative of Cesaro or Poisson–Abel type. The proof is given in [98] for $f(x)$, with $x \in \mathbb{R}$. Note that this statement can be proved by the direct summation of series for entire functions. This property means that lattice derivative of first order is the exact discrete analog of the derivative of first order. As a result, this lattice operator has the same algebraic properties as the operator of differentiation on the space $A(\mathbb{Z})$ of discrete functions.

Using Equations 5.4.2.6 and 5.4.2.8 (or 5.4.2.10, 5.4.2.12, 5.4.2.13) of [67], we obtain [97] expression of lattice derivatives of sine and cosine functions. The lattice derivative of the discrete sine function is equal to the cosine function

$$^{\mathrm{T}}\boldsymbol{\Delta}^1 \sin(k\,n) = k \cos(k\,n). \tag{58}$$

The lattice derivative of the discrete cosine function is equal to the sine function with minus:

$$^{\mathrm{T}}\boldsymbol{\Delta}^1 \cos(k\,n) = -k \sin(k\,n), \tag{59}$$

where $k \in \mathbb{R}$, $n \in \mathbb{Z}$.

The Leibniz rule, which is also called the product rule, is a characteristic property of the derivative operator of first order [76, 95]. Using the principle of algebraic correspondence, we postulate that the exact discretization of the derivative of first order should satisfy the Leibniz rule. In this case, the lattice derivatives of integer orders can be considered as a derivative operator on a space of discrete functions. For lattice derivative of first order, the Leibniz rule

$$^{\mathrm{T}}\boldsymbol{\Delta}^1(f[n]\,g[n]) = (^{\mathrm{T}}\boldsymbol{\Delta}^1 f[n])\,g[n] + f[n]\,(^{\mathrm{T}}\boldsymbol{\Delta}^1 g[n]) \tag{60}$$

holds for discrete functions $f[n]$ and $g[n]$ belonging to the space $A(\mathbb{Z})$, where $^{\mathrm{T}}\boldsymbol{\Delta}^1$ is the lattice derivative of Cesaro or Poisson–Abel type. The proof is given in [98], where we restrict ourselves to the space $A(\mathbb{Z})$ of discrete functions.

Let us give some examples of lattice (difference) equations that are exact discretization of ordinary differential equations of integer orders. These lattice (discrete) equations have solutions that exactly correspond to the solutions of the corresponding differential equations.

The differential equation

$$D^1 \frac{df(x)}{dx} = -\lambda f(x) \tag{61}$$

has the solution

$$f(x) = f(0)\, e^{-\lambda x}. \tag{62}$$

Note that the standard forward-difference equation

$$^{f}\Delta^1 f[n] = -\lambda f[n], \tag{63}$$

where $^{f}\Delta^1 f[n] := f[n+1] - f[n]$, has the solution

$$f[n] = f[0]\,(1-\lambda)^n, \tag{64}$$

which cannot be considered as an exact discretization of the solution (62), since $(1-\lambda)^n \neq e^{-\lambda n}$. The lattice (T-difference) equation

$$^{\mathrm{T}}\boldsymbol{\Delta}^1 f[n] = -\lambda f[n] \tag{65}$$

has the solution

$$f[n] = f[0]\, e^{-\lambda n}, \tag{66}$$

which coincides with (62) in the sense that $f[n] = f(n)$ for all $n \in \mathbb{Z}$.

Analogously, the lattice (T-difference) equation

$$^{\mathrm{T}}\boldsymbol{\Delta}^2 f[n] + \lambda^2 f[n] = 0 \tag{67}$$

has the solution

$$f[n] = A\,\cos(\lambda n + \phi), \tag{68}$$

which exactly corresponds to the solutions of the corresponding differential equations.

The nonlinear differential and lattice equations

$$\frac{df(x)}{dx} - a^2 (f(x))^{k/(k+1)} = 0, \tag{69}$$

$$^{\mathrm{T}}\boldsymbol{\Delta}^1 f[n] - a^2 (f[n])^{k/(k+1)} = 0 \tag{70}$$

have the solutions

$$f(x) = \left(\frac{a^2}{k+1}x + C\right)^{k+1}, \tag{71}$$

$$f[n] = \left(\frac{a^2}{k+1}n + C\right)^{k+1}, \tag{72}$$

which are connected exactly by $f[n] = f(n)$ for all $n \in \mathbb{Z}$.

Some other examples of lattice differential equations, which are exact discretization of continuous differential equations, are given in [98, 106, 107, 113, 114].

An important property of lattice derivatives of integer orders is the semigroup property. In particular, this property implies that the repeated action of lattice derivative of first-order is equal to action of lattice derivative of the second order [98]. If a discrete function $f[n]$, with $n \in \mathbb{Z}$, belongs to the space $A(\mathbb{Z})$, then the repeated action of the lattice derivative of first order is given [98] by the equation ${}^{\mathrm{T}}\mathbf{\Delta}^{1\,\mathrm{T}}\mathbf{\Delta}^1 f[n] = {}^{\mathrm{T}}\mathbf{\Delta}^2 f[n]$, which holds for all functions $f[n]$ belonging to the space $A(\mathbb{Z})$.

As a result, we see that the lattice integro-differential operators satisfy the same algebraic characteristic relations as the continuum integro-differential operators. Consequently, the suggested lattice operators of integer orders can be considered as an exact discretization of standard derivatives of the same orders.

8 Asymptotic lattice fractional derivatives

Let us give examples of kernels of lattice integro-differentiation that satisfy the asymptotic conditions in the form

$$\hat{K}_\alpha^+(k) - \hat{K}_\alpha^+(0) = |k|^\alpha + o(|k|^\alpha), \tag{73}$$

$$\hat{K}_\alpha^-(k) = i\ \mathrm{sgn}(k)\,|k|^\alpha + o(|k|^\alpha) \quad (k \to 0). \tag{74}$$

It is obvious that the corresponding asymptotic lattice operators cannot be considered as an exact discretization of the standard continuum operators.

As a primal example of asymptotic kernel, let us give the expression of the kernel $K_\alpha^+(n)$ in the form

$$K_\alpha^+(n) = \frac{(-1)^n\,\Gamma(\alpha+1)}{\Gamma(\alpha/2+1+n)\,\Gamma(\alpha/2+1-n)}, \tag{75}$$

where $\alpha > -1$. This expression of kernel has been suggested in [71, 72] (see also [79, 87]) to describe long-range interactions of the lattice particles for noninteger values of α. Using equation 5.4.8.12 of [67] for the kernel (75), we get asymptotic expression (73). We see that the kernel (75) is the even function $K_\alpha^+(-n) = K_\alpha^+(n)$. Note that $K_\alpha^+(0)$ is not equal to zero for the kernel (75). The term $K_\alpha^+(0)$ describes a self-interaction of lattice particles. The interaction of different particles is described by $K^+_\alpha(n-m)$, with $n-m \neq 0$. For integer values of $\alpha \in \mathbb{N}$, we have $K^+_\alpha(n-m) = 0$ for $|n-m| \geq \alpha/2+1$. Using $K^+_{2j}(n-m) = 0$ for all $|n-m| \geq j+1$, we see that the kernel $K^+_{2j}(n-m)$ describes an interaction of the n-particle, with $2j$ particles, with numbers $n \pm 1, \dots, n \pm j$. Note that the asymptotic kernel (75) can be applied for integer and noninteger values of $\alpha > -1$.

As a second example of asymptotic kernel, we present the asymptotic kernel $K^-_\alpha(n)$ in the form

$$K^-_\alpha(n) = \frac{(-1)^{(n+1)/2}\,(2[(n+1)/2]-n)\,\Gamma(\alpha+1)}{2^\alpha\,\Gamma((\alpha+n)/2+1)\Gamma((\alpha-n)/2+1)}, \quad (\alpha > -1), \tag{76}$$

where $n \in \mathbb{Z}$, and the brackets [ ] mean the integral part (the floor function). Expression (76) can be written in the form

$$K^-_\alpha(n) = \begin{cases} \frac{(-1)^{m+1}\,\Gamma(\alpha+1)}{2^\alpha\,\Gamma(\alpha/2+1/2+m)\Gamma(\alpha/2+3/2-m)}, & n = 2m-1, \\ 0, & n = 2m, \end{cases} \tag{77}$$

where $m \in \mathbb{N}$. The kernel (76) has been suggested in [79, 87]. Using Equation (5.4.8.13) of [67] for the kernel (76) and (77), we get asymptotic expression (74). Note that Equation (76) gives (77), since $(2[(n+1)/2]-n)$ is equal to zero for $n = 2m$, and it is equal to 1 for $n = 2m-1$. Expression (76) combines two cases of (77), for even and odd values of $n \in \mathbb{N}$. Note that the kernel (76) is real valued function, since we have zero when the expression $(-1)^{(n+1)/2}$ becomes a complex number. It is easy to see that the kernel $K^-_\alpha(n)$ is the odd function ($K^-_\alpha(-n) = -K^-_\alpha(n)$), and $K^-_\alpha(0) = 0$.

As a third example of asymptotic kernel, we give the power-law kernel

$$K^+_\alpha(n) = |n|^{-(s+1)}, \quad s > 0, \tag{78}$$

which describes a lattice derivative of the order

$$\alpha = \begin{cases} s, & 0 < s < 2, \\ 2, & s > 2. \end{cases} \tag{79}$$

For $0 < s < 2$ ($s \neq 1$), we have

$$\hat{K}^+_\alpha(k) - \hat{K}^+_\alpha(0) = 2\Gamma(-s)\cos(\pi s/2)\,|k|^s. \tag{80}$$

For $s = 1$, Equation (78) gives

$$\hat{K}^+_\alpha(k) - \hat{K}^+_\alpha(0) = -\frac{\pi}{2}\,k. \tag{81}$$

For noninteger $s > 2$,

$$\frac{\hat{K}^+_\alpha(k) - \hat{K}^+_\alpha(0)}{k^2} = -\zeta(s-1), \tag{82}$$

where $\zeta(z)$ is the Riemann zeta-function. As a result, the kernel (78) defines a lattice derivative of integer and noninteger orders.

The other examples of asymptotic kernels of lattice fractions derivatives are given in [71, 72, 74, 79, 87, 97].

9 Lattice fractional derivatives of the Grünwald–Letnikov type

Let us define a lattice fractional partial derivative of the Grünwald–Letnikov type.

Lattice fractional partial derivatives ${}^{\mathrm{GL}}\mathbb{D}^{\pm}\left[\begin{smallmatrix}\alpha\\ j\end{smallmatrix}\right]$ of the Grünwald–Letnikov type, in the direction $\mathbf{e}_i = \mathbf{a}_i/|\mathbf{a}_i|$, are the operators

$$ {}^{\mathrm{GL}}\mathbb{D}^{\pm}_L\begin{bmatrix}\alpha\\ j\end{bmatrix} f(\mathbf{m}) = \frac{1}{a_j^{\alpha}} \sum_{m_j=-\infty}^{+\infty} {}^{\mathrm{GL}}K^{\pm}_{\alpha}(n_j - m_j) f(\mathbf{m}), \tag{83} $$

where $j = 1, 2, 3$; the kernels ${}^{\mathrm{GL}}K^{\pm}_{\alpha}(n)$ are defined by the equations

$$ {}^{\mathrm{GL}}K^{\pm}_{\alpha}(n) = \frac{(-1)^n\,\Gamma(\alpha+1)}{2\,\Gamma(|n|+1)}\,\frac{\Gamma(\alpha+n+1) \pm \Gamma(\alpha-n+1)}{\Gamma(\alpha+n+1)\,\Gamma(\alpha-n+1)}, \tag{84} $$

and α is the order of these derivatives.

It should be noted that lattice models with the long-range interaction of the form ${}^{\mathrm{GL}}K^{+}_{\alpha}(n)$, and correspondent fractional nonlocal continuum models, have been suggested in [83, 85], (see also [74, 90, 97, 101]).

Note that the kernels ${}^{\mathrm{GL}}K^{\pm}_{\alpha}(n)$ are even and odd functions, ${}^{\mathrm{GL}}K^{\pm}_{\alpha}(-n) = \pm{}^{\mathrm{GL}}K^{\pm}_{\alpha}(n)$. The form of these lattice fractional derivatives are defined by the addition and subtraction of the Grünwald–Letnikov fractional differences. These differences of order $\alpha \in \mathbb{R}_+$ are defined in the form of the infinite series

$$ \nabla^{\alpha}_{a,\pm} f(x) = \sum_{n=0}^{\infty} \frac{(-1)^n\,\Gamma(\alpha+1)}{\Gamma(n+1)\Gamma(\alpha-n+1)} f(x \mp na), \tag{85} $$

where $a > 0$. The series in (85) converges absolutely and uniformly for every bounded function $f(x)$ and $\alpha > 0$ (see Section 20 in [69]). The difference $\nabla^{\alpha}_{a,+}$ is the called left-sided fractional difference, and $\nabla^{\alpha}_{a,-}$ is called the right-sided fractional difference.

The Fourier transforms of the Grünwald–Letnikov fractional differences are given by the equation

$$ \mathrm{F}\{\nabla^{\alpha}_{a,\pm} f(x)\}(k) = (1 - \exp\{\pm ika\})^{\alpha}\,\mathrm{F}\{f(x)\}(k) $$

for any function $f(x) \in L_1(\mathbb{R})$ (see Property 2.30 in [35]). Therefore, the Grünwald–Letnikov fractional differences can be considered as an asymptotic lattice derivative only.

The left- and right-sided partial Grünwald–Letnikov fractional derivatives of order $\alpha > 0$ are defined by

$$ {}^{\mathrm{GL}}D^{\alpha}_{x_j,\pm} f(\mathbf{r}) = \lim_{a_j \to 0+} \frac{\nabla^{\alpha}_{a_j,\pm} f(\mathbf{r})}{|a_j|^{\alpha}}. \tag{86} $$

Substitution of (85) into (86) gives

$$ {}^{\mathrm{GL}}D^{\alpha}_{x_j,\pm} f(\mathbf{r}) = \lim_{a_j \to 0+} \frac{1}{|a_j|^{\alpha}} \sum_{n_j=0}^{\infty} \frac{(-1)^{n_j}\,\Gamma(\alpha+1)}{\Gamma(n_j+1)\Gamma(\alpha-n_j+1)} f(\mathbf{r} \mp n_j \mathbf{a}_j). \tag{87} $$

Note that the Grünwald–Letnikov derivatives (87) of integer orders $\alpha = n \in \mathbb{N}$ are connected with the standard partial derivatives by the equations

$$ {}^{\mathrm{GL}}D^{n}_{x_j,\pm} f(\mathbf{r}) = (\pm 1)^n \frac{\partial^n f(\mathbf{r})}{\partial x_j^n}. \tag{88} $$

The asymptotic lattice fractional derivatives (83) are transformed by the continuous limit operation ($a_j \to 0$) into the fractional partial derivatives of Grünwald–Letnikov type of order α with respect to coordinate x_j in the form

$$ \mathrm{Lim}\left({}^{\mathrm{GL}}\mathbb{D}^{\pm}_{L} \begin{bmatrix} \alpha \\ j \end{bmatrix} f(\mathbf{m}) \right) = {}^{\mathrm{GL}}\mathrm{D}^{\alpha,\pm}_{j} f(\mathbf{r}), \tag{89} $$

where ${}^{\mathrm{GL}}D^{\alpha,\pm}_{j}$ are the fractional derivatives of the Grünwals–Letnikov type

$$ {}^{\mathrm{GL}}\mathrm{D}^{\alpha,\pm}_{j} = \frac{1}{2}\left({}^{\mathrm{GL}}D^{\alpha}_{x_j,+} \pm {}^{\mathrm{GL}}D^{\alpha}_{x_j,-} \right), \tag{90} $$

which contain the Grünwald–Letnikov fractional derivatives (87).

Using (88), the derivatives (90) for integer orders $\alpha = n \in \mathbb{N}$ have the forms

$$ {}^{\mathrm{GL}}\mathrm{D}^{n,+}_{j} = \frac{1}{2}\left(\frac{\partial^n}{\partial x_j^n} + (-1)^n \frac{\partial^n}{\partial x_j^n} \right) = \begin{cases} 0, & n = 2m-1, \\ \frac{\partial^n}{\partial x_j^n}, & n = 2m, \end{cases} \tag{91} $$

$$ {}^{\mathrm{GL}}\mathrm{D}^{n,-}_{j} = \frac{1}{2}\left(\frac{\partial^n}{\partial x_j^n} - (-1)^n \frac{\partial^n}{\partial x_j^n} \right) = \begin{cases} \frac{\partial^n}{\partial x_j^n}, & n = 2m-1, \\ 0, & n = 2m, \end{cases} \tag{92} $$

where $m \in \mathbb{N}$. Therefore, ${}^{\mathrm{GL}}\mathrm{D}^{n,+}_{j}$ is the standard derivative of integer order n for even values α only, and ${}^{\mathrm{GL}}\mathrm{D}^{n,-}_{j}$ is the derivative of integer order n for odd values α only.

The lattice fractional integral operations of Grünwald–Letnikov type can be defined by expression (83) for $\alpha < 0$. This possibility is based on the fact that the series (85) can be used for $\alpha < 0$ (see Section 20 in [69]). Equation (87) defines the Grünwald–Letnikov fractional integration if

$$ |f(x)| < c(1+|x|)^{-\mu}, \quad \mu > |\alpha|. \tag{93} $$

The existence of the Grünwald–Letnikov fractional integration means that we have a possibility to define a lattice fractional integration.

The suggested lattice derivatives are represented by infinite series, instead of finite series, which are the usually used in standard and nonstandard (Mickens) differences. Infinite series of the fractional-order differences and the corresponding derivatives of noninteger orders were first proposed by Grünwald [29] and by Letnikov [41]. Now, there are other types of fractional differences, which are proposed by Kuttner [39], Cargo and Shisha [10], Diaz and Osler [14], Ortigueira and Coito [60], Ortigueira [58, 59], Tarasov [71, 72, 74], Ortigueira, Rivero, Trujillo [61, 63], and others. The Grünwald–Letnikov differences [29, 41, 69] and other types of fractional differences, which are proposed in [10, 14, 39, 58–61, 63, 71, 72, 74], cannot be considered an exact discretization of corresponding derivatives. These differences have algebraic properties, which do not coincide with the characteristic properties of differential operators. It is easy to see that these differences for integer values of orders do not have the same algebraic properties as the integer-order derivatives. For example, the Grünwald–Letnikov fractional differences are defined (see Section 20 in [69]) by the infinite series

$$({}^{\mathrm{GL}}\nabla^{\alpha}_{1,+}f)[n] = \sum_{j=0}^{\infty} \frac{(-1)^j\,\Gamma(\alpha+1)}{\Gamma(j+1)\Gamma(\alpha-j+1)} f[n-j]. \tag{94}$$

For integer values of $\alpha = m \in \mathbb{N}$, the differences (94) are equal to the backward finite difference

$$({}^{\mathrm{GL}}\nabla^{m}_{1,+}f)[n] = \sum_{j=0}^{m} \frac{(-1)^m m!}{j!\,(m-j)!} f(n-j) = ({}^{b}\Delta^m f)[n]. \tag{95}$$

The properties of this operator for $m = 1$ are the following:

(1) It is well known that the standard Leibniz rule is not satisfied by the backward finite difference of first order:

$$({}^{b}\Delta^1 f\,g)[n] \neq ({}^{b}\Delta^1 f)[n]\,g[n] + f[n]\,({}^{b}\Delta^1 g)[n]. \tag{96}$$

(2) The backward finite difference of power functions does not give standard expressions. For example,

$${}^{b}\Delta^1 n^3 = n^3 - (n-1)^3 = 3n^2 - 3n + 1 \neq 3n^2. \tag{97}$$

(3) The backward finite difference of entire functions does not give standard expressions. For example,

$$\begin{aligned} {}^{b}\Delta^1 \cos(k\,n) &= \cos(k\,n) - \cos(k\,(n-1)) \\ &= -2\sin(k/2)\,\sin(k\,n - k/2) \neq -k\,\sin(k\,n). \end{aligned} \tag{98}$$

It should be emphasized that the lattice derivatives, which are proposed in [79, 87, 97], and the exact fractional differences, which are proposed in [88, 98, 99, 105], give the standard expressions ${}^{\mathrm{T}}\boldsymbol{\Delta}^1 n^3 = 3n^2$, ${}^{\mathrm{T}}\boldsymbol{\Delta}^1 \cos(k\,n) = -k \sin(k\,n)$, and the standard Leibniz holds.

The Fourier series transform $\mathrm{F}_{a,\Delta}$ of the Grünwald–Letnikov fractional differences has the form

$$\mathrm{F}_{a,\Delta}\{({}^{\mathrm{GL}}\nabla^{\alpha}_{a,\mp}f)[n]\}(k) = (1 - \exp\{\mp i\,k\,a\})^{\alpha}\,\mathrm{F}_{a,\Delta}\{f[n]\}(k) \tag{99}$$

(see Property 2.30 of [35]). Using (99), we can see that these differences cannot be considered as an exact discretization of the derivative of integer order m, since necessary condition (36) is not satisfied,

$$\frac{\mathrm{F}^{-1}\{\mathrm{F}_{a,\Delta}\{{}^{\mathrm{GL}}\nabla^{m}_{a,+}\}\}}{a^m} \neq \frac{d^m}{dx^m} \quad (m \in \mathbb{N}), \tag{100}$$

and we have inequalities (96)–(98). By analogy with the Grünwald–Letnikov differences, we have the same situation with all other types of fractional differences, which are proposed in [10, 14, 39, 58–61, 63, 71, 72, 74] and other. All these fractional differences cannot be considered as lattice operators that are exact discrete analogs of the derivative of integer and noninteger orders.

10 Conclusion

In this chapter, we demonstrate that long-range interactions are directly related with the fractional calculus. The fractional lattice operators and exact fractional differences allow us to describe long-range interactions in the physical lattices. From the mathematical point of view, the fractional lattice operators describe an exact discretization of the standard integro-differentiation of integer and noninteger orders. The lattice derivatives have the same algebraic properties as the standard (continuum) partial derivatives. The action of the lattice operators (exact differences) on entire functions gives the standard expressions as an action of corresponding continuum derivatives. The lattice integro-differentiation allows us to obtain an exact discretization of wide class of differential equations. As a result, the lattice analogs of solutions of differential equations are solutions of the corresponding equations with lattice integro-differentiation. The considered lattice approach, which is based on fractional lattice integro-differentiation and exact fractional differences, can be used to get structural models of objects, media, and systems with nonlocality, where the long-range interactions are crucial in determining nonlocal properties. This approach has wide applications in various fields of physics, including the continuum mechanics, the statistical physics, the nonlinear dynamics, the quantum mechanics, and the quantum field theory.

Bibliography

[1] G. L. Alfimov, V. M. Eleonsky, and L. M. Lerman, Solitary wave solutions of nonlocal sine-Gordon equations, *Chaos*, **8** (1998), 257–271.

[2] G. L. Alfimov and V. G. Korolev, On multikink states described by the nonlocal sine-Gordon equation, *Physics Letters A*, **246** (1998), 429–435.

[3] R. Bachelard, C. Chandre, D. Fanelli, X. Leoncini, and S. Ruffo, Abundance of regular orbits and nonequilibrium phase transitions in the thermodynamic limit for long-range systems, *Physics Review Letters*, **101**(26) (2008), 260603.

[4] C. Baesens and R. S. MacKay, Algebraic localisation of linear response in networks with algebraically decaying interaction, and application to discrete breathers in dipole–dipole systems, *Helvetica Physica Acta*, **72** (1999), 23–32.

[5] J. Barre, F. Bouchet, T. Dauxois, and S. Ruffo, Large deviation techniques applied to systems with long-range interactions, *Journal of Statistical Physics*, **119** (2005), 677–713.

[6] O. M. Braun and Y. S. Kivshar, Nonlinear dynamics of the Frenkel–Kontorova model, *Physics Reports*, **306** (1998), 2–108.

[7] O. M. Braun and Y. S. Kivshar, *The Frenkel–Kontorova Model: Concepts, Methods, and Applications*, Springer, Belin, New York, 2004, 472 pages.

[8] O. M. Braun, Y. S. Kivshar, and I. I. Zelenskaya, Kinks in the Frenkel–Kontorova model with long-range interparticle interactions, *Physical Review B*, **41** (1990), 7118–7138.

[9] A. Campa, T. Dauxois, and S. Ruffo, Statistical mechanics and dynamics of solvable models with long-range interactions, *Physics Reports*, **480**(3–6) (2009), 57–159.

[10] G. T. Cargo and O. Shisha, Zeros of polynomials and fractional order differences of their coefficients, *Journal of Mathematical Analysis and Applications*, **7**(2) (1963), 176–182, DOI:10.1016/0022-247X(63)90046-5.

[11] T. Dauxois, S. Ruffo, E. Arimondo, M. Wilkens, eds. *Dynamics and Thermodynamics of Systems With Long-Range Interactions*, Springer-Verlag, Berlin, Heidelberg, New York, 2002, ISBN 3-540-44315-0.

[12] M. Di Paola, F. Marino, and M. Zingales, A generalized model of elastic foundation based on long-range interactions: integral and fractional model, *International Journal of Solids and Structures* **46** (2009), 3124–3137.

[13] M. Di Paola and M. Zingales, Long-range cohesive interactions of non-local continuum faced by fractional calculus, *International Journal of Solids and Structures*, **45** (2008), 5642–5659.

[14] J. B. Diaz and T. J. Osler, Differences of fractional order, *Mathematics of Computation*, **28**(125) (1974), 185–202, DOI:10.1090/S0025-5718-1974-0346352-5.

[15] P. A. M. Dirac, The fundamental equations of quantum mechanics, *Proceedings of the Royal Society London. Series A*, **109**(752) (1925), 642–653, DOI:10.1098/rspa.1925.0150.

[16] F. J. Dyson, Existence of a phase-transition in a one-dimensional Ising ferromagnet, *Communications in Mathematical Physics*, **12**(2) (1969), 91–107.

[17] F. J. Dyson, Non-existence of spontaneous magnetization in a one-dimensional Ising ferromagnet, *Communications in Mathematical Physics*, **12** (1969), 212–215.

[18] F. J. Dyson, An Ising ferromagnet with discontinuous long-range order, *Communications in Mathematical Physics*, **21** (1971), 269–283.

[19] A. Erdelyi (ed.), *Tables of Integral Transforms. Volume II*, McGraw Hill, New York, London, Toronto, 1954, 451 pages.

[20] G. M. Fichtenholz, *Differential and Integral Calculus. Volume 2*, 7th Ed., Moscow, Nauka, 1969, 800 pages, Chapters 11 and 12 [in Russian].

[21] G. M. Fichtenholz, *Infinite series: ramifications*, Gordon and Breach, New York, 1970, 139 pages, ISBN-10: 0677209401, ISBN-13: 978-0677209401.

[22] G. M. Fichtenholz, *Functional Series*, Gordon and Breach Science, New York, 1970, 176 pages, ISBN-10: 0677209509, ISBN-13: 978-0677209500.
[23] S. Flach, Breathers on lattices with long-range interaction, *Physical Review E*, **58** (1998), R4116–R4119.
[24] S. Flach and C. R. Willis, Discrete breathers, *Physics Reports*, **295** (1998), 181–264.
[25] J. Frohlich, R. Israel, E. H. Lieb, and B. Simon, Phase transitions and reflection positivity I. General theory and long-range lattice model, *Communications in Mathematical Physics*, **62** (1978), 1–34.
[26] Yu. Gaididei, N. Flytzanis, A. Neuper, and F. G. Mertens, Effect of nonlocal interactions on soliton dynamics in anharmonic lattices, *Physical Review Letters*, **75** (1995), 2240–2243.
[27] Yu. B. Gaididei, S. F. Mingaleev, P. L. Christiansen, and K. O. Rasmussen, Effects of nonlocal dispersive interactions on self-trapping excitations, *Physical Review E*, **55** (1997), 6141–6150.
[28] A. V. Gorbach and S. Flach, Compactlike discrete breathers in systems with nonlinear and nonlocal dispersive terms, *Physical Review E*, **72** (2005), 056607.
[29] A. K. Grünwald, About "limited" derivations their application, *Zeitschrift für Angewandte Mathematik und Physik*, **12** (1897), 441–480 [in German].
[30] G. H. Hardy, *Divergent Series*, 2nd Ed., American Mathematical Society, New York, 2000, 396 pages.
[31] H. Hinrichsen, Non-equilibrium phase transitions with long-range interactions, *Journal of Statistical Mechanics: Theory and Experiment*, **2007** (2007), P07006. DOI:10.1088/1742-5468/2007/07/P07006 (arxiv:cond-mat/0702169).
[32] Y. Ishimori, Solitons in a one-dimensional Lennard–Jones lattice, *Progress of Theoretical Physics*, **68** (1982), 402–410.
[33] R. Ishiwata and Y. Sugiyama, Relationships between power-law long-range interactions and fractional mechanics, *Physica A*, **391** (2012), 5827–5838, DOI:10.1016/j.physa.2012.06.055.
[34] G. S. Joyce, Absence of ferromagnetism or antiferromagnetism in the isotropic Heisenberg model with long-range interactions, *Journal of Physics C*, **2** (1969), 1531–1533.
[35] A. A. Kilbas, H. M. Srivastava, and J. J. Trujillo, *Theory and Applications of Fractional Differential Equations*, Elsevier, Amsterdam, 2006, 353 pages.
[36] V. Kiryakova, *Generalized Fractional Calculus and Applications*, Longman, Harlow and Wiley, New York, 1994.
[37] N. Korabel and G. M. Zaslavsky, Transition to chaos in discrete nonlinear Schrödinger equation with long-range interaction, *Physica A*, **378**(2) (2007), 223–237.
[38] N. Korabel, G. M. Zaslavsky, and V. E. Tarasov, Coupled oscillators with power-law interaction and their fractional dynamics analogues, *Communications in Nonlinear Science and Numerical Simulation*, **12**(8) (2007), 1405–1417 (arXiv:math-ph/0603074).
[39] B. Kuttner, On differences of fractional order, *Proceedings of the London Mathematical Society*, **3–7**(1) (1957), 453–466, DOI:10.1112/plms/s3-7.1.453.
[40] N. Laskin and G. M. Zaslavsky, Nonlinear fractional dynamics on a lattice with long-range interactions, *Physica A*, **368**(1) (2006), 38–54. (arXiv:nlin.SI/0512010).
[41] A. V. Letnikov, Theory of differentiation with arbitrary pointer, *Matematicheskii Sbornik*, **3** (1868), 1–68 [in Russian].
[42] Y. Levin, R. Pakter, F. B. Rizzato, T. N. Teles, and F. P. C. Benetti, Nonequilibrium statistical mechanics of systems with long-range interactions, *Physics Reports*, **535**(1) (2014), 1–60, DOI:10.1016/j.physrep.2013.10.001.
[43] Y. L. Luke, *The Special Functions and Their Approximations: Volume 1 I*, Academic Press, New York, London, 1969, 349 pages (Academic Press, 2012, ISBN: 978-0124110373).

[44] T. M. Michelitsch, B. A. Collet, A. P. Riascos, A. F. Nowakowski, and F. C. G. A. Nicolleau, A fractional generalization of the classical lattice dynamics approach, *Chaos, Solitons and Fractals*, **92** (2016), 43–50, DOI:10.1016/j.chaos.2016.09.009 (arXiv:1610.03744).
[45] R. E. Mickens, Difference equation models of differential equations, *Mathematical and Computer Modelling*, **11** (1988), 528–530, DOI:10.1016/0895-7177(88)90549-3.
[46] R. E. Mickens, *Nonstandard Finite Difference Models of Differential Equations*, World Scientific, Singapore, 1994.
[47] R. E. Mickens, Discretizations of nonlinear differential equations using explicit nonstandard methods, *Journal of Computational and Applied Mathematics*, **110**(1) (1999), 181–185, DOI:10.1016/S0377-0427(99)00233-2.
[48] R. E. Mickens (ed.), *Applications of Nonstandard Finite Difference Schemes*, World Scientific, Singapore, 2000, 264 pages.
[49] R. E. Mickens, Nonstandard finite difference schemes for differential equations, *Journal of Difference Equations and Applications*, **8**(9) (2002), 823–847, DOI:10.1080/1023619021000000807.
[50] R. E. Mickens (ed.), *Advances in the Applications of Nonstandard Finite Difference Schemes*, World Scientific, Singapore, 2005, 664 pages.
[51] R. E. Mickens, J. Munyakazi, and T. M. Washington, A note on the exact discretization for a Cauchy–Euler equation: application to the Black–Scholes equation, *Journal of Difference Equations and Applications*, **21**(7) (2015), 547–552, DOI:10.1080/10236198.2015.1034118.
[52] S. F. Mingaleev, Y. B. Gaididei, and F. G. Mertens, Solitons in anharmonic chains with power-law long-range interactions, *Physical Review E*, **58** (1998), 3833–3842.
[53] S. F. Mingaleev, Y. B. Gaididei, and F. G. Mertens, Solitons in anharmonic chains with ultra-long-range interatomic interactions, *Physical Review E*, **61** (2000), R1044–R1047. (arXiv:patt-sol/9910005).
[54] H. Nakano and M. Takahashi, Quantum Heisenberg chain with long-range ferromagnetic interactions at low temperatures, *Journal of the Physical Society of Japan*, **63** (1994), 926–933. (arXiv:cond-mat/9311034).
[55] H. Nakano and M. Takahashi, Quantum Heisenberg model with long-range ferromagnetic interactions, *Physical Review B*, **50** (1994), 10331–10334.
[56] H. Nakano and M. Takahashi, Quantum Heisenberg ferromagnets with long-range interactions, *Journal of the Physical Society of Japan*, **63** (1994), 4256–4257.
[57] H. Nakano and M. Takahashi, Magnetic properties of quantum Heisenberg ferromagnets with long-range interactions, *Physical Review B*, **52** (1995), 6606–6610.
[58] M. D. Ortigueira, Riesz potential operators and inverses via fractional centred derivatives, *International Journal of Mathematics and Mathematical Sciences*, **2006** (2006), 48391, DOI:10.1155/IJMMS/2006/48391.
[59] M. D. Ortigueira, Fractional central differences and derivatives, *Journal of Vibration and Control*, **14**(9–10) (2008), 1255–1266, DOI:10.1177/1077546307087453.
[60] M. D. Ortigueira and F. Coito, From differences to derivatives, *Fractional Calculus & Applied Analysis. An International Journal for Theory and Applications*, **7**(4) (2004), 459–471.
[61] M. D. Ortigueira, M. Rivero, and J. J. Trujillo, The incremental ratio based causal fractional calculus, *International Journal of Bifurcation and Chaos in Applied Sciences and Engineering*, **22**(4) (2012), 1250078, 9 pages, DOI:10.1142/S0218127412500782.
[62] M. D. Ortigueira and J. A. Tenreiro Machado, What is a fractional derivative?, *Journal of Computational Physics*, **293** (2015), 4–13, DOI:10.1016/j.jcp.2014.07.019.
[63] M. D. Ortigueira and J. J. Trujillo, A unified approach to fractional derivatives, *Communications in Nonlinear Science and Numerical Simulation*, **17**(12) (2012), 5151–5157, DOI:10.1016/j.cnsns.2012.04.021.

[64] V. L. Pokrovsky and A. Virosztek, Long-range interactions in commensurate-incommensurate phase transition, *Journal of Physics C*, **16** (1983), 4513–4525.

[65] R. B. Potts, Differential and difference equations, *The American Mathematical Monthly*, **89**(6) (1982), 402–407, DOI:10.2307/2321656.

[66] R. B. Potts, Ordinary and partial difference equations, *Journal of the Australian Mathematical Society. Series B. Applied Mathematics*, **27**(6) (1986), 488–501, DOI:10.1515/fca-2015-0020.

[67] A. P. Prudnikov, Yu. A. Brychkov, and O. I. Marichev, *Integrals and Series, Vol. 1: Elementary Functions*, Gordon and Breach, New York, 1986.

[68] K. O. Rasmussen, P. L. Christiansen, M. Johansson, Yu. B. Gaididei, and S. F. Mingaleev, Localized excitations in discrete nonlinear Schröedinger systems: effects of nonlocal dispersive interactions and noise, *Physica D*, **113** (1998), 134–151.

[69] S. G. Samko, A. A. Kilbas, and O. I. Marichev, *Fractional Integrals and Derivatives Theory and Applications*, Gordon and Breach, New York, 1993, 1006 pages.

[70] J. R. Sousa, Phase diagram in the quantum XY model with long-range interactions, *The European Physical Journal. B, Condensed Matter Physics*, **43** (2005), 93–96.

[71] V. E. Tarasov, Map of discrete system into continuous, *Journal of Mathematical Physics*, **47**(9) (2006), 092901, 24 pages, DOI:10.1063/1.2337852 (arXiv:0711.2612).

[72] V. E. Tarasov, Continuous limit of discrete systems with long-range interaction, *Journal of Physics A*, **39**(48) (2006), 14895–14910, DOI:10.1088/0305-4470/39/48/005. (arXiv:0711.0826).

[73] V. E. Tarasov, *Quantum Mechanics of Non-Hamiltonian and Dissipative Systems*, Elsevier Science, Amsterdam, Boston, 2008, 540 pages.

[74] V. E. Tarasov, Fractional dynamics of media with long-range interaction, in *Fractional Dynamics: Applications of Fractional Calculus to Dynamics of Particles, Fields and Media*, pp. 153–214, Springer, Berlin, Heidelberg, 2010, DOI:10.1007/978-3-642-14003-7_8.

[75] V. E. Tarasov, *Fractional Dynamics: Applications of Fractional Calculus to Dynamics of Particles, Fields and Media*, Springer, New York, 2010, 450 pages, DOI:10.1007/978-3-642-14003-7.

[76] V. E. Tarasov, No violation of the Leibniz rule. No fractional derivative, *Communications in Nonlinear Science and Numerical Simulation*, **18**(11) (2013), 2945–2948, DOI:10.1016/j.cnsns.2013.04.001.

[77] V. E. Tarasov, Lattice model with power-law spatial dispersion for fractional elasticity, *Central European Journal of Physics*, **11**(11) (2013), 1580–1588 (arXiv:1501.01201).

[78] V. E. Tarasov, Review of some promising fractional physical models, *International Journal of Modern Physics B*, **27**(9) (2013), 1330005, DOI:10.1142/S0217979213300053.

[79] V. E. Tarasov, Toward lattice fractional vector calculus, *Journal of Physics A: Mathematical and Theoretical*, **47**(35) (2014), 355204, 51 pages, DOI:10.1088/1751-8113/47/35/355204.

[80] V. E. Tarasov, Fractional gradient elasticity from spatial dispersion law, *ISRN Condensed Matter Physics*, **2014** (2014), 794097, DOI:10.1155/2014/794097.

[81] V. E. Tarasov, Lattice with long-range interaction of power-law type for fractional non-local elasticity, *International Journal of Solids and Structures* **51**(15–16) (2014), 2900–2907, DOI:10.1016/j.ijsolstr.2014.04.014 (arXiv:1502.05492).

[82] V. E. Tarasov, General lattice model of gradient elasticity, *Modern Physics Letters B*, **28**(7) (2014), 1450054, 17 pages, DOI:10.1142/S0217984914500547 (arXiv:1501.01435).

[83] V. E. Tarasov, Lattice model of fractional gradient and integral elasticity: long-range interaction of Grunwald–Letnikov–Riesz type, *Mechanics of Materials*, **70**(1) (2014), 106–114, DOI:10.1016/j.mechmat.2013.12.004 (arXiv:1502.06268).

[84] V. E. Tarasov, Large lattice fractional Fokker–Planck equation, *Journal of Statistical Mechanics: Theory and Experiment*, **2014**(9) (2014), P09036, DOI:10.1088/1742-5468/2014/09/P09036 (arXiv:1503.03636).

[85] V. E. Tarasov, Fractional diffusion equations for lattice and continuum: Grunwald–Letnikov differences and derivatives approach, *International Journal of Statistical Mechanics*, **2014** (2014), Article ID 873529, 7 pages, DOI:10.1155/2014/873529 (arXiv:1503.03201).

[86] V. E. Tarasov, Fractional quantum field theory: from lattice to continuum, *Advances in High Energy Physics*, **2014** (2014), 957863, DOI:10.1155/2014/957863.

[87] V. E. Tarasov, Lattice fractional calculus, *Applied Mathematics and Computation*, **257** (2015), 12–33, DOI:10.1016/j.amc.2014.11.033.

[88] V. E. Tarasov, Exact discrete analogs of derivatives of integer orders: differences as infinite series, *Journal of Mathematics*, **2015** (2015), 134842, DOI:10.1155/2015/134842.

[89] V. E. Tarasov, Three-dimensional lattice approach to fractional generalization of continuum gradient elasticity, *Progress in Fractional Differentiation and Applications*, **1**(4) (2015), 243–258, DOI:10.12785/pfda/010402.

[90] V. E. Tarasov, Fractional-order difference equations for physical lattices and some applications, *Journal of Mathematical Physics*, **56**(10) (2015), 103506, DOI:10.1063/1.4933028.

[91] V. E. Tarasov, Discretely and continuously distributed dynamical systems with fractional nonlocality, In: *Fractional Dynamics*, C. Cattani, H. M. Srivastava, X.-J. Yang (Eds.), pp. 31–49, De Gruyter Open, Warsaw, Berlin, Chapter 3, 2015, ISBN (Online): 9783110472097, DOI (Chapter): 10.1515/9783110472097-003.

[92] V. E. Tarasov, Lattice model with nearest-neighbor and next-nearest-neighbor interactions for gradient elasticity, *Discontinuity, Nonlinearity, and Complexity*, **4**(1) (2015), 11–23, DOI:10.5890/DNC.2015.03.002 (arXiv:1503.03633).

[93] V. E. Tarasov, Non-linear fractional field equations: weak non-linearity at power-law non-locality, *Nonlinear Dynamics*, **80**(4) (2015), 1665–1672, DOI:10.1007/s11071-014-1342-0.

[94] V. E. Tarasov, Fractional Liouville equation on lattice phase-space, *Physica. A, Statistical Mechanics and Its Applications*, **421** (2015), 330–342, DOI:10.1016/j.physa.2014.11.031. (arXiv:1503.04351).

[95] V. E. Tarasov, Leibniz rule and fractional derivatives of power functions, *Journal of Computational and Nonlinear Dynamics*, **11**(3) (2016), 031014, DOI:10.1115/1.4031364.

[96] V. E. Tarasov, On chain rule for fractional derivatives, *Communications in Nonlinear Science and Numerical Simulation*, **30**(1–3) (2016), 1–4, DOI:10.1016/j.cnsns.2015.06.007.

[97] V. E. Tarasov, United lattice fractional integro-differentiation, *Fractional Calculus and Applied Analysis*, **19**(3) (2016), 625–664, DOI:10.1515/fca-2016-0034.

[98] V. E. Tarasov, Exact discretization by Fourier transforms, *Communications in Nonlinear Science and Numerical Simulation*, **37** (2016), 31–61, DOI:10.1016/j.cnsns.2016.01.006.

[99] V. E. Tarasov, Exact finite differences: a brief overview, *Almanac of Modern Science and Education (Almanah Sovremennoj Nauki i Obrazovaniya)*, **7**(109) (2016), 105–108, [in Russian] ISSN 1993-5552.

[100] V. E. Tarasov, Discrete model of dislocations in fractional nonlocal elasticity, *Journal of King Saud University—Science*, **28**(1) (2016), 33–36, DOI:10.1016/j.jksus.2015.04.001.

[101] V. E. Tarasov, Three-dimensional lattice models with long-range interactions of Grunwald–Letnikov type for fractional generalization of gradient elasticity, *Meccanica*, **51**(1) (2016), 125–138, DOI:10.1007/s11012-015-0190-4.

[102] V. E. Tarasov, What discrete model corresponds exactly to gradient elasticity equation?, *Journal of Mechanics of Materials and Structures*, **11**(4) (2016), 329–343, DOI:10.2140/jomms.2016.11.329.

[103] V. E. Tarasov, Exact discretization of Schrodinger equation, *Physics Letters A*, **380**(1–2) (2016), 68–75, DOI:10.1016/j.physleta.2015.10.039.

[104] V. E. Tarasov, Exact discrete analogs of canonical commutation and uncertainty relations, *Mathematics*, **4**(3) (2016), 44, DOI:10.3390/math4030044.

[105] V. E. Tarasov, Exact discretization of fractional Laplacian, *Computers & Mathematics with Applications*, **73**(5) (2017), 855–863, DOI:10.1016/j.camwa.2017.01.012.

[106] V. E. Tarasov, Discrete wave equation with infinite differences, *Applied Mathematics and Information Sciences Letters*, **5**(2) (2017), 41–44, DOI:10.18576/amisl/050201.

[107] V. E. Tarasov, Exact solution of T-difference radial Schrodinger equation, *International Journal of Applied and Computational Mathematics*, **3**(4) (2017), 2779–2784, DOI:10.1007/s40819-016-0270-8.

[108] V. E. Tarasov, No nonlocality. No fractional derivative, *Communications in Nonlinear Science and Numerical Simulation*, **62** (2018), 157–163, DOI:10.1016/j.cnsns.2018.02.019. (arXiv:1803.00750).

[109] V. E. Tarasov and G. M. Zaslavsky, Fractional dynamics of coupled oscillators with long-range interaction, *Chaos*, **16**(2) (2006), 023110. (arXiv:nlin.PS/0512013).

[110] V. E. Tarasov and G. M. Zaslavsky, Fractional dynamics of systems with long-range interaction, *Communications in Nonlinear Science and Numerical Simulation*, **11**(8) (2006), 885–898.

[111] V. E. Tarasov and G. M. Zaslavsky, Fractional dynamics of systems with long-range space interaction and temporal memory, *Physica A*, **383**(2) (2007), 291–308 (arXiv:math-ph/0702065).

[112] V. V. Tarasova and V. E. Tarasov, Exact discretization of economic accelerators and multipliers with memory, *Journal of Economy and Entrepreneurship [Ekonomika i Predprinimatelstvo]*, **7**(84) (2017), 1063–1069 [in Russian].

[113] V. V. Tarasova and V. E. Tarasov, Exact discretization of economic accelerator and multiplier with memory, *Fractal and Fractional*, **1**(1) (2017), Article ID 6. DOI:10.3390/fractalfract1010006.

[114] V. V. Tarasova and V. E. Tarasov, Accelerators in macroeconomics: a comparison of discrete and continuous approaches, *Scientific Journal [Nauchnyy Zhurnal]*, **8**(21) (2017), 4–14 [in Russian].

[115] V. V. Tarasova and V. E. Tarasov, Accelerators in macroeconomics: comparison of discrete and continuous approaches, *American Journal of Economics and Business Administration*, **9**(3) (2017), 47–55. DOI: 10.3844/ajebasp.2017.47.55.

[116] C. J. Tessone, M. Cencini, and A. Torcini, Synchronization of extended chaotic systems with long-range interactions: an analogy to Levy-flight spreading of epidemics, *Physical Review Letters*, **97**(22) (2006), 224101.

[117] T. L. Van Den Berg, D. Fanelli, and X. Leoncini, Stationary states and fractional dynamics in systems with long-range interactions, *Europhysics Letters*, **89**(5) (2010), 50010, DOI:10.1209/0295-5075/89/50010. E-print: arXiv:0912.3060.

[118] D. B. Vasconcelos, R. L. Viana, S. R. Lopes, A. M. Batista, and S. E. de S. Pinto, Spatial correlations and synchronization in coupled map lattices with long-range interactions, *Physica A*, **343** (2004), 201–218, DOI:10.1016/j.physa.2004.06.063.

[119] G. M. Zaslavsky, M. Edelman, and V. E. Tarasov, Dynamics of the chain of oscillators with long-range interaction: from synchronization to chaos, *Chaos*, **17**(4) (2007), 043124 (arXiv:0707.3941).

Mark Edelman

Dynamics of nonlinear systems with power-law memory

Abstract: Dynamics of fractional (with power-law memory) nonlinear systems may demonstrate features, which are fundamentally different from behavior of regular (memoryless) nonlinear systems. The new features of fractional dynamics include self-intersection of trajectories in dimensions α, less than two, and overlapping of chaotic attractors; existence of periodic points only in an asymptotic sense, and a cascade of bifurcations-type attracting trajectories. Fractional nonlinear systems demonstrate the standard scenario of the transition to chaos, through a cascade of period-doubling bifurcations, which depends on the fractional dimension (memory parameter) α. Two-dimensional bifurcation diagrams, limiting the value of the coordinate x_{lim} as a function of the nonlinearity parameter K and the memory parameter α, possess some universal properties. Bifurcations with the change in α, (α, x)-bifurcation diagrams, allow control of fractional systems by changing the memory parameter.

Keywords: Power-law memory, fractional maps, attractors, bifurcations, chaos

MSC 2000: 47H99, 60G99, 34A99, 39A70

1 Introduction

The author's motivation in research on fractional dynamics comes from his work with George Zaslavsky on chaos in Hamiltonian systems. The phase space of a typical chaotic nonlinear Hamiltonian system is not purely ergodic; it is fractal with a hierarchy of islands (structure of higher-order islands around lower-order islands) of regular motion. The dynamics in ergodic areas near islands' borders is almost integrable. A system stays in a thin layer near the border of an island for a long time; the higher the order of an island, the longer the exit time from its boundary layer. The Fokker–Plank–Kolmogorov equation, with a fractional time derivative (FFPK), provides an accurate description of transport in the presence of the hierarchy of the exit times, corresponding to Hamiltonian dynamics (a fractional space derivative in FFPK results from the fractal structure of the phase space) [61–64, 66, 72]. Dynamics

Acknowledgement: The author expresses his gratitude to the administration of the Courant Institute for the opportunity to complete this work at Courant and to Virginia Donnelly for technical help.

Mark Edelman, Department of Physics, Stern College at Yeshiva University, 245 Lexington Ave, New York, NY 10016, USA; and Courant Institute of Mathematical Sciences, New York University, 251 Mercer St., New York, NY 10012, USA; and Department of Mathematics, BCC, CUNY, 2155 University Avenue, Bronx, New York, NY 10453, USA, e-mail: edelman@cims.nyu.edu

https://doi.org/10.1515/9783110571707-005

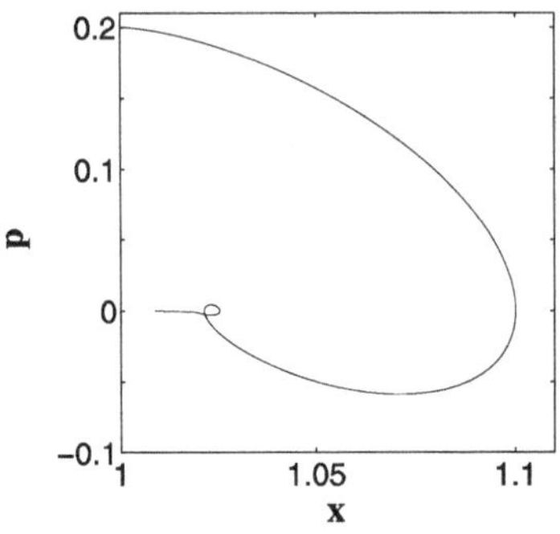

Figure 1: An example of a self-intersecting phase space trajectory for the fractional Caputo Duffing equation ${}^{C}D_{0}^{1.5}x(t) = x(1 - x^{2})$ on $t \in [0, 40]$. The initial conditions are $x(0) = 1$ and $p = dx/dt(0) = 0.2$. Here the standard definition of the fractional Caputo derivative, assumed in this book, is used.

of systems with fractional time derivative is called fractional dynamics. The ubiquity of the problems of transport in Hamiltonian systems provides a strong argument in support of the investigation of fractional dynamics. The author's contributions to the research on fractional transport in Hamiltonian systems and billiards include papers [5, 40, 65–75].

Various aspects of the application of fractional dynamics in physics are reflected in the fourth and fifth volumes of the present book [51]. An intriguing fundamental problem is the origin of the Universe, and a related problem of the origin of memory in living species. Were there seeds of memory present at the origin of the Universe? Are the fundamental laws of nature memoryless, or do they have some form of memory? One of the approaches is to assume that on the time and length scales, smaller than Planck time and length, the fundamental laws should have some memory, and a feedback mechanism to manage the evolution of the Universe. This is a purely philosophical question, unless we show that the presence of memory may lead to a fundamentally different behavior of the Universe on the large scales, and compare it with observations. This is yet another motivation to investigate the very basic properties of systems with memory.

The fundamental difference in behavior of fractional and regular systems becomes obvious at the very beginning, as one starts numerical simulations of fractional differential/difference equations. One of the effects of memory illustrated in Figure 1 is a possibility of self-intersections of trajectories in systems of less than two-dimensions. As a result, the Poincaré–Bendixson theorem does not apply to fractional systems, and in continuous systems of orders $\alpha < 2$, the nonexistence of chaos is only a conjecture (see [10, 18]). Another effect, noticed by many researchers, and illustrated in Figure 2 as an example of the fractional Duffing oscillator driven by a periodic force, is that nondissipative fractional systems behave similar to the higher-order integer systems with dissipation. The example in Figure 2 also demonstrates that chaos may exist in fractional systems of lower than three (2.9 in Figure 2(b)) dimensions.

It is much easier to investigate general properties of discrete systems with a power-law memory than properties of fractional differential equations with power-law ker-

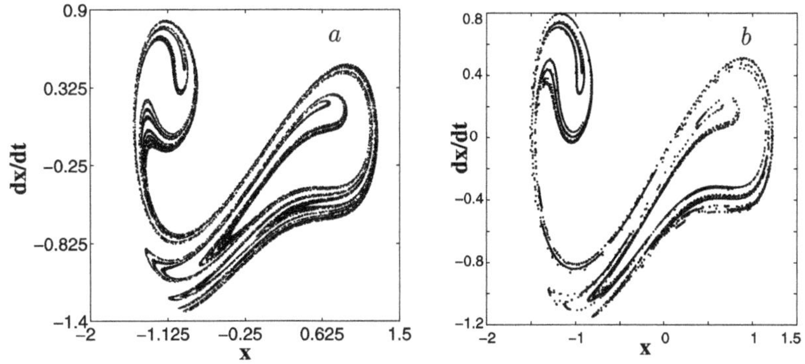

Figure 2: Poincaré maps: (a) for the ordinary Duffing equation driven by periodic force: $\ddot{x} + 0.1172\dot{x} - x + x^3 = 0.279\sin(1.0149t)$; (b) for the fractional Duffing equation ${}^{C}D_0^{1.9}x(t) - x + x^3 = 0.3\sin(t)$, $t \in [0, 16\pi]$, driven by a periodic force. Number of trajectories in (b) is 2000. This figure is reprinted from [76], with the permission of AIP Publishing.

nels. One property, which has been noticed in discrete fractional systems and not yet reported in continuous fractional systems, is the existence of cascade of bifurcations-type trajectories (CBTT). The periodicity of such trajectories changes with time: a trajectory may start converging to a period $T = 2^n$ sink, but then bifurcates and starts converging to a period $T = 2^{2n+1}$ sink, and so on. Examples of CBTT for $0 < \alpha < 1$ are given in Figure 3(a); the trajectory that starts as $T = 4$ trajectory ends as a $T = 16$ sink, and in Figure 3(b), a trajectory ends in chaos. Figure 4 contains examples of phase spaces ((a), (c), and (d)), and a time evolution (b), for a cascade of bifurcations and intermittent cascade of bifurcations-type ((a), (b), and (c)) trajectories in the Riemann–Liouville standard α-family of maps, Equations (35) and (36) (see [12]). In the Caputo logistic α-family of maps, with $1 < \alpha < 2$, Equations (44) and (45) numerical simulations [14] demonstrate trajectories on which mergers (inverse bifurcations) occur. A trajectory may initially converge to a period $T = 2^{2n+1}$ sink, which then evolve into a period $T = 2^n$ sink, and so on. An example of an inverse CBTT, a phase space and time evolution on a single trajectory for the member of the Caputo logistic α-family of maps, with the order $\alpha = 1.2$ and the nonlinearity parameter $K = 3.45$, is presented in Figure 5.

Another phenomenon, which appears in systems with memory, is overlapping of attractors. In Figure 6, a chaotic attractor in the Caputo standard α-family of maps, Equations (38) and (39) [12], of the order $\alpha = 1.02$, is overlapping with a CBTT (which is an attracting trajectory).

In this paper, we review the research on general properties of fractional systems, based on the investigation of the fractional/fractional difference maps introduced in [13–15, 18, 19, 47–49, 53, 57, 58], and reviewed in the author's paper in volume two of the present book [22].

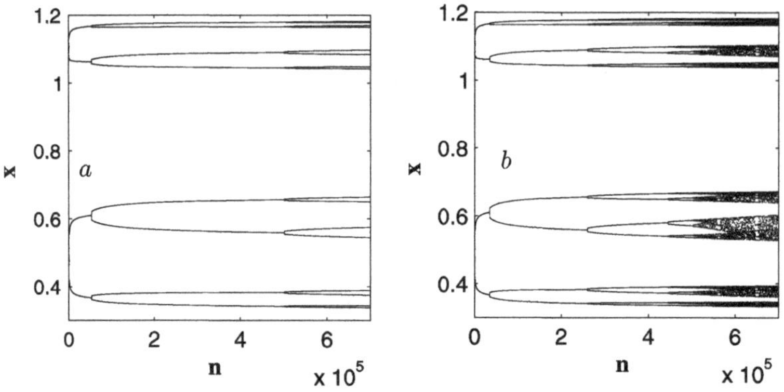

Figure 3: Two examples of cascade of bifurcations-type trajectories (CBTT) in the Caputo logistic α-family of maps; Equation (29) (see [22]) with the order $\alpha = 0.1$ and the initial condition $x_0 = 0.001$: (a) for the nonlinearity parameter $K = 22.37$, the last bifurcation from the period $T = 8$ to the period $T = 16$ occurs after approximately 5×10^5 iterations; (b) when $K = 22.423$, the trajectory becomes chaotic after approximately 5×10^5 iterations. This figure is reprinted from [21] with the permission of AIP Publishing.

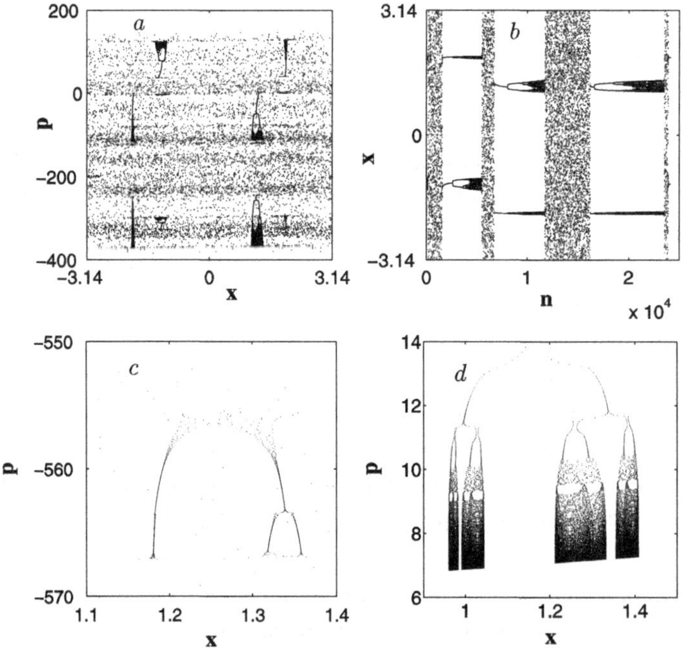

Figure 4: Cascade of bifurcations-type trajectories in the Riemann–Liouville standard α-family of maps, Equation (35) and (36): (a) $\alpha = 1.65$, $K = 4.5$; one intermittent trajectory in the phase space; (b) The time dependence of the coordinate (x of n) for the case (a); (c) $\alpha = 1.98$, $K = 6.46$; a zoom of a small feature for a single intermittent trajectory in phase space; (d) $\alpha = 1.1$, $K = 3.5$; a single trajectory enters the cascade after a few iterations and stays there during 500 000 iterations. This figure is reprinted from [13] with the permission of L&H Scientific Publishing.

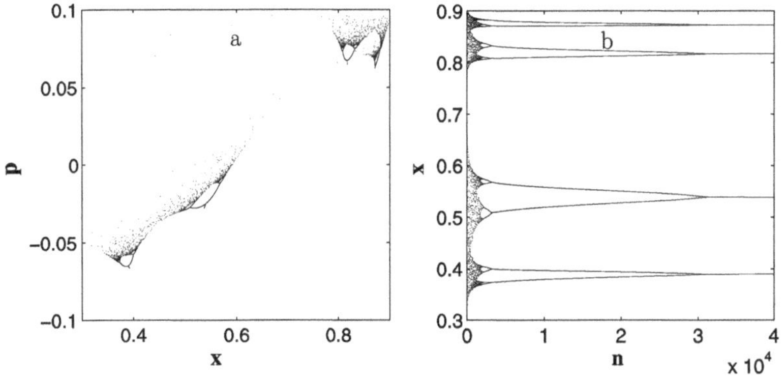

Figure 5: An inverse CBTT in the Caputo logistic α-family of maps, Equations (44) and (45), with α = 1.2 and K = 3.45. 40 000 iterations on a trajectory with the initial conditions x_0 = 0.01 and p_0 = 0.1. (a) Phase space. (b) $x - n$ graph. This figure is reprinted from [14] with the permission of AIP Publishing.

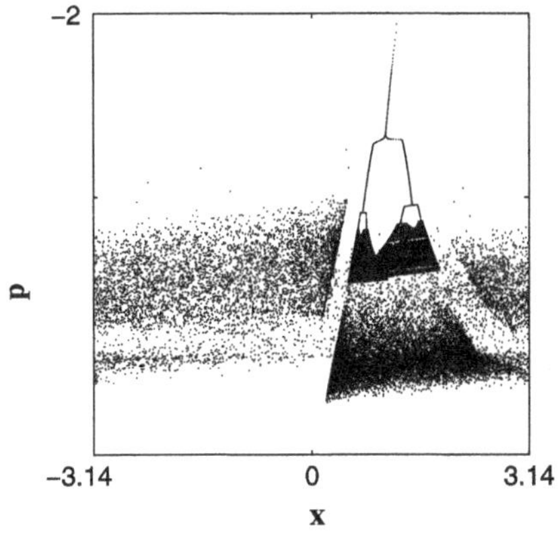

Figure 6: 20 000 iterations on each of two overlapping independent attractors in the Caputo standard α-family of maps, Equation (38) and (39) with K = 4.5, α = 1.02, and the initial coordinate x_0 = 0. The CBTT has the initial momentum p_0 = −1.8855, and the chaotic attractor p_0 = −2.5135. This figure is reprinted from [12] with the permission of Elsevier.

2 Fractional/fractional difference maps

For additional references, in this section we will list the maps introduced in volume two of this book [22], which will be used to illustrate properties of fractional systems.

2.1 Fractional maps

The Riemann–Liouville universal map, which is a solution at time $t_n = nh$ ($t_0 = 0$) of the differential equation, for the x-coordinate of a particle experiencing periodic, with the period h, delta-function kicks, driven by a force defined by a nonlinear function $G_K(x)$ (where K is a nonlinearity parameter, defining nonlinear properties of $G_K(x)$), with the Riemann–Liouville fractional derivative, is defined as

$$x_{n+1} = \sum_{k=1}^{m-1} \frac{c_k}{\Gamma(\alpha-k+1)} h^{\alpha-k}(n+1)^{\alpha-k} + \frac{h^{\alpha}}{\Gamma(\alpha)} \sum_{k=0}^{n} (n+1-k)^{\alpha-1} G_K(x_k), \tag{1}$$

where $n \in \mathbb{Z}$, $n > 0$, $\Gamma()$ is the Gamma function, and $x_n = x(t_n)$. This map can be written as the m-dimensional map

$$x_{n+1} = \sum_{k=1}^{m-1} \frac{c_k}{\Gamma(\alpha-k+1)} h^{\alpha-k}(n+1)^{\alpha-k} - \frac{h^{\alpha}}{\Gamma(\alpha)} \sum_{k=0}^{n} G_K(x_k)(n-k+1)^{\alpha-1}, \tag{2}$$

$$p^s_{n+1} = \sum_{k=1}^{m-s-1} \frac{c_k}{(m-s-1-k)!} h^{m-s-1-k}(n+1)^{m-s-1-k} - \frac{h^{m-s-1}}{(m-s-2)!} \sum_{k=0}^{n} G_K(x_k)(n-k+1)^{m-s-2}, \tag{3}$$

where $s = 0, 1, 2, \ldots, m-2$, $p^s_k = p^{(s)}(t_k)$, $p^{(0)}(t) = {}^{RL}D_0^{\alpha-m+1}x(t)$, and $p^{(s)}(t) = D^s p^{(0)}(t)$, for $s = 1, 2, \ldots, m-2$.

Likewise, the Caputo universal map is defined as

$$x_{n+1} = \sum_{k=0}^{m-1} \frac{b_k}{k!} h^k (n+1)^k - \frac{h^{\alpha}}{\Gamma(\alpha)} \sum_{k=0}^{n} G_K(x_k)(n-k+1)^{\alpha-1}, \tag{4}$$

where $n \in \mathbb{Z}$, $n > 0$. This map can be written as the m-dimensional map

$$x^{(s)}_{n+1} = \sum_{k=0}^{m-s-1} \frac{x_0^{(k+s)}}{k!} h^k (n+1)^k - \frac{h^{\alpha-s}}{\Gamma(\alpha-s)} \sum_{k=0}^{n} G_K(x_k)(n-k+1)^{\alpha-s-1}, \tag{5}$$

where $x^{(s)}(t) = D^s x(t)$, $s = 0, 1, \ldots, m-1$.

2.2 Fractional difference maps

The only numerically investigated form of fractional difference maps is the h-difference Caputo universal α-family of maps

$$x_{n+1} = \sum_{k=0}^{m-1} \frac{c_k}{k!} ((n+1)h)_h^{(k)} - \frac{h^{\alpha}}{\Gamma(\alpha)} \sum_{s=0}^{n+1-m} (n-s-m+\alpha)^{(\alpha-1)} G_K(x_{s+m-1}), \tag{6}$$

which describes solutions of the Caputo-like h-difference equation (see [22])

$$({}_0\Delta^{\alpha}_{h,*}x)(t) = -G_K(x(t+(\alpha-1)h)), \tag{7}$$

where $t \in (h\mathbb{N})_m = \{m, m+h, m+2h, \dots\}$, with the initial conditions

$$({}_0\Delta^{k}_{h}x)(0) = c_k, \quad k = 0, 1, \dots, m-1, \quad m = \lceil \alpha \rceil. \tag{8}$$

The h-factorial $t_h^{(\alpha)}$ is defined as

$$t_h^{(\alpha)} = h^{\alpha}\frac{\Gamma(\frac{t}{h}+1)}{\Gamma(\frac{t}{h}+1-\alpha)} = h^{\alpha}\left(\frac{t}{h}\right)^{(\alpha)}, \quad \frac{t}{h} \neq -1, -2, -3, \dots. \tag{9}$$

The falling factorial

$$t^{(\alpha)} = \frac{\Gamma(t+1)}{\Gamma(t+1-\alpha)}, \quad t \neq -1, -2, -3 \dots. \tag{10}$$

is asymptotically a power function:

$$\lim_{t\to\infty} \frac{\Gamma(t+1)}{\Gamma(t+1-\alpha)t^{\alpha}} = 1, \quad \alpha \in \mathbb{R}. \tag{11}$$

2.3 Standard and logistic α-families of maps

Various forms of fractional standard and logistic maps were used in numerical simulations to investigate general properties of fractional systems.

The α-families of maps with $G_K(x) = K\sin x$, which in the case $\alpha = 2$ and $h = 1$, yield the regular standard map, called the standard α-families of maps. The α-families of maps, with $G_K(x) = x - Kx(1-x)$, which in the case $\alpha = 1$ and $h = 1$ yield the regular logistic map, are called the logistic α-families of maps.

2.3.1 Integer orders

For $\alpha = 0$, all fractional maps are identically zero: $x_n = 0$. The Caputo fractional difference standard map is the sine map:

$$x_{n+1} = -K\sin x_n. \tag{12}$$

The Caputo fractional difference logistic map is a quadratic map:

$$x_{n+1} = -x_n + Kx_n(1-x_n). \tag{13}$$

For $\alpha = 1$, all universal fractional and fractional difference maps can be written as

$$x_{n+1} = x_n - hG_K(x_n). \tag{14}$$

With $h = 1$, all forms of the standard α-families of maps turn into the circle map with zero driving phase:

$$x_{n+1} = x_n - K\sin(x_n), \tag{15}$$

and all forms of the logistic α-families of maps turn into the regular logistic map:

$$x_{n+1} = Kx_n(1 - x_n). \tag{16}$$

For $\alpha = 2$, all universal fractional maps assume the form [21]:

$$\begin{cases} p_{n+1} = p_n - hG_K(x_n), & n \geq 0, \\ x_{n+1} = x_n + hp_{n+1}, & n \geq 0, \end{cases} \tag{17}$$

and all universal fractional difference maps assume the form

$$\begin{cases} p_{n+1} = p_n - hG_K(x_n), & n \geq 1, \quad p_1 = p_0, \\ x_{n+1} = x_n + hp_{n+1}, & n \geq 0. \end{cases} \tag{18}$$

With $h = 1$, all forms of the standard α-families of maps turn into the regular standard map [8]:

$$\begin{cases} p_{n+1} = p_n - K\sin(x_n) \pmod{2\pi}, \\ x_{n+1} = x_n + p_{n+1} \pmod{2\pi}, \end{cases} \tag{19}$$

and all forms of the logistic α-families of maps turn into a quadratic map, which we call the 2D logistic map [22]:

$$\begin{cases} p_{n+1} = p_n + Kx_n(1 - x_n) - x_n, \\ x_{n+1} = x_n + p_{n+1}. \end{cases} \tag{20}$$

The N-dimensional ($N \in \mathbb{Z}$, $N > 1$, $h = 1$) members of the fractional [14]

$$p^s_{n+1} = p^s_n + \sum_{k=0}^{N-s-3} \frac{p^{k+s+1}_n}{(k+1)!} - \frac{G_K(x_n)}{(N-s-2)!}, \quad s = 0, 1, \ldots, N-2, \tag{21}$$

$$x_{n+1} = x_n + \sum_{k=0}^{N-2} \frac{p^k_n}{(k+1)!} - \frac{G_K(x_n)}{(N-1)!}, \tag{22}$$

and fractional difference [15]

$$x^s_{n+1} = \sum_{k=s}^{N-1} x^k_n - G_K(x^0_n), \quad s = 0, 1, \ldots, N-1, \tag{23}$$

universal α-families of maps are volume preserving.

2.3.2 $0 < \alpha < 1$

In this case, all forms of the universal α-families of maps and hence Equations (1), (4), and (6) can be written as

$$x_{n+1} = x_0 - \sum_{k=0}^{n} \tilde{G}(x_k)U_\alpha(n-k+1), \tag{24}$$

where $\tilde{G}(x) = h^\alpha G_K(x)/\Gamma(\alpha)$, and x_0 is the initial condition ($x_0 = 0$ in Equation (1), and the corresponding map is identically zero [14]). In fractional maps, Equations (1) and (4),

$$U_\alpha(n) = n^{\alpha-1}, \quad U_\alpha(1) = 1, \tag{25}$$

and in fractional difference maps, Equation (6),

$$\begin{aligned} U_\alpha(n) &= (n+\alpha-2)^{(\alpha-1)}, \\ U_\alpha(1) &= (\alpha-1)^{(\alpha-1)} = \Gamma(\alpha). \end{aligned} \tag{26}$$

The Caputo standard α-family of maps can be written as [14]

$$x_n = x_0 - \frac{K}{\Gamma(\alpha)} \sum_{k=0}^{n-1} \frac{\sin(x_k)}{(n-k)^{1-\alpha}} \pmod{2\pi}, \tag{27}$$

and the fractional difference Caputo standard α-family of maps is [15]

$$x_{n+1} = x_0 - \frac{K}{\Gamma(\alpha)} \sum_{s=0}^{n} \frac{\Gamma(n-s+\alpha)}{\Gamma(n-s+1)} \sin(x_s) \pmod{2\pi}. \tag{28}$$

The Caputo logistic α-family of maps can be written as [14]

$$x_n = x_0 - \frac{1}{\Gamma(\alpha)} \sum_{k=0}^{n-1} \frac{x_k - Kx_k(1-x_k)}{(n-k)^{1-\alpha}}, \tag{29}$$

and the fractional difference Caputo logistic α-family of maps is [19]

$$x_{n+1} = x_0 - \frac{1}{\Gamma(\alpha)} \sum_{s=0}^{n} \frac{\Gamma(n-s+\alpha)}{\Gamma(n-s+1)} [x_s - Kx_s(1-x_s)]. \tag{30}$$

2.3.3 $1 < \alpha < 2$

In this case the map Equations (1), (4), and (6) can be written as

$$x_{n+1} = x_0 + f(\alpha)[h(n+1)]^\beta p_0 - h \sum_{k=0}^{n} \tilde{G}(x_k)U_\alpha(n-k+1) + hf_1(n). \tag{31}$$

Here, $\tilde{G}(x) = h^{\alpha-1}G_K(x)/\Gamma(\alpha)$, x_0, and $U(n)$ are defined the same way as in Equations (24), (25), and (26); p_0 is the initial momentum (b_k or c_k in corresponding formulae); β is equal to 1 in Equations (4) and (6), and $\alpha - 1$ in Equation (1); $f(\alpha)$ is 1 in Equations (4) and (6), $1/\Gamma(\alpha)$ in Equation (1), and $f_1(n) = 0$ in Equations (1) and (4), and $f_1(n) = h^{\alpha-1}G(x_0)(n-1+\alpha)^{(\alpha-1)}/\Gamma(\alpha) \sim n^{\alpha-1}$ in Equation (6).

Let us define the momentum

$$p_{n+1} = \frac{x_{n+1} - x_n}{h}. \tag{32}$$

Then, taking into account that $U_\alpha(0) = 0$, from Equation (31) it follows

$$p_{n+1} = \tilde{f}(n)p_0 - \sum_{k=0}^{n} \tilde{G}(x_k)\tilde{U}_\alpha(n-k+1) + f_1(n) - f_1(n-1), \tag{33}$$

where

$$\begin{aligned} \tilde{U}_\alpha(n) &= U_\alpha(n) - U_\alpha(n-1) \\ &= \begin{cases} n^{\alpha-1} - (n-1)^{\alpha-1} \sim n^{\alpha-2} \quad \text{and} \quad \tilde{U}_\alpha(1) = 1 \quad \text{in Eqs. (1), (4);} \\ (n+\alpha-2)^{(\alpha-1)} - (n+\alpha-3)^{(\alpha-1)} \\ \quad = (\alpha-1)(n+\alpha-3)^{(\alpha-2)} \\ \quad = (\alpha-1)U_{\alpha-1}(n) \sim n^{\alpha-2} \quad \text{and} \quad \tilde{U}_\alpha(1) = \Gamma(\alpha) \quad \text{in Eq. (6),} \end{cases} \end{aligned} \tag{34}$$

$f_1(n) - f_1(n-1) = 0$ in Equations (1) and (4), and $f_1(n) - f_1(n-1) \sim n^{\alpha-1}$ in Equation (6); $\tilde{f}(n) = 1$ in Equations (4) and (6), and $\tilde{f}(n) \sim n^{\alpha-2}$ in Equation (1). Note, that the definitions of $\tilde{U}_\alpha(1)$ in Equation (34) and $U_\alpha(1)$ in Equations (25) and (26) are identical.

The Riemann–Liouville standard α-family of maps is defined on a cylinder and can be written as (see [12, 24, 25])

$$p_{n+1} = p_n - K \sin x_n, \tag{35}$$

$$x_{n+1} = \frac{1}{\Gamma(\alpha)} \sum_{i=0}^{n} p_{i+1} V_\alpha^1(n-i+1) \pmod{2\pi}, \tag{36}$$

where

$$V_\alpha^n(m) = m^{\alpha-n} - (m-1)^{\alpha-n}. \tag{37}$$

The defined on a torus Caputo standard α-family of maps is [14]

$$p_{n+1} = p_n - \frac{K}{\Gamma(\alpha-1)} \left[\sum_{i=0}^{n-1} V_\alpha^2(n-i+1)\sin(x_i) + \sin(x_n) \right] \pmod{2\pi}, \tag{38}$$

$$x_{n+1} = x_n + p_0 - \frac{K}{\Gamma(\alpha)} \sum_{i=0}^{n} V_\alpha^1(n-i+1)\sin(x_i) \pmod{2\pi}. \tag{39}$$

The fractional difference Caputo standard α-family of maps can be defined on a torus as

$$p_n = p_1 - \frac{K}{\Gamma(\alpha-1)} \times \sum_{s=2}^{n} \frac{\Gamma(n-s+\alpha-1)}{\Gamma(n-s+1)} \sin(x_{s-1}) \pmod{2\pi}, \tag{40}$$

$$x_n = x_{n-1} + p_n \pmod{2\pi}, \quad n \geq 1, \tag{41}$$

The Riemann–Liouville logistic α-family of maps can be written as [13, 16]

$$p_{n+1} = p_n - Kx_n(1-x_n) - x_n, \tag{42}$$

$$x_{n+1} = \frac{1}{\Gamma(\alpha)} \sum_{i=0}^{n} p_{i+1} V_\alpha^1(n-i+1), \tag{43}$$

and requires the initial condition $x_0 = 0$. The Caputo Logistic αFM is [13, 16]

$$x_{n+1} = x_0 + p_0(n+1) - \frac{1}{\Gamma(\alpha)} \sum_{k=0}^{n} [x_k - Kx_k(1-x_k)](n-k+1)^{\alpha-1}, \tag{44}$$

$$p_{n+1} = p_0 - \frac{1}{\Gamma(\alpha-1)} \sum_{k=0}^{n} [x_k - Kx_k(1-x_k)](n-k+1)^{\alpha-2}, \tag{45}$$

and the fractional difference Caputo logistic α-family of maps is [19]

$$p_n = p_1 - \frac{K}{\Gamma(\alpha-1)} \times \sum_{s=2}^{n} \frac{\Gamma(n-s+\alpha-1)}{\Gamma(n-s+1)} [x_{s-1} - Kx_{s-1}(1-x_{s-1})], \tag{46}$$

$$x_n = x_{n-1} + p_n, \quad n \geq 1. \tag{47}$$

3 Periodic points

Even regular nonlinear systems demonstrate quite complicated regular and chaotic behavior. Transition to chaos in nonlinear systems occurs through period-doubling cascades of bifurcations, with the change in nonlinearity parameters of these systems. The situation becomes even more complicated in fractional systems, in which cascades of bifurcations may occur on single trajectories. Stability of fractional systems is a hot topic, investigated in numerous papers by various methods, including Lyapunov's direct and indirect methods, Lyapunov function, and Routh–Hurwitz criterion. The most cited article on stability of linear fractional differential equations is, probably, Matignon's paper [41]. Some of the papers in which stability of nonlinear fractional differential equations was investigated are [1, 2, 26, 29, 35, 37, 38, 54]. A remarkable result, showing that for a wide class of continuous fractional systems, stability of an integer system implies stability of the corresponding fractional system of a lower-order. This was obtained in [7]. Some of the papers in which stability of discrete fractional systems was investigated are [4, 6, 32, 33, 46, 59]. The reviews on stability of fractional systems include papers [36, 43, 45] and books [44, 77]. Most of the cited papers investigate sufficient conditions of stability and do not allow calculation of the ranges of nonlinearity parameters and orders of derivatives, for which fixed points are stable. In this section we will present methods, which allow calculations of the critical values of parameters, at which the bifurcation from a stable fixed point to a stable asymptotically period-two point occurs.

3.1 Fixed points

In the case, $0 < \alpha < 2$, fixed points x_{fp} in Equations (24) and (31) are the solutions of the equation $G_K(x) = 0$. In addition, for $1 < \alpha < 2$, p_0 also must be zero. Here, following papers [12, 25], we will present an example of the investigation of stability of the fixed point $(0, 0)$ in the Riemann–Liouville and Caputo standard α-families of maps, Equations (35) and (36) (RLSM), and Equations (38) and (39) (CSM) for $1 < \alpha < 2$.

The regular standard map, the RLSM, and the CSM have the same fixed point $(0, 0)$. In the case of the RLSM, the evolution of perturbations near the fixed point is described by the following system of equations:

$$\delta p_{n+1} = \delta p_n - K\delta x_n, \tag{48}$$

$$\delta x_{n+1} = \frac{1}{\Gamma(\alpha)} \sum_{i=0}^{n} \delta p_{i+1} V_\alpha(n - i + 1). \tag{49}$$

The system describing the evolution of trajectories near the fixed point $(0, 0)$ in the CSM is

$$\delta p_{n+1} = \delta p_n - \frac{K}{\Gamma(\alpha - 1)} \left[\sum_{i=0}^{n-1} V_\alpha^2(n - i + 1)\delta x_i + \delta x_n \right], \tag{50}$$

$$\delta x_{n+1} = \delta x_n + \delta p_0 - \frac{K}{\Gamma(\alpha)} \sum_{i=0}^{n} V_\alpha^1(n - i + 1)\delta x_i. \tag{51}$$

Direct computations in [12], using Equations (38) and (39), show that the critical curve $K_{cr}(\alpha)$ (see Fig. 7(a)) on which the transition from the stable fixed point $(0, 0)$ ($K < K_{cr}$) to a stable period-two ($T = 2$) point ($K > K_{cr}$) occurs, in the case of the CSM, is the same as the critical curve obtained from the semi-analytic stability analysis of the RLSM in [25]. The equation of this curve is

$$\frac{V_{\alpha l} K_{cr}}{2\Gamma(\alpha)} = 1, \tag{52}$$

where

$$V_{\alpha l} = \sum_{i=1}^{\infty} (-1)^{i+1} V_\alpha^1(i). \tag{53}$$

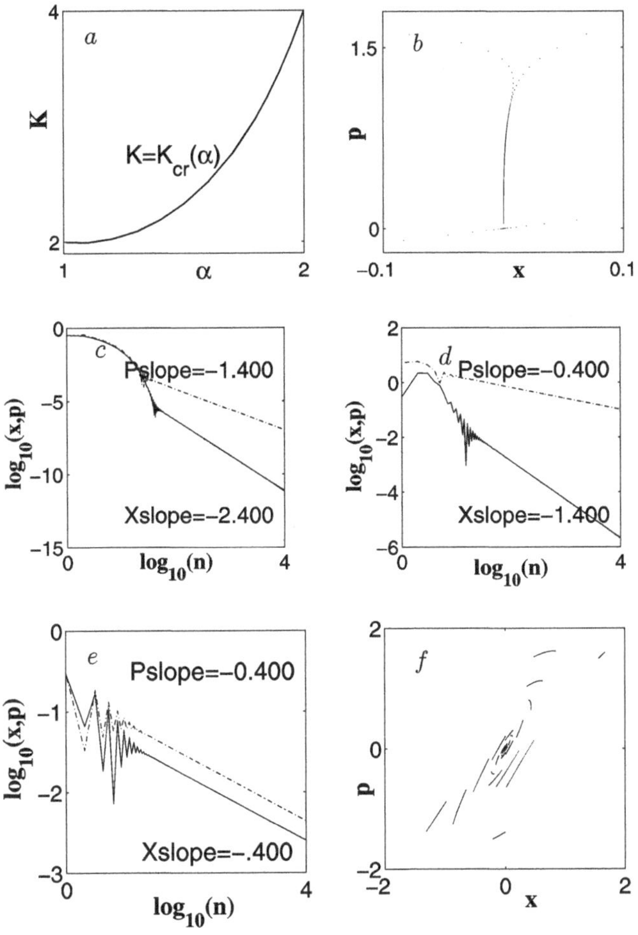

Figure 7: The RLSM and CSM stable fixed point (0, 0): (a) The fixed point (0, 0) for both the RLSM and the CSM is stable below the curve $K = K_{cr}(\alpha)$; (b) Two RLSM trajectories, with $K = 2$, $\alpha = 1.4$, and 10^5 iterations on each trajectory. The bottom one ($p_0 = 0.3$) is a fast converging trajectory. The upper trajectory ($p_0 = 5.3$) is an example of the RLSM's attracting slow converging trajectory (ASCT). The momentum on this trajectory after 10^5 iterations is $p \approx 0.042$; (c) Time dependence of the coordinate and momentum for the fast converging trajectory from (b); (d) x and p time dependence for the ASCT from (b); (e) x and p time dependence for the CSM with $K = 2$, $\alpha = 1.4$, and $p_0 = 0.3$; (f) Evolution of the CSM trajectories with $p_0 = 1.6 + 0.002i$, $0 \le i < 50$, for the case $K = 3$, $\alpha = 1.9$. The line segments correspond to the nth iteration on the set of trajectories with close initial conditions. The evolution of the trajectories with smaller p_0, and for the RLSM with the same K, α, and $p_0 < 1.6$ is similar. This figure is reprinted from [12] with the permission of Elsevier.

Because not only the stability problems (Equations (48)–(51)), but also the original maps (Equations (24) and (31)), contain convolutions, the use of generating functions [28], which allow transformations of sums of products into products of sums, may be utilized in the investigation of the fractional standard maps, and other maps with memory. After the introduction of the generating functions

$$\tilde{X}(t) = \sum_{i=0}^{\infty} \delta x_i t^i \quad \text{and} \quad \tilde{P}(t) = \sum_{i=0}^{\infty} \delta p_i t^I, \tag{54}$$

$$\tilde{W}_\alpha^1(t) = \frac{K}{\Gamma(\alpha)} \sum_{i=0}^{\infty} [(i+1)^{\alpha-1} - i^{\alpha-1}] t^i, \tag{55}$$

$$\tilde{W}_\alpha^2(t) = \frac{K}{\Gamma(\alpha-1)} \left(1 + \sum_{i=1}^{\infty} [(i+1)^{\alpha-2} - i^{\alpha-2}] t^i \right), \tag{56}$$

the stability analysis can be reduced to the analysis of the asymptotic behavior at $t = 0$ of the derivatives of the analytic functions

$$\tilde{X}(t) = \frac{p_0 \tilde{W}_\alpha^1(t)}{K} \frac{t}{1 - t(1 - \tilde{W}_\alpha^1(t))}, \tag{57}$$

$$\tilde{P}(t) = p_0 \frac{1 + \tilde{W}_\alpha^1(t)}{1 - t(1 - \tilde{W}_\alpha^1(t))} \tag{58}$$

for the RLSM and

$$\tilde{X}(t) = \frac{tp_0 + (1-t)x_0}{(1-t)(1-t+t\tilde{W}_\alpha^1(t))}, \tag{59}$$

$$\tilde{P}(t) = \frac{1}{1-t} \left[p_0 - t \frac{tp_0 + (1-t)x_0}{(1-t)(1-t+t\tilde{W}_\alpha^1(t))} \tilde{W}_\alpha^2(t) \right] \tag{60}$$

for the CSM.

The following results are conjectures, based on the large number of numerical simulations for some values of parameters K and α for which the fixed point $(0, 0)$ is stable. In the standard map this point is stable for $K < K_{cr} = 4$, and all stable periodic points are elliptic points with zero Lyapunov exponent. They are surrounded by islands of regular motion. As can be seen in Figure 8, in the case of fractional standard maps with $\alpha < 2$, elliptic points turn into sinks or, as it relates to RLSM (see Figure 8(a), (c), and (e)), attracting slowly diverging trajectories (ASDT). The islands turn into the basins of attraction associated with these attractors. For $K < K_{cr}$ $(1 < \alpha < 2)$, there are no chaotic or periodic (except the fixed point) trajectories. Two initially close trajectories that start in the area between basins of attraction, at first diverge, but then converge to the same or different attractors.

There is more than one way to approach an attracting fixed point of the RLSM. Two RLSM trajectories for $K = 2$ and $\alpha = 1.4$ are shown in Figure 7(b). The bottom one is a fast converging trajectory that starts in the basin of attraction, in which $x_n \sim n^{-1-\alpha}$, and $p_n \sim n^{-\alpha}$ (see Figure 7(c)). The upper trajectory is an example of the ASCT, introduced

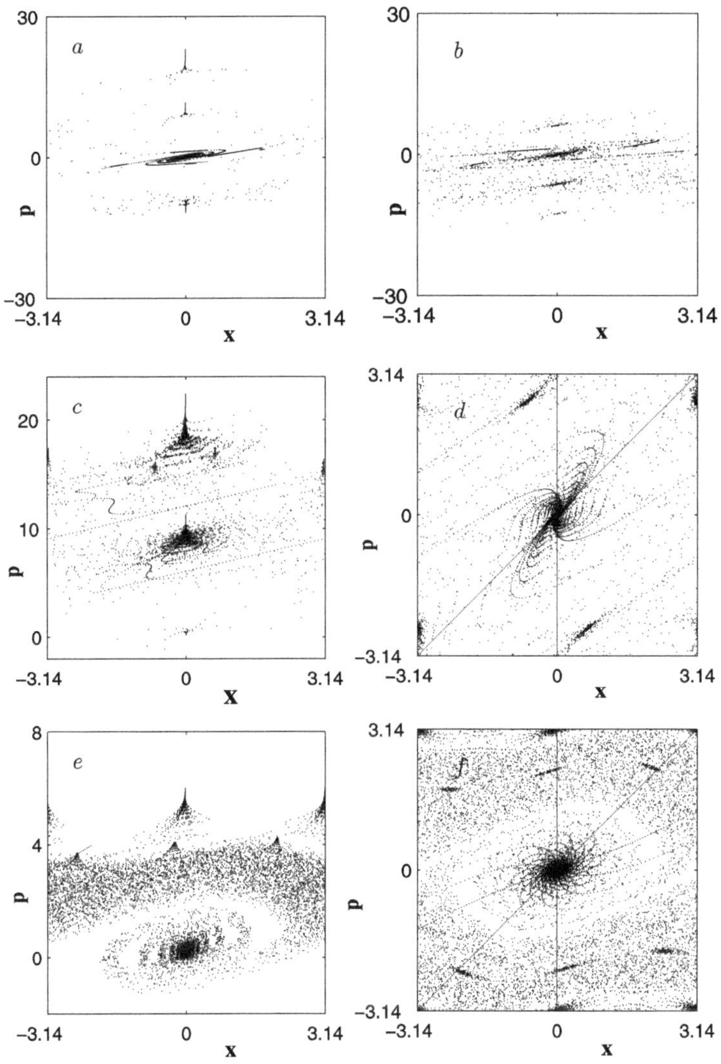

Figure 8: The RLSM and CSM phase space for $K < K_{cr}$: (a) The RLSM with the same values of parameters as in Figure 7(f), but $p_0 = 1.6 + 0.002i$, $0 \leq i < 50$; (b) The CSM with the same values of parameters as in (a) but $p_0 = 1.7 + 0.002i$, $0 \leq i < 50$; (c) 400 iterations on the RLSM trajectories with $p_0 = 4 + 0.08i$, $0 \leq i < 125$ for the case $K = 2$, $\alpha = 1.9$. Trajectories converging to the fixed point; ASDTs with $x = 0$, and period 4 ASDTs are present; (d) 100 iterations on the CSM trajectories with $p_0 = -3.14 + 0.0314i$, $0 \leq i < 200$ for the same case as in (c) ($K = 2$, $\alpha = 1.9$), but considered on a torus. In this case all trajectories converge to the fixed point or period four stable attracting points; (e) 400 iterations on trajectories with $p_0 = 2 + 0.04i$, $0 \leq i < 50$, for the RLSM case $K = 0.6$, $\alpha = 1.9$. Trajectories converging to the fixed point and ASDTs of period 2 and 3 are present; (f) 100 iterations on the CSM trajectories with $p_0 = -3.14 + 0.0314i$, $0 \leq i < 200$, for the same case as in (e) ($K = 0.6$, $\alpha = 1.9$) considered on a torus. In this case all trajectories converge to the fixed point, period-two, and period-three stable attracting points. This figure is reprinted from [12] with the permission of Elsevier.

in [25], on which $x_n \sim n^{-\alpha}$ and $p_n \sim n^{1-\alpha}$ (see Figure 7(d)). Trajectories converging to the fixed point, which start outside of the basin of attraction, are ASCTs. In the case of the CSM, all trajectories converging to the fixed point have the same asymptotic behavior: $x_n \sim n^{1-\alpha}$ and $p_n \sim n^{1-\alpha}$ (see Figure 7(e)).

In the RLSM and the CSM, considered on a cylinder, the stable fixed point $(0,0)$ is surrounded by a finite basin of attraction, whose width W depends on the values of K and α. For the RLSM, with $K = 3$ and $\alpha = 1.9$ (see Figs. 7(f) and 8(a)), the width of the basin of attraction is $1.6 < W < 1.7$ ([25]). In CSM with the same parameters, Figure 8(b), the width is $1.7 < W < 1.8$. Simulations of thousands of trajectories with $p_0 < 1.6$, performed by the author of [12], of which only 50 (with $1.6 \le p_0 < 1.7$) are presented in Figure 7(f), show only converging trajectories, whereas among 50 trajectories with $1.7 \le p_0 < 1.8$ in Figure 8(b), there are trajectories converging to the fixed point $(0,0)$, and some trajectories converging to the fixed points $(0,2\pi)$, $(0,-2\pi)$, and $(0,-4\pi)$.

As it can be seen from Figure 8, when the order of a map is reducing from $\alpha = 2$, elliptic points of the regular standard map evolve into fractional attractors, and islands of various orders (periodicity) evolve into attractors of the same orders, inheriting some of the structural features of the map's phase space. In the RLSM, with the exception of the $(0,0)$ fixed point, elliptic points evolve into periodic, attracting slowly diverging trajectories (see Figures 8(a), (c), and (e)), whereas in CSM they evolve into the corresponding sinks (see Figures 8(b), (d), and (f) for CSM considered on a torus). Presence of only period-one structures in the phase spaces of fractional standard maps, in the case where $K = 3$, $\alpha = 1.9$ (Figures 8(a) and (b)), corresponds to the fact that the phase space of the standard map with $K = 3$ has only one central island. Period $T = 4$ structures in Figures 8(c) and (d), and period $T = 2$ and $T = 3$ in Figures 8(e) and (f) correspond to the phase spaces of the standard map with $K = 2$ and $K = 0.6$. Numerical computations ([25]) show that ASDTs, which converge to trajectories along the p-axis ($x \to x_{\lim} = 0$), have the following asymptotic behavior: $x_n \sim n^{1-\alpha}$ and $p_n \sim n^{2-\alpha}$.

3.2 Period-two sinks

In [25], the condition Equation (52) for the stability of the standard families of maps' fixed point $(0,0)$ was numerically confirmed, but this condition can be obtained analytically as the condition of the appearance of the period two ($T = 2$) point. Here, we will follow paper [21] to investigate period-two sinks.

Continuous and discrete fractional systems may not have periodic solutions except fixed points (see, for example, [3, 30, 31, 34, 55, 56, 60]). Periodic points may exist only in the asymptotic sense. We will consider the asymptotic stability of periodic points in discrete fractional systems. A periodic point is asymptotically stable

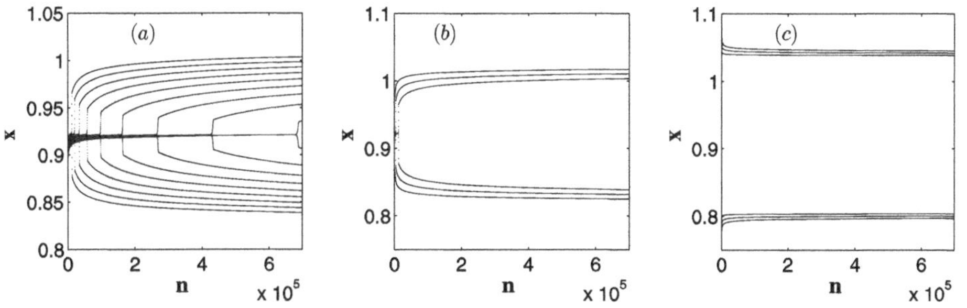

Figure 9: Asymptotically period-two trajectories for the Caputo logistic α-family of maps, Equation (29), with $\alpha = 0.1$ and $K = 15.5$: (a) nine trajectories with the initial conditions $x_0 = 0.29 + 0.04i$, $i = 0, 1, \ldots, 8$ ($i = 0$ corresponds to the rightmost bifurcation); (b) $x_0 = 0.61 + 0.06i$, $i = 1, 2, 3$; (c) $x_0 = 0.95 + 0.04i$, $i = 1, 2, 3$. As $n \to \infty$, all trajectories converge to the limiting values $x_{\lim 1} = 0.80629$ and $x_{\lim 2} = 1.036030$ (see Equation (85)). The unstable fixed point is $x_{\lim 0} = (K-1)/K = 0.93548$. This figure is reprinted from [21] with the permission of AIP Publishing.

if there exists an open set such that all trajectories with initial conditions from this set converge to this point, as $t \to \infty$. It is known that as a nonlinearity parameters of ordinary nonlinear dynamical systems change, the systems bifurcate. At the value of the parameter at which the period 2^{n+1}, ($T = 2^{n+1}$) sink is born, and the $T = 2^n$ sink becomes unstable. We will present results of investigation of the $T = 2$ sinks of discrete fractional systems and their application to the problem of stability of fixed points.

In discrete fractional systems, not only the speed of convergence of trajectories to the periodic sinks (as can be seen from Figure 7(b)), but also the way in which convergence occurs depends on initial conditions. When $n \to \infty$, all trajectories in Figure 9 converge to the same, as in Figure 9(c), $T = 2$ sink, but for small initial conditions x_0, all trajectories first converge to the $T = 1$ trajectory, which then bifurcates and becomes the $T = 2$ sink, which converges to its asymptotic value. When x_0 increases, the bifurcation point n_{bif} gradually moves from the right to the left (Figure 9(a)). Ignoring this feature may result in very messy bifurcation diagrams.

3.2.1 $0 < \alpha < 1$

For $n = 2N$, after subtracting x_{2N}, Equation (24) can be written as

$$x_{2N+1} = x_{2N} - \tilde{G}(x_{2N})U_\alpha(1) + \sum_{n=1}^{N} \tilde{G}(x_{2N-2n+1})\big(U_\alpha(2n-1) - U_\alpha(2n)\big)$$
$$+ \sum_{n=1}^{N} \tilde{G}(x_{2N-2n})\big(U_\alpha(2n) - U_\alpha(2n+1)\big). \tag{61}$$

The terms $U_\alpha(2n-1) - U_\alpha(2n)$ are of the order $n^{\alpha-2}$. If an asymptotic ($n \to \infty$) period $T = 2$ sink exists:

$$x_o = \lim_{n\to\infty} x_{2n+1}, \quad x_e = \lim_{n\to\infty} x_{2n}, \tag{62}$$

then the series in Equation (61) converges absolutely. In the limit $n \to \infty$, Equation (61) converges to

$$x_o - x_e = [\tilde{G}(x_o) - \tilde{G}(x_e)]W_\alpha, \tag{63}$$

where W_α is a converging series

$$W_\alpha = \sum_{n=1}^{\infty}[U_\alpha(2n-1) - U_\alpha(2n)], \tag{64}$$

which can be computed numerically with $U_\alpha(n)$ defined either by Equation (25) or by Equation (26).

Now, let us add x_{2N} to x_{2N+1}:

$$\begin{aligned} x_{2N+1} + x_{2N} = 2x_0 - \sum_{n=1}^{2N}[\tilde{G}(x_{2N-n+1}) \\ + \tilde{G}(x_{2N-n})]U_\alpha(n) - \tilde{G}(x_0)U_\alpha(2N+1). \end{aligned} \tag{65}$$

If $T = 2$ sink exists, then, in the limit $n \to \infty$, the left-hand side of Equation (65), the first term on the right-hand side, and the last term of this equation are finite. Expressions in the brackets in Equation (65) tend to the limit $\tilde{G}(x_o) + \tilde{G}(x_e)$. Because the series $\sum_{n=1}^{\infty} U_\alpha(n)$ is diverging, the equality

$$\tilde{G}(x_o) + \tilde{G}(x_e) = 0 \tag{66}$$

presents the only possibility for Equation (65) to be true. The equations defining the value (and existence) of the asymptotic $T = 2$ sink can be written as

$$\begin{cases} G_K(x_o) + G_K(x_e) = 0, \\ x_o - x_e = \frac{W_\alpha}{\Gamma(\alpha)}h^\alpha[G_K(x_o) - G_K(x_e)]. \end{cases} \tag{67}$$

3.2.2 $1 < \alpha < 2$

If we assume that the $T = 2$ sink exists, and limits x_o and x_e are defined by Equation (62), then the limiting values for p (see Equation (32)) are defined by

$$\begin{aligned} p_o = \lim_{n\to\infty} p_{2n+1} = \lim_{n\to\infty} \frac{x_{2n+1} - x_{2n}}{h} = \frac{x_o - x_e}{h} \\ \text{and} \quad p_e = \lim_{n\to\infty} p_{2n} = -p_o. \end{aligned} \tag{68}$$

Similar to the derivation of Equations (63) and (66), addition and subtraction of p_{2N+1} and p_{2N} yields

$$p_o - p_e = [\tilde{G}(x_o) - \tilde{G}(x_e)]\tilde{W}_\alpha \tag{69}$$

and

$$\tilde{G}(x_o) + \tilde{G}(x_e) = 0, \tag{70}$$

where $\tilde{W}_\alpha$ is the converging series

$$\tilde{W}_\alpha = \sum_{n=1}^{\infty}[\tilde{U}_\alpha(2n-1) - \tilde{U}_\alpha(2n)]. \tag{71}$$

Let us note that if, as defined in Equation (25), $U_\alpha(n) = n^{\alpha-1}$, then $\tilde{W}$ is identical to the $V_{\alpha l}$ defined by Equation (53). A high-accuracy algorithm for calculating $V_{\alpha l}$ can be found in Appendix [14]. For $U(n)$, defined by Equation (26), $\tilde{W}$ was calculated in [15]. Taking into account that converging series Equation (71) can be written as

$$\tilde{W}_\alpha = \tilde{U}_{\alpha 1} - \sum_{n=1}^{\infty}[\tilde{U}_\alpha(2n) - \tilde{U}_\alpha(2n+1)], \tag{72}$$

where

$$\tilde{U}_{\alpha 1} = \begin{cases} 1 & \text{in Eqs. (1), (4),} \\ \Gamma(\alpha) & \text{in Eq. (6),} \end{cases} \tag{73}$$

and using the absolute convergence of series Equation (64) (and, correspondingly, the series on the first line of Equation (74) below), for $0 < \alpha < 1$, we may write:

$$\begin{aligned} \tilde{W}_\alpha &= \tilde{U}_{\alpha 1} - \sum_{n=1}^{\infty}\{[U_\alpha(2n) - U_\alpha(2n-1)] - [U_\alpha(2n+1) - U_\alpha(2n)]\} \\ &= \tilde{U}_{\alpha 1} + \sum_{n=1}^{\infty}[U_\alpha(2n-1) - U_\alpha(2n)] - \sum_{n=1}^{\infty}[U_\alpha(2n) - U_\alpha(2n+1)] \\ &= W_\alpha + \tilde{U}_{\alpha 1} - U_\alpha(2) + U_\alpha(3) - U_\alpha(4) + U_\alpha(5) - \cdots = 2W_\alpha. \end{aligned} \tag{74}$$

Note that, for fractional difference maps, from Equation (72) it follows that

$$\begin{aligned} \tilde{W}_\alpha &= (\alpha-1)\Gamma(\alpha-1) - (\alpha-1)\sum_{n=1}^{\infty}[U_{\alpha-1}(2n) - U_{\alpha-1}(2n+1)] \\ &= (\alpha-1)W_{\alpha-1} = \frac{\alpha-1}{2}\tilde{W}_{\alpha-1}. \end{aligned} \tag{75}$$

3.2.3 $0 < \alpha < 2$

Based on the results for $0 < \alpha < 1$ and $1 < \alpha < 2$ from the previous two sections, we write the equations of the asymptotic $T = 2$ sink for $0 < \alpha < 2$ as

$$\begin{cases} G_K(x_o) + G_K(x_e) = 0, \\ x_o - x_e = \frac{\tilde{W}_\alpha}{2\Gamma(\alpha)} h^\alpha [G_K(x_o) - G_K(x_e)], \end{cases} \tag{76}$$

where $\tilde{W}_\alpha$ is defined by Equations (72) and (73). Notice that, according to Equation (72), $\tilde{W}_1 = 1$.

- For $0 < \alpha < 2$, the fixed point, defined by the equation $G_K(x_o) = 0$, is a solution of the equation system (76).
- When $h \to 0$, fractional difference equations converge to the corresponding fractional differential equations, and $x_o - x_e \to 0$. This implies that in fractional differential equations of the order $0 < \alpha < 2$, transition from a fixed point to a periodic trajectory will never happen. This can be considered as an argument in support of what is mentioned in Section 1 conjecture:

Conjecture 1. *Chaos does not exist in continuous fractional systems of the orders* $0 < \alpha < 2$.

3.3 $T = 2$ points in standard and logistic families of maps

Here we will apply Equation (76) to investigate fixed and $T = 2$ points of the standard and logistic α-families of maps introduced in Section 2.3.

3.3.1 Standard α-families of maps

When $G_K(x) = K\sin(x)$, the first equation of the equation system (76) yields

$$\sin\frac{x_o + x_e}{2}\cos\frac{x_o - x_e}{2} = 0, \tag{77}$$

which on $x \in [-\pi, \pi]$ has two solutions:

$$\begin{aligned} &\text{symmetric point} \quad x_{\text{osy}} = -x_{\text{esy}} \quad \text{and} \\ &\text{shift} - \pi \text{ point} \quad x_{\text{osh}} = x_{\text{esh}} - \pi. \end{aligned} \tag{78}$$

Then, the second equation in (76), for $0 < \alpha < 2$, allows us to calculate two $T = 2$ sinks defined by the equations

$$\sin x_{\text{osy}} = \frac{2\Gamma(\alpha)}{\tilde{W}_\alpha h^\alpha K} x_{\text{osy}} \tag{79}$$

and

$$\sin x_{\text{osh}} = \frac{\pi\Gamma(\alpha)}{\tilde{W}_\alpha h^\alpha K}. \tag{80}$$

The symmetric $T = 2$ sink appears when

$$h^\alpha|K| > h^\alpha|K_{C1s}| = \frac{2\Gamma(\alpha)}{\tilde{W}_\alpha}, \tag{81}$$

and the shift-π $T = 2$ sink appears when

$$h^\alpha|K| > \frac{\pi}{2}h^\alpha|K_{C1s}|. \tag{82}$$

The condition of the fixed-point stability for $h = 1$ from Equation (52) is opposite to the condition of the appearance of the symmetric $T = 2$ sink (see Equation (81)).

3.3.2 Logistic α-families of maps

When $G_K(x) = x - Kx(1-x)$, the equation system (76) becomes

$$\begin{cases} (1-K)(x_o + x_e) + K(x_o^2 + x_e^2) = 0, \\ x_o - x_e = \frac{\tilde{W}_\alpha}{2\Gamma(\alpha)}h^\alpha(x_o - x_e)[1 - K + (x_o + x_e)]. \end{cases} \tag{83}$$

Two fixed-point solutions, with $x_o = x_e$, are $x_o = 0$, stable for $K < 1$, and $x_o = (K-1)/K$.

The $T = 2$ sink is defined by the equation

$$x_o^2 - \left(\frac{2\Gamma(\alpha)}{\tilde{W}Kh^\alpha} + \frac{K-1}{K}\right)x_o + \frac{2\Gamma^2(\alpha)}{(\tilde{W}Kh^\alpha)^2} + \frac{(K-1)\Gamma(\alpha)}{\tilde{W}K^2h^\alpha} = 0, \tag{84}$$

which has the solutions

$$x_o = \frac{K_{C1s} + K - 1 \pm \sqrt{(K-1)^2 - K_{C1s}^2}}{2K}, \tag{85}$$

defined when

$$K \geq 1 + \frac{2\Gamma(\alpha)}{\tilde{W}h^\alpha} = 1 + K_{C1s} \quad \text{or} \quad K \leq 1 - \frac{2\Gamma(\alpha)}{\tilde{W}h^\alpha} = 1 - K_{C1s}. \tag{86}$$

Conditions (86) are valid for all forms of the logistic α-families of maps considered in this review for $0 < \alpha < 2$. For the Riemann–Liouville logistic α-family of maps, with $h = 1$ and $1 < \alpha < 2$, the first inequality in (86) was derived in [14]. Here, we will consider $K > 0$ and $h \leq 1$. From the definition (72) it follows that $\tilde{W}$ is less than $\tilde{U}_1$, which is either 1 or $\Gamma(\alpha)$ (also $\Gamma(\alpha) > 0.885$ for $\alpha > 0$). Then, $2\Gamma(\alpha)/(\tilde{W}h^\alpha) > 1$, and we may ignore the second of the inequalities in (86). Note that the fixed point $x = (K-1)/K$ is stable when

$$1 \leq K < K_{C1l} = 1 + \frac{2\Gamma(\alpha)}{\tilde{W}h^\alpha} = 1 + K_{C1s}. \tag{87}$$

4 Universality in fractional dynamics

Universality in regular nonlinear dynamics is frequently associated with the universal scenario of transition to chaos through the period-doubling cascade of bifurcations. The major steps in the development of the notion of universality in nonlinear dynamics are gathered in the reprinted selection of papers compiled by Predrag Cvitanovic [9]. This book, among other works, contains one of the most cited publication by Robert M. May [42] and Mitchell J. Feigenbaum's paper [27], in which the Feigenbaum numbers and function are introduced. Applications of universality encompass all areas of science.

4.1 Bifurcation diagrams

Evolution of fractional systems with changes in the nonlinearity (K) and memory (the order α of a map) parameters was investigated on the examples of the fractional/fractional difference standard and logistic maps [13–17, 19–21, 23]. The universal behavior of transition to chaos through the period-doubling cascade of bifurcations persists in fractional systems, but the corresponding constants (similar to the Feigenbaum constants) are not identified yet, because of the complexity of systems with memory. Numerical simulations of the cascades of bifurcations are quite complicated, because of the slow convergence of trajectories to periodic sinks, and existence of CBBT. Comparison of bifurcation diagrams for fractional and fractional difference maps is given in Figure 10 (for the Caputo standard α-families of maps) and Figure 11 (for the Caputo logistic α-families of maps). The obvious difference between fractional and fractional difference maps is that, as α decreases towards zero, (K, x)-bifurcation diagrams of the fractional difference maps, Figures 10(a), (c), (e), and Figure 11(b) contract along the K-axis, whereas the bifurcation diagrams of the fractional maps, Figures 10(b), (d), (f) and Figure 11(a), expand along the K-axis.

Bifurcations with the changes in the memory parameter (when K is fixed), Figure 12, is a new type of the universal behavior, which appears in fractional/fractional difference systems.

4.2 2D bifurcation diagrams

The existence of the (K, x)- and (α, x)-bifurcation diagrams is reflected in the universal 2D-bifurcation diagrams in Figure 13. For two families of maps (standard and logistic) investigated so far, the 2D bifurcation diagrams look similar. Here, we must note that the bifurcation diagrams in Figure 13 follow the bifurcations, which originate from the stable (for small values of the nonlinearity parameter) $(0, 0)$ fixed point.

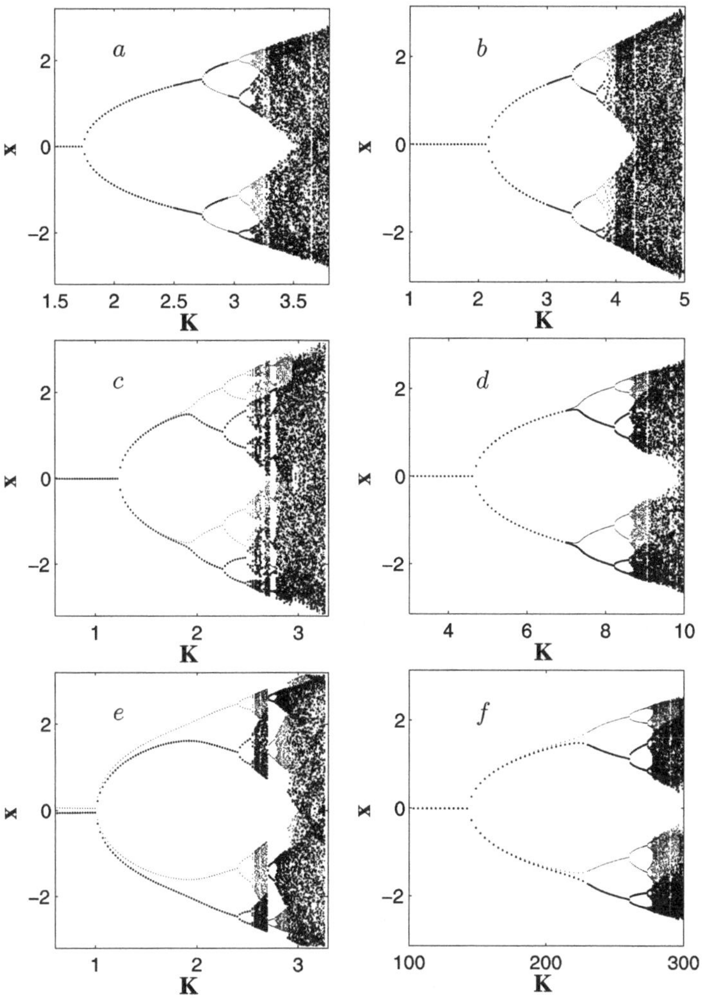

Figure 10: Bifurcation diagrams for the fractional difference Caputo standard α-families of maps, Equation (28) ((a), (c), and (e)), and the fractional Caputo standard α-families of maps, Equation (27) ((b), (d), and (f)). The diagrams were obtained after 5000 iterations with the initial condition $x_0 = 0.1$ (regular points) and $x_0 = -0.1$ (bold points). $\alpha = 0.8$ in a and b; $\alpha = 0.3$ in (c) and (d); $\alpha = 0.01$ in (e) and (f). This figure is reprinted from [15] with the permission of AIP Publishing.

5 Conclusion

We conclude this chapter by listing the main results and the perspectives of the research on nonlinear fractional dynamics.

- Phase space trajectories in continuous fractional systems of orders less than two may intersect, and chaotic attractors may overlap. At present, there is no strict proof that chaos in these systems is impossible, and there is no counterexample.

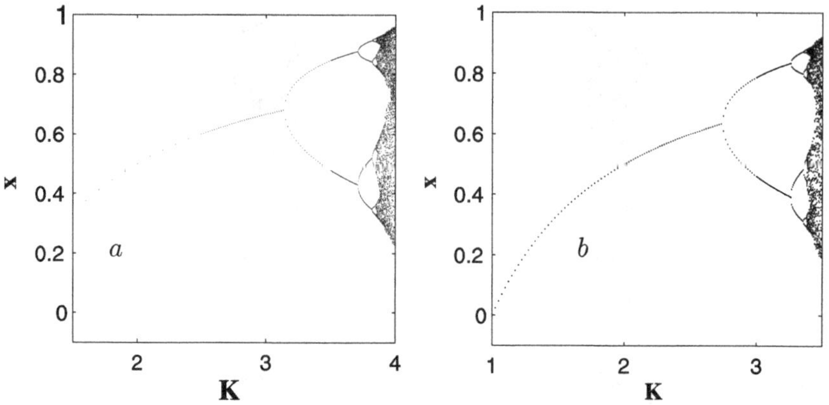

Figure 11: Bifurcation diagrams for the Caputo, Equation (29) and (a) fractional difference Caputo, Equation (30); (b) logistic α-families of maps with $\alpha = 0.8$ obtained after 1000 iterations with the initial condition $x_0 = 0.01$. This figure is reprinted from [19] with the permission of L&H Scientific Publishing.

- Periodic sinks (except the fixed point) may exist only in an asymptotic sense, and asymptotically attracting points may not belong to their own basins of attraction.
- The speed with which a trajectory approaches an attracting point depends on its origin. Trajectories originating from the basin of attraction may converge faster than trajectories originating from the chaotic sea.
- Cascade of bifurcations-type trajectories (CBTT) are general features of discrete fractional systems. One of the problems is to find if they exist in continuous fractional systems. Another problem is to find symmetries related to cascades of bifurcations on single trajectories.
- Fractional systems of the orders, less than the orders of the corresponding integer volume preserving systems, behave as integer system, with dissipation. Correspondingly, the types of attractors, which may exist in fractional systems, include sinks, limiting cycles, and chaotic attractors.
- General properties of discrete fractional systems of orders more than two are not well investigated (see [13, 16]), and their connections to the three-dimensional discrete systems [11, 39] are not established.

Most of the listed properties of fractional systems were derived from the investigation of the fractional generalizations of the simplest quadratic (logistic) and harmonic (standard) maps. Investigations of more complex fractional systems (see, for example, [50, 52] and Figure 8 from [24] for the fractional dissipative standard map) give examples of more complicated behaviors, whose studies should be a subject of the future research.

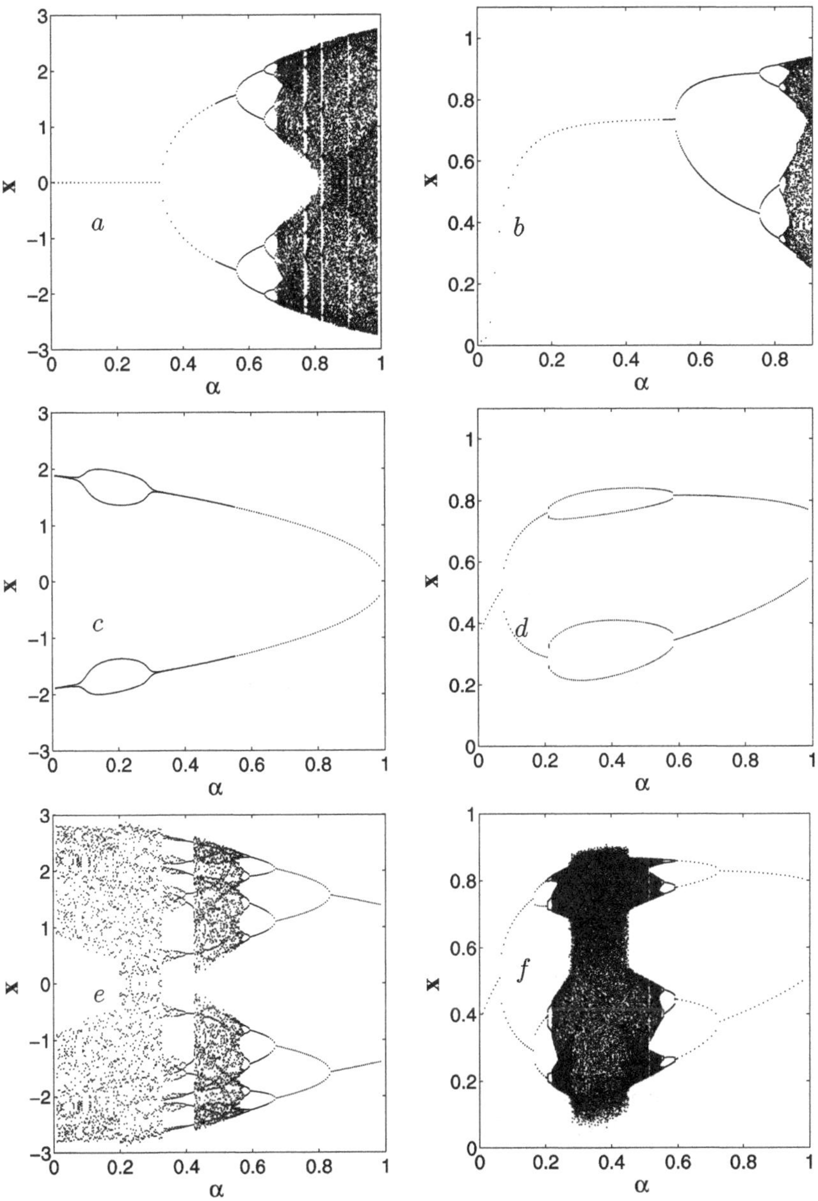

Figure 12: The memory (α,x)-bifurcation diagrams for fractional Caputo standard and (a) logistic; (b) maps and for fractional difference Caputo standard; (c) and (e) logistic; (d) and (f) maps obtained after 5000 iterations. $K = 4.2$ in (a), $K = 3.8$ in (b), $K = 2.0$ in (c), $K = 3.1$ in (d), $K = 2.8$ in (e), and $K = 3.2$ in (f). This figure is reprinted from [23] with the permission of Springer.

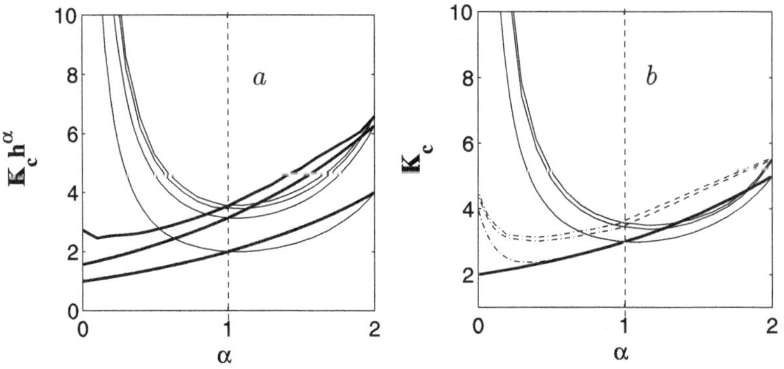

Figure 13: 2D bifurcation diagrams for fractional (solid thing lines) and fractional difference (bold and dashed lines) Caputo standard and (a) $h = 1$ logistic; (b) maps. The first bifurcation, transition from the stable fixed point to the stable period two ($T = 2$) sink, occurs on the bottom curves. $T = 2$ sink (in the case of standard α-families of maps antisymmetric $T = 2$ sink with $x_{n+1} = -x_n$) is stable between the bottom and middle curves. Transition to chaos occurs on the top curves. For the standard fractional map, transition from $T = 2$ to $T = 4$ sink occurs on the line below the top line (the third from the bottom line). Period-doubling bifurcations leading to chaos occur in the narrow band between the two top curves. All bottom curves, and the next to the bottom in (a), are obtained using Equations (81), (82), and (86). Two dashed lines for $1 < \alpha < 2$ in (b) are obtained by interpolation. The remaining lines are results of the direct numerical simulations. Stability of the fixed point for the fractional difference logistic α-family of maps is calculated using both Equation (86) (bold solid line) and direct numerical simulations (a dashed line branching from the solid line). The difference is due to the slow, as $n^{-\alpha}$ (see [15]), convergence of trajectories to the $T = 2$ sink for small α (x vs. K, fixed α; bifurcation diagrams used to find the first bifurcation were calculated on trajectories after 5000 iterations. This figure is reprinted from [21] with the permission of AIP Publishing.)

Bibliography

[1] N. Aguila-Camacho, M. A. Duarte-Mermoud, and J. A. Gallegos, Lyapunov functions for fractional order systems, *Communications in Nonlinear Science and Numerical Simulation*, **19** (2014), 2951–2957.

[2] E. Ahmed, A. M. A. El-Sayed, and Hala A. A. El-Saka, On some Routh–Hurwitz conditions for fractional order differential equations and their applications in Lorenz, Rössler, Chua and Chen systems, *Physics Letters A*, **358** (2006), 1–4.

[3] I. Area, J. Losada, and J. J. Nieto, On fractional derivatives and primitives of periodic functions, *Abstract and Applied Analysis*, **2014** (2014), 392598.

[4] D. Baleanu, G.-C. Wu, Y.-R. Bai, and F.-L. Chen, Stability analysis of Caputo-like discrete fractional systems, *Communications in Nonlinear Science and Numerical Simulation*, **48** (2017), 520–530.

[5] B. A. Carreras, V. E. Lynch, L. Garcia, M. Edelman, and G. M. Zaslavsky, Topological instability along filamented invariant surfaces, *Chaos*, **13** (2003), 1175–1187.

[6] F. Chen and Z. Liu, Asymptotic stability results for nonlinear fractional difference equations, *Journal of Applied Mathematics*, **2012** (2012), 879657.

[7] Y. Chen, Y. Wei, X. Zhou, and Y. Wang, Stability for nonlinear fractional order systems: an indirect approach, *Nonlinear Dynamics*, **89** (2017), 1011–1018.

[8] B. V. Chirikov, A universal instability of many dimensional oscillator systems, *Physics Reports*, **52** (1979), 263–379.

[9] P. Cvitanovic, *Universality in Chaos*, Adam Hilger, Bristol and New York, 1989.

[10] A. S. Deshpande and V. Daftardar-Gejji, On disappearance of chaos in fractional systems, *Chaos, Solitons and Fractals*, **102** (2017), 119–126.

[11] H. R. Dullin and J. D. Meiss, Resonances and twist in volume-preserving maps, *SIAM Journal on Applied Dynamical Systems*, **11** (2012), 319–359.

[12] M. Edelman, Fractional standard map: Riemann–Liouville vs. Caputo, *Communications in Nonlinear Science and Numerical Simulation*, **16** (2011), 4573–4580.

[13] M. Edelman, Fractional maps and fractional attractors. Part I: α-families of maps, *Discontinuity, Nonlinearity, and Complexity*, **1** (2013), 305–324.

[14] M. Edelman, Universal fractional map and cascade of bifurcations type attractors, *Chaos*, **23** (2013), 033127.

[15] M. Edelman, Caputo standard α-family of maps: fractional difference vs. fractional, *Chaos*, **24** (2014), 023137.

[16] M. Edelman, Fractional maps as maps with power-law memory, in V. Afraimovich, A. C. J. Luo, and X. Fu (eds.) *Nonlinear Dynamics and Complexity*, Nonlinear Systems and Complexity, pp. 79–120, Springer, New York, 2014.

[17] M. Edelman, Universality in fractional dynamics, in *International Conference on Fractional Differentiation and Its Applications (ICFDA), 2014*, pp. 1–6, 2014. DOI:10.1109/ICFDA.2014.6967376.

[18] M. Edelman, On the fractional Eulerian numbers and equivalence of maps with long term power-law memory (integral Volterra equations of the second kind) to Grünvald–Letnikov fractional difference (differential) equations, *Chaos*, **25** (2015), 073103.

[19] M. Edelman, Fractional maps and fractional attractors. Part II: fractional difference α-families of maps, *Discontinuity, Nonlinearity, and Complexity*, **4** (2015), 391–402.

[20] M. Edelman, On nonlinear fractional maps: nonlinear maps with power-law memory, in X. Leoncini, C. Eloy, and G. Boedec (eds.) *Chaos, Complexity and Transport, Proceedings of the CCT '15, Conference on Chaos, Complexity, and Transport 2015*, Marseilles, France, 1–5 June 2015, pp. 119–130, World Scientific, Singapore, 2017.

[21] M. Edelman, On stability of fixed points and chaos in fractional systems, *Chaos*, **28** (2018), 023112.

[22] M. Edelman, Maps with power-law memory: direct introduction and Eulerian numbers, fractional maps, and fractional difference maps, in A. Kochubei and Y. Luchko (eds.) *Handbook of Fractional Calculus With Applications, Volume 2, Theory*, De Gruyter, Berlin, 2018.

[23] M. Edelman, Universality in systems with power-law memory and fractional dynamics, in M. Edelman, E. Macau, and M. A. F. Sanjuan (eds.) *Chaotic, Fractional, and Complex Dynamics: New Insights and Perspectives*, Understanding Complex Systems, pp. 147–171, Springer, eBook, 2018.

[24] M. Edelman and L. A. Taieb, New types of solutions of non-linear fractional differential equations, in A. Almeida, L. Castro, and T.-O. Speck (eds.) *Advances in Harmonic Analysis and Operator Theory*, Operator Theory: Advances and Applications, vol. 229, pp. 139–155, Springer, Basel, 2013.

[25] M. Edelman and V. E. Tarasov, Fractional standard map, *Physics Letters A*, **374** (2009), 279–285.

[26] H. A. El-Saka, E. Ahmed, M. I. Shehata, and A. M. A. El-Sayed, On stability, persistence, and Hopf bifurcation in fractional order dynamical systems, *Nonlinear Dynamics*, **56** (2009), 121–126.

[27] M. J. Feigenbaum, Universal behavior in nonlinear systems, *Los Alamos Science*, **1** (1980), 4–27.

[28] W. Feller, *An Introduction to Probability Theory and Its Applications*, Wiley, New York, 1968.

[29] I. Grigorenko and E. Grigorenko, Chaotic dynamics of the fractional Lorenz system, *Physical Review Letters*, **91** (2003), 034101.

[30] J. Jagan Mohan, Periodic solutions of fractional nabla difference equations, *Communications in Applied Analysis*, **20** (2016), 585–609.

[31] J. Jagan Mohan, Quasi-periodic solutions of fractional nabla difference systems, *Fractional Differential Calculus*, **7** (2017), 339–355.

[32] J. Jagan Mohan, N. Shobanadevi, and G. V. S. R. Deekshitulu, Stability of nonlinear nabla fractional difference equations using fixed point theorems, *Italian Journal of Pure and Applied Mathematics*, **32** (2014), 165–184.

[33] F. Jarad, T. Abdeljawad, D. Baleanu, and K. Bicen, On the stability of some discrete fractional nonautonomous systems, *Abstract and Applied Analysis*, **2012** (2012), 476581.

[34] E. Kaslik and S. Sivasundaram, Non-existence of periodic solutions in fractional-order dynamical systems and a remarkable difference between integer and fractional-order derivatives of periodic functions, *Nonlinear Analysis: Real World Applications*, **13** (2012), 1489–1497.

[35] B. K. Lenka and S. Banerjee, Asymptotic stability and stabilization of a class of nonautonomous fractional order systems, *Nonlinear Dynamics*, **85** (2016), 167–177.

[36] C. P. Li and F. R. Zhang, A survey on the stability of fractional differential equations, *The European Physical Journal Special Topics*, **193** (2011), 27–47.

[37] T. Li and Y. Wang, Stability of a class of fractional-order nonlinear systems, *Discrete Dynamics in Nature and Society*, **2014** (2014), 724270.

[38] Y. Li, Y. Q. Chen, and I. Podlubny, Stability of fractional-order nonlinear dynamic systems: Lyapunov direct method and generalized Mittag-Leffler stability, *Computers & Mathematics with Applications*, **59** (2010), 1810–1821.

[39] H. E. Lomeli and J. D. Meiss, Quadratic volume-preserving maps, *Nonlinearity*, **11** (1998), 557–574.

[40] O. Lyubomudrov, M. Edelman, and G. M. Zaslavsky, Pseudochaotic systems and their fractional kinetics, *International Journal of Modern Physics B*, **17** (2003), 4149–4167.

[41] D. Matignon, Stability result on fractional differential equations with applications to control processing, in *Proceedings of the International Meeting on Automated Compliance Systems and the International Conference on Systems, Man, and Cybernetics. (IMACS-SMC 96)*, pp. 963–968, Lille, France, 1996.

[42] R. M. May, Simple mathematical models with very complicated dynamics, *Nature*, **261** (1976), 459–467.

[43] I. Petras, Stability of fractional order systems with rational order: a survey, *Fractional Calculus & Applied Analysis. An International Journal for Theory and Applications*, **12** (2009), 269–298.

[44] I. Petras, *Fractional-Order Nonlinear Systems*, Springer, Berlin, 2011.

[45] M. Rivero, S. V. Rogozin, J. A. T. Machado, and J. J. Trujilo, Stability of fractional order systems, *Mathematical Problems in Engineering*, **2013** (2013), 356215.

[46] A. A. Stanislavsky, Long-term memory contribution as applied to the motion of discrete dynamical system, *Chaos*, **16** (2006), 043105.

[47] V. E. Tarasov, Discrete map with memory from fractional differential equation of arbitrary positive order, *Journal of Mathematical Physics*, **50** (2009), 122703.

[48] V. E. Tarasov, Differential equations with fractional derivative and universal map with memory, *Journal of Physics A: Mathematical and Theoretical*, **42** (2009), 465102.

[49] V. E. Tarasov, *Fractional Dynamics: Application of Fractional Calculus to Dynamics of Particles, Fields and Media*, HEP, Springer, Beijing, Berlin, Heidelberg, 2010.

[50] V. E. Tarasov, Fractional Zaslavsky and Henon map, in A. C. J. Luo and V. Afraimovich (eds.) *Long-Range Interactions, Stochasticity and Fractional Dynamics*, pp. 1–26, Springer and HEP, Berlin, 2010.
[51] V. E. Tarasov (ed.), *Handbook of Fractional Calculus With Applications*, Applications in Physics, Volumes 4 and 5, De Gruyter, Berlin, 2018.
[52] V. E. Tarasov and M. Edelman, Fractional dissipative standard map, *Chaos*, **20** (2010), 023127.
[53] V. E. Tarasov and G. M. Zaslavsky, Fractional equations of kicked systems and discrete maps, *Journal of Physics A: Mathematical and Theoretical*, **41** (2008), 435101.
[54] M. S. Tavazoei and M. Haeri, A necessary condition for double scroll attractor existence in fractional order systems, *Physics Letters A*, **367** (2007), 102–113.
[55] M. S. Tavazoei and M. Haeri, A proof for non existence of periodic solutions in time invariant fractional order systems, *Automatica*, **45** (2009), 1886–1890.
[56] J. Wang, M. Feckan, and Y. Zhou, Nonexistence of periodic solutions and asymptotically periodic solutions for fractional differential equations, *Communications in Nonlinear Science and Numerical Simulation*, **18** (2013), 246–256.
[57] G.-C. Wu and D. Baleanu, Discrete fractional logistic map and its chaos, *Nonlinear Dynamics*, **75** (2014), 283–287.
[58] G.-C. Wu, D. Baleanu, and S.-D. Zeng, Discrete chaos in fractional sine and standard maps, *Physics Letters A*, **378** (2014), 484–487.
[59] M. Wyrwas, E. Pawluszewicz, and E. Girejko, Stability of nonlinear h-difference systems with N fractional orders, *Kybernetika*, **15** (2015), 112–136.
[60] M. Yazdani and H. Salarieh, On the existence of periodic solutions in time-invariant fractional order systems, *Automatica*, **47** (2011), 1834–1837.
[61] G. M. Zaslavsky, Fractional kinetic equation for Hamiltonian chaos, *Physica D*, **76** (1994), 110–122.
[62] G. M. Zaslavsky, Renormalization group theory of anomalous transport in systems with Hamiltonian chaos, *Chaos*, **4** (1994), 25–33.
[63] G. M. Zaslavsky, Chaos, fractional kinetics, and anomalous transport, *Physics Reports*, **371** (2002), 461–580.
[64] G. M. Zaslavsky, *Hamiltonian Chaos and Fractional Dynamics*, Oxford University Press, Oxford, 2008.
[65] G. M. Zaslavsky, B. A. Carreras, V. E. Lynch, L. Garcia, and M. Edelman, Topological instability along invariant surfaces and pseudochaotic transport, *Physical Review E*, **72** (2005), 026227.
[66] G. M. Zaslavsky and M. Edelman, Hierarchical structures in the phase space and fractional kinetics: I. Classical systems, *Chaos*, **10** (2000), 135–146.
[67] G. M. Zaslavsky and M. Edelman, Weak mixing and anomalous kinetics along filamented surfaces, *Chaos*, **11** (2001), 295–305.
[68] G. M. Zaslavsky and M. Edelman, Pseudochaos, in E. Kaplan, J. E. Marsden, and R. S. Katepalli (eds.) *Perspectives and Problems in Nonlinear Science*, pp. 421–443, Springer, New York, 2003.
[69] G. M. Zaslavsky and M. Edelman, Fractional dynamics of pseudochaotic systems, in *First Workshop on Fractional Differentiation and Its Applications of the International Federation of Automatic Control*, Bordeaux, France, July 19–21, 2004, 2004.
[70] G. M. Zaslavsky and M. Edelman, Fractional kinetics: from pseudochaotic dynamics to Maxwell's Demon, *Physica D*, **193** (2004), 128–147.
[71] G. M. Zaslavsky and M. Edelman, Fractional kinetics of pseudochaotic dynamics, in A. Le Mechaute, J. A. Tenreiro Machado, J. C. Trigeassou, and J. Sabatier (eds.) *Fractional Differentiation and Its Applications*, pp. 183–193, Books on Demand, Germany, 2005.

[72] G. M. Zaslavsky, M. Edelman, and B. A. Niyazov, Self-similarity, renormalization, and phase space nonuniformity of Hamiltonian chaotic dynamics, *Chaos*, **7** (1997), 159–181.
[73] G. M. Zaslavsky, M. Edelman, H. Weitzner, B. A. Carreras, G. McKee, R. Bravenec, and R. Fonk, Large-scale behavior of tokamak density fluctuations, *Physics of Plasmas*, **7** (2000), 3691–3698.
[74] G. M. Zaslavsky, P. N. Guzdar, M. Edelman, M. I. Sitnov, and A. S. Sharma, Selfsimilarity and fractional kinetics of solar wind–magnetosphere coupling, *Physica A*, **373** (2007), 11–20.
[75] G. M. Zaslavsky, P. N. Guzdar, M. Edelnman, M. I. Sitnov, and A. S. Sharma, Multiscale behavior and fractional kinetics from the data of solar wind–magnetosphere coupling, *Communications in Nonlinear Science and Numerical Simulation*, **13** (2008), 314–330.
[76] G. M. Zaslavsky, A. A. Stanislavsky, and M. Edelman, Chaotic and pseudochaotic attractors of perturbed fractional oscillator, *Chaos*, **16** (2006), 013102.
[77] Y. Zhou, *Basic Theory of Fractional Differential Equations*, World Scientific, Singapore, 2014.

Aleksander Stanislavsky

Fractional oscillator basics

Abstract: In fractional dynamics the oscillator model plays a key role, just as in physics generally. Being a natural generalization of the conventional harmonic oscillator, the main feature of the fractional oscillator is an intrinsic damping, which vanishes in the classical limit. The fractional oscillator clearly demonstrates an algebraic decay together with harmonic motion. The model is introduced in this chapter to present its basic properties in free, forced, and coupled cases. Interpreting the fractional oscillator as an ensemble average of ordinary harmonic oscillators, governed by a time arrow with irregular steps, the intrinsic absorption results from the full contribution of the harmonic oscillator ensemble in which each oscillator differs a little from others in frequency, so that its response is compensated by an anti-phase response of another harmonic oscillator, allows one to find a logical relationship between the fractional oscillator and fractional diffusive waves.

Keywords: Power-law memory, fractional derivative, oscillator

MSC 2000: 35R11, 26A33

1 Introduction

Considering physical systems that oscillate and observable everywhere, the harmonic oscillator serves as a desirable prototype in the mathematical treatment of different phenomena. It is no wonder that the paradigm has many important and useful applications in both classical and quantum mechanics. A simple realization of the harmonic oscillator in classical mechanics is the particle motion, under the action of a restoring force proportional to its displacement from its equilibrium position. However, the applicability of this model has its strictly defined boundaries. As the idea, more than any other, is fundamental to our understanding of physical phenomena, the simple harmonic oscillator as a concept was generalized. Here, we are talking not only about friction damping, driven, or coupled systems, but also about the quantum harmonic oscillator as a quantum-mechanical analog of the classical oscillator. In this case an arbitrary potential is approximated as a harmonic potential at the vicinity of a stable equilibrium point, being one of the most important model systems in quantum mechanics (for example, in analyzing the spectra of diatomic molecules). Note also, classical harmonic oscillators are defined in integer-order derivatives. The mathemat-

Aleksander Stanislavsky, Institute of Radio Astronomy, National Academy of Sciences of Ukraine, Mystetstv St., 4, 61002 Kharkiv, Ukraine; and V. N. Karazin Kharkiv National University, Svobody Sq., 4, 61022 Kharkiv, Ukraine, e-mail: a.a.stanislavsky@rian.kharkov.ua

https://doi.org/10.1515/9783110571707-006

ical tools are extended to fractional calculus [12, 51], with the help of fractional operators and—in some circumstances—one may build a new oscillator model (so-called the fractional oscillator). Thanks to Mainardi [37], this idea was embodied in physics: "By replacing the time derivative ... with a fractional derivative of order α with $0 < \alpha < 2$, we expect to obtain processes which interpolate (if $1 < \alpha < 2$) or extrapolate (if $0 < \alpha < 1$) the classical phenomena; ...".

Today, fractional calculus, as a branch of mathematics, has been investigated well from mathematical points of view (with the theorems, solving of equations in fractional operators, analysis of mathematical properties of the operators, and so on) [46]. Naturally, the question arises regarding what physics corresponds to the mathematics. The question is nontrivial. In fact, it is well known that the equations of classic mechanics, field theory, hydrodynamics, quantum mechanics, and so on, are described mainly by means of ordinary derivatives of integer order, despite their opportunity of generalization as derivatives of fractional order. How can fractional calculus appear in the description of physical processes? What physical interpretation stand behind fractional calculus? These and other questions are of a large scientific and practical interest. In this connection, it is appropriate to make mention of the insightful words of Dirac [13] regarding the extraordinary efficiency of mathematics in physical researches (and it is enough to recall here, as an example, quantum mechanics and the theory of self-conjugate operators in the Hilbert space). All this supports an argument that every mathematical concept plays a constitutive role in physics, and it is worthy of note.

The classical Hamiltonian (or Lagrangian) mechanics, formulated with derivatives of integer order, suggests effective methods for the analysis of conservative systems, although the real physical world is rather nonconservative because of friction. Accounting for friction forces in physical models increases complexity of the mathematical apparatus necessary for their descriptions. Hamiltonian (and Lagrangian) equations of motion in fractional derivatives, describing nonconservative systems, were introduced by Riewe [47, 48]. His approach is based on a simple observation. If the frictional force is proportional to the velocity, the functional form of a classical Lagrangian can be supplemented by a term with a fractional derivative of half order. After applying the variational principle, the resulting equations of motion for such nonconservative systems contain the contribution of friction forces. A generalization of the variational problem with fractional derivatives to the case of systems with constraints (motion constraints) was proposed in the works [1, 41]. However, as it was noted by Dreisigmeier and Young [14], this concept does not lead to the strict causal equations. The reason is that the variational principle uses integration by parts. By virtue of this left Riemann–Liouville derivative is replaced by a right-hand derivative, in which the temporal variable is directed from the future into the past. Unfortunately, the attempt to overcome this problem by interpreting the action in the form of the Volterra series was not successful [15]. An alternative approach for deriving the equations of motion in fractional derivatives was recently developed in the papers [59, 60]. The equations of motion are found using the fractional normalization condition. This condition assumes that the normalization for the distribution function takes place in a space of fractional dimension. The volume element of such a fractional phase space is expressed in terms of external derivatives of fractional order. In this case, dynamical systems are described by power functions, with an arbitrary exponent of coordinates and impulses. Thus, such systems prove to be essentially nonlinear.

Describing specific properties of the fractal oscillator and neighboring fractional models, the purpose of this chapter is to present a current state of researches in the field. It should be recalled that the penetration of fractional calculus into physics was quickly accelerated after the establishment of its close connection with stable distributions from the theory of probability [24]. Perhaps the understanding of the fractional oscillator also lies hereabouts. In this context, we discuss basic properties of the fractional concept. Starting with the free fractional oscillator presentation in Section 2, its equation can be derived from the Hamiltonian formalism (Section 3). Subsequent sections describe a wide range of fractional systems related to the fractional oscillator, from two-term fractional oscillation equation (Section 4), and excitation of the fractional oscillator by an externally applied force (Section 5), to coupled fractional oscillators (Section 6), and forced oscillations of a multiple fractional system (Section 7). The behavior of continuous fractional systems will be previewed in Section 8. Finally, we sum up our consideration in Section 9.

2 Free fractional oscillator

Although damping can be introduced in the classical harmonic oscillator through a variety of physical mechanisms, the fractional derivative is not the most common of them, but rather a very original one. Usually, the damping force is modeled proportional to the velocity of the oscillating object. Adding the friction force to the undamped harmonic oscillator equation, we obtain the second-order differential equation with constant coefficients, where the damping contribution is contained in one of equation terms. The equation of the free fractional oscillator does have the term, but in contrast to the classical case, the amplitude of oscillation does not remain unchanged forever.

The integral equation of motion of the fractional oscillator is written [45, 66] as

$$x(t) = x(0) + \dot{x}(0)t - \frac{\omega^\alpha}{\Gamma(\alpha)} \int_0^t (t-\tau)^{\alpha-1} x(\tau)\, d\tau,$$

with $1 < \alpha \leq 2$, where $x(0)$ and $\dot{x}(0)$ are the displacement and velocity of the oscillator at $t = 0$, and ω denotes the circular frequency. Solving the equation by taking the Laplace transforms on both sides, we obtain

$$\widetilde{x}(s) = \frac{x(0)s^{\alpha-1}}{s^\alpha + \omega^\alpha} + \frac{\dot{x}(0)s^{\alpha-2}}{s^\alpha + \omega^\alpha},$$

where $\widetilde{x}(t) = \int_0^\infty x(t)\, e^{-st}\, dt$. The inverse Laplace transform gives

$$x(t) = x(0)E_{\alpha,1}(-\omega^\alpha t^\alpha) + \dot{x}(0)\, tE_{\alpha,2}(-\omega^\alpha t^\alpha), \tag{1}$$

expressed in terms of the two-parameter Mittag Leffler function defined in [39], for example. Without loss of generality, the initial conditions can be taken by $x(0) = x_0$ and $\dot{x}(0) = 0$. The solution, corresponding to the initial conditions, explicitly reads: $x(t) = x_0 E_{\alpha,1}(-\omega^\alpha t^\alpha)$. From the asymptotic behavior of the Mittag-Leffler function, the solution has the so-called algebraic decay:

$$x(t) \sim x_0 \frac{(\omega t)^{-\alpha}}{\Gamma(1-\alpha)} \quad \text{as } \omega t \to \infty.$$

By analogy with the harmonic oscillator [45], the generalized momentum of the fractional oscillator is defined by

$$p = m\left(\frac{d^{\alpha/2}x}{dt^{\alpha/2}}\right) = -mx_0\omega^\alpha t^{\alpha/2} E_{\alpha,1+\alpha/2}(-\omega^\alpha t^\alpha). \tag{2}$$

Here, the dimension of the parameter m is $MT^{\alpha-2}$ (not having the dimension of a mass, as in the case of a simple harmonic oscillator), and for ω^α, it is T^{-1}. Then, the total energy yields

$$E_F = \frac{p^2}{2m} + \frac{kx^2}{2} = \frac{1}{2}kx_0^2\big(E_{\alpha,1}(-\omega^\alpha t^\alpha)\big)^2 + \frac{1}{2}mx_0^2\omega^{2\alpha}t^\alpha\big(E_{\alpha,1+\alpha/2}(-\omega^\alpha t^\alpha)\big)^2,$$

where the parameter k is present as one in $E_H = \frac{1}{2}m\dot{x}^2 + \frac{1}{2}kx^2$ for the harmonic oscillator. Note that the derivative in Equation (2) is the Caputo-type fractional derivative of order $\alpha/2$ [7]. Then, the fractional differential equation of the system is

$$\frac{d^\alpha x}{dt^\alpha} + \omega^\alpha x = 0, \quad 1 < \alpha \le 2, \tag{3}$$

where $\omega^\alpha = k/m$. Another important feature of the fractional oscillations is a finite number of damped oscillations with an algebraic decay [18, 22, 23, 27, 56]. The point is that such oscillations can be represented in the form of a simple sum of two parts. One of them has asymptotically an algebraic (monotonic) decay, and another term is an exponentially damped harmonic oscillation. The second term decreases faster than the first. Thus, the fractional oscillations possess a finite number of zeros. A laboratory component for undergraduates studying the fractional harmonic oscillator in theoretical, experimental, and numerical ways, was presented in [6].

In Equation (3), instead of the Caputo fractional derivative, it would be possible to use the Riemann–Liouville fractional differentiation operator [51], but this case requires the initial conditions in fractional derivatives. Although such initial value problems can successfully be solved mathematically, their solutions are fruitless in physics. The reason is that there is no clear physical interpretation of this type of initial conditions [46]. The Caputo fractional derivative definition is based on standard initial conditions in terms of derivatives of integer order. These initial conditions have clear physical interpretation as an initial position $x(a)$ at the point a, the initial velocity $\dot{x}(a)$, initial acceleration $\ddot{x}(a)$, and so on. On the other hand, the Caputo fractional derivative is more restrictive, as it requires the existence of the n-th derivative of the solution. Fortunately, most fractional models that appear in applications fulfill this requirement. Later, whenever the Caputo operator is used, it is assumed that this condition is satisfied.

If the order of fractional derivatives α slightly deviates from an integer value, then there is a possibility to use an expansion over the small parameter $\varepsilon = n - \alpha$. The approach is called the ε-expansion by Tarasov and Zaslavsky [64], motivated by a lack of explicit methods for the analysis of equations with fractional derivatives. For the linear fractional oscillator, in particular, the obtained ε-expansion allows one to compare with the exact one. When $\varepsilon \ll 1$, the Caputo fractional derivative is written as

$$^{C}D_t^{2-\varepsilon} f(t) = \ddot{f}(t) + \varepsilon D_1^2 f(t) + \cdots,$$

where D_1^2 is defined in the form

$$D_1^2 f(t) = \dot{f}(0)\ln(t) + \gamma \ddot{f}(t) + \int_0^t \dddot{f}(\tau)\ln(t-\tau)\,d\tau.$$

In the limit $\varepsilon \to 0$, there follows the correct expansion $\lim_{\varepsilon\to 0} D^{2-\varepsilon} f(t) = \ddot{f}(t)$. For the fractional oscillator, the expansion takes the form

$$x(t) = x_0(t) + \varepsilon x_1(t) + \cdots.$$

The equation for $x_0(t)$ satisfies the harmonic oscillator equation with the initial conditions $x_0(0) = x_0$ and $\dot{x}_0(0) = \dot{x}(0)$ and has the conventional solution $x_0(t) = x_0\cos(\omega t) + (\dot{x}(0)/\omega)\sin(\omega t)$. The equation for $x_1(t)$ and the initial conditions are

$$\ddot{x}_1(t) + \omega^\alpha x_1(t) + D_1^2 x_0(t) = 0, \quad x_1(0) = 0, \quad \dot{x}_1(0) = 0.$$

For simplicity in calculations, if one puts $x(0) = 1$, $\dot{x}(0) = 0$, and $\omega = 1$, then the solution of this equation is

$$x_1(t) = -\frac{t}{2}[\sin(t)\mathrm{Ci}(t) + \cos(t)\mathrm{Si}(t)] - \frac{1}{2} + \frac{1}{2}\cos(t),$$

where $\mathrm{Ci}(t)$ and $\mathrm{Si}(t)$ are the sine and cosine integral functions, respectively, of the form

$$\mathrm{Si}(t) = \int_0^t \frac{\sin(x)}{x}dx, \quad \mathrm{Ci}(t) = -\int_t^\infty \frac{\cos(x)}{x}dx.$$

Finally, the solution in the ε-expansion is

$$x(t) = \cos(t) - \frac{\varepsilon}{2}(1 - \cos(t) + t[\sin(t)\mathrm{Ci}(t) + \cos(t)\mathrm{Si}(t)]) + \cdots.$$

The term, proportional to ε, may be considered as a correction to the solution $x(t) = \cos(t)$, due to the fractional derivative of order $\alpha = 2-\varepsilon$. Comparing this result with the exact solution (1), its decomposition [23], when $\alpha = 2 - \varepsilon$ and $\varepsilon t \ll 1$, gives the same result [64].

The phenomena of amplified oscillations corresponding to the case $2 < \alpha < 3$ were considered in [67]. Contrasting with the common dissipation interpretation for fractional derivative of order $\alpha = 2 - \varepsilon$ ($\varepsilon \ll 1$), the time rate of change of the energy is always positive for $\alpha = 2 + \varepsilon$. Hence, the total energy of the system is a monotonous increasing function of time, which can mean, for example, that the system absorbs energy from the environment.

3 Hamiltonian under subordination

If the time variable represents a sum of random temporal intervals T_i, being nonnegative, independent, and identically distributed, and if the waiting times T_i belong to an α-stable distribution ($0 < \alpha < 1$), then their sum $n^{-1/\alpha}(T_1 + T_2 + \cdots + T_n)$, $n \in \mathbf{N}$ converges in distribution to a stable law, with the same index α [40]. To determine a walker position at the true time t, one needs to find the number of jumps up to time t. This discrete counting process is $\{N_t\}_{t\geq 0} = \max\{n \in \mathbf{N} \mid \sum_{i=1}^{n} T_i \leq t\}$. The continuous limit of $\{N_t\}_{t\geq 0}$ is conventionally denoted by $S(t)$. For a fixed time, it represents the first passage of the stochastic time evolution above that time level. The random process is nondecreasing, and it can be chosen as a new time clock (time arrow with irregular steps) [53]. The probability density of the process $S(t)$ has the following Laplace image:

$$p^S(t,\tau) = \frac{1}{2\pi j}\int_{\mathrm{Br}} e^{ut-\tau u^\alpha}\, u^{\alpha-1}\, du = t^{-\alpha}F_\alpha(\tau/t^\alpha), \tag{4}$$

where Br denotes the Bromwich path. This probability density determines the probability to be at the internal time (or so-called operational time) τ on the real time t [20]. The function $F_\alpha(z)$ (called also the Mainardi function [35, 38]) can be expanded as a Taylor series. Besides, it has the Fox' H-function representation:

$$F_\alpha(z) = H_{11}^{10}\left(z \,\middle|\, \begin{matrix}(1-\alpha,\alpha)\\(0,1)\end{matrix}\right) = \sum_{k=0}^{\infty}\frac{(-z)^k}{k!\Gamma(1-\alpha(1+k))},$$

where $\Gamma(x)$ is the Euler gamma function. In the theory of anomalous diffusion, the random process $S(t)$ is applied for the subordination of purely random processes [40, 54], for example, Gaussian or Lévy ones. In our case, we are going to use it for subordinating a deterministic motion (harmonic oscillation in perspective).

Let the temporal evolution of a Hamiltonian system depend on the operational time τ. Then, the equation of motion takes the form

$$\frac{dq}{d\tau} = \frac{\partial \mathcal{H}}{\partial p}, \quad \frac{dp}{d\tau} = -\frac{\partial \mathcal{H}}{\partial q}. \tag{5}$$

Consider such a dynamical system under subordination [57]. Then, one can write the momentum and the coordinate in the form of subordination relations:

$$p_\alpha(t) = \int_0^\infty p^S(t,\tau)\, p(\tau)\, d\tau, \quad q_\alpha(t) = \int_0^\infty p^S(t,\tau)\, q(\tau)\, d\tau.$$

However, the values $\partial \mathcal{H}/\partial p$ and $\partial \mathcal{H}/\partial q$ are also expressed in terms of operational time. Therefore, we suppose that

$$\frac{\partial \mathcal{H}_\alpha}{\partial p_\alpha} = \int_0^\infty p^S(t,\tau)\, \frac{\partial \mathcal{H}}{\partial p}\, d\tau, \quad \frac{\partial \mathcal{H}_\alpha}{\partial q_\alpha} = \int_0^\infty p^S(t,\tau)\, \frac{\partial \mathcal{H}}{\partial q}\, d\tau.$$

In this case, Equations (5) transform into the fractional form

$$\frac{d^\alpha q_\alpha}{dt^\alpha} = \frac{\partial \mathcal{H}_\alpha}{\partial p_\alpha} = p_\alpha, \quad \frac{d^\alpha p_\alpha}{dt^\alpha} = -\frac{\partial \mathcal{H}_\alpha}{\partial q_\alpha}. \tag{6}$$

Here, we apply the Caputo derivative [7, 23], namely

$$\frac{d^\alpha x(t)}{dt^\alpha} = {}^{C}D_t^\alpha x(t) = \frac{1}{\Gamma(n-\alpha)} \int_0^t \frac{x^{(n)}(\tau)}{(t-\tau)^{\alpha+1-n}}\, d\tau, \quad n-1 < \alpha < n,$$

where $x^{(n)}(t) = D^n x(t)$ means the n-derivative of $x(t)$.

One of the simplest (but nontrivial) physical models supported by the method is the fractional oscillator. Probably, the fractional model was first studied by Mainardi [37]. In this case, the generalized Hamiltonian takes the form

$$H_\alpha = (p_\alpha^2 + \omega^2 q_\alpha^2)/2, \tag{7}$$

where ω is the circular frequency, and q_α and p_α are the displacement and momentum, respectively. This magnitude describes the total energy for this dynamical system [45, 66]. It should be noted that in this section (in comparison with the previous section) our notations suppose $0 < \alpha \le 1$. Then, the fractional Hamiltonian equations for the oscillator are written as

$${}^{C}D_t^\alpha q_\alpha = \frac{d^\alpha q_\alpha}{dt^\alpha} = \frac{\partial H_\alpha}{\partial p_\alpha} = p_\alpha, \tag{8}$$

$${}^{C}D_t^\alpha p_\alpha = \frac{d^\alpha p_\alpha}{dt^\alpha} = -\frac{\partial H_\alpha}{\partial q_\alpha} = -\omega^2 q_\alpha. \tag{9}$$

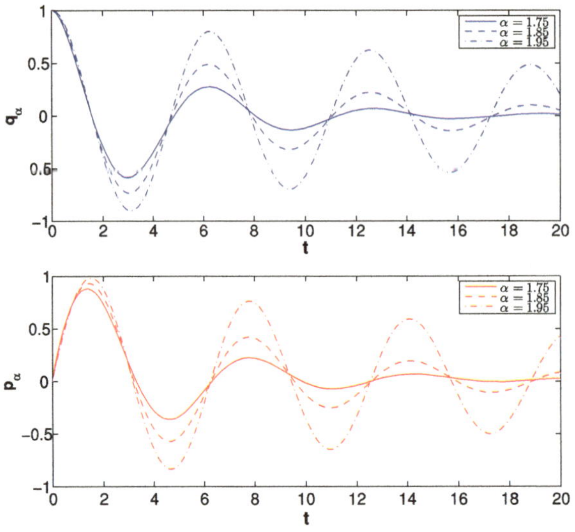

Figure 1: Fractional oscillations of q_α and p_α for different values α.

Following this way, we come to the equations of motion:

$$ {}^{C}\mathcal{D}_t^{2\alpha} q_\alpha + \omega^2 q_\alpha = 0 \quad \text{or} \quad {}^{C}\mathcal{D}_t^{2\alpha} p_\alpha + \omega^2 p_\alpha = 0. \tag{10} $$

Each of the equations has two independent solutions [57]. It suffices to solve one of these equations. For example, the coordinate is determined by the expression

$$ q_\alpha(t) = A\,E_{2\alpha,1}(-\omega^2 t^{2\alpha}) + B\,\omega t^\alpha E_{2\alpha,1+\alpha}(-\omega^2 t^{2\alpha}), \tag{11} $$

where A and B are constants, and

$$ E_{\mu,\nu}(z) = \sum_{k=0}^{\infty} \frac{z^k}{\Gamma(\mu k + \nu)}, \qquad \mu, \nu > 0, $$

is the two-parameter Mittag-Leffler function [39]. The momentum of the fractional oscillator is expressed in terms of a fractional time derivative on the coordinate

$$ p_\alpha(t) = m\, {}^{C}\mathcal{D}_t^{\alpha} q_\alpha(t), $$

where m is the generalized mass (Figure 1). The phase portrait of the fractional oscillator is a spiral. The intrinsic dissipation in the fractional oscillator can be treated as an ensemble average of ordinary harmonic oscillators [56]. In this system, the oscillators differ from each other in frequency randomly. Even if they start in phase, after a time the oscillators will be allocated uniformly up to the clock-face. Although each oscillator is conservative, the system of such oscillators—with the dynamics like a fractional oscillator—demonstrates a dissipative core.

As has been shown in [11, 26], the formalism given above can be retrieved from the fractional embedding theory, based on the least action principle [10]. In particular, this means that fractional Hamiltonian systems can be described by a variational principle. Moreover, this fractional formalism is coherent, meaning that there exists a commutative diagram for obtaining the fractional equations.

In the paper [62], the linear fractional oscillator is considered as an open (non-isolated) system with memory. It describes the interaction between simple classical systems and some environments. The system consists of independent harmonic oscillators not interacting with each other, but power-law spectral densities for the environment generate a memory function with power laws. Equations of motion with this memory are differential equations with fractional time derivatives.

4 Two-term fractional oscillation equation

Now we take notice of the two-term fractional relaxation-oscillation equation with the Caputo derivatives under the initial conditions $y(0) = 1$, and $\dot{y}(0) = 0$ [5]. Let the derivative orders be $1 < \alpha \leq 2$ and $0 < \beta \leq \alpha$. Then, the equation is written as

$$ {}^{C}D_t^{\alpha} y(t) + c\, {}^{C}D_t^{\beta} y(t) = \omega y(t) \tag{12} $$

with $c > 0$. By applying Laplace transform it follows that

$$ \tilde{y}(s) = \frac{s^{\alpha-1} + cs^{\beta-1}}{s^{\alpha} + cs^{\beta} + \omega}. \tag{13} $$

The inverse Laplace transform gives the solution in terms of the Taylor series,

$$ y(t) = 1 - \sum_{n=0}^{\infty} \sum_{p=0}^{\infty} (-1)^n \binom{n}{p} c^p \omega^{n-p+1} \frac{t^{\alpha n - \beta p + \alpha}}{\Gamma(\alpha n - \beta p + \alpha + 1)}, \tag{14} $$

and the asymptotic behavior

$$ y(t) \sim \frac{t^{-\alpha}}{\omega \Gamma(1-\alpha)} + \frac{ct^{-\beta}}{\omega \Gamma(1-\beta)} \quad \text{as } t \to \infty. \tag{15} $$

The result shows a similar behavior in comparison with the simple fractional oscillator.

The friction force, proportional to the fractional velocity (derivative), can also be inserted in the fractional oscillator model [8]. The corresponding equation is

$$ {}^{C}D_t^{2\nu} z(t) = -w^2 z(t) - 2\gamma\, {}^{C}D_t^{\nu} z(t), \quad 0 < \nu < 1. $$

Thus, three cases exist: overdamped oscillator ($\gamma^2 > w^2$), critically damped oscillator ($\gamma^2 = w^2$), and underdamped oscillator ($\gamma^2 < w^2$). They are consistent with the classical result as ν goes to 1.

In this regard, we cannot fail to mention the linear harmonic oscillator, with the fractional derivative interpreted as "viscoelastic" friction (standard oscillator with nonstandard friction). The problem has a long history and remains under scrutiny in our days [4, 49]. The corresponding homogeneous equation reads:

$$\ddot{x}(t) + b\,{}^{C}\!D_t^{\alpha} x(t) + c\,x(t) = 0,$$

where α satisfies $0 < \alpha < 1$ or $1 < \alpha < 2$, with the initial conditions $x(0) = x_0$ and $\dot{x}(0) = x_1$. Applying the Laplace transform, we obtain the characteristic equation

$$s^2 + b\,s^{\alpha} + c = 0.$$

According to [42], this equation always has a couple of conjugate complex roots, the simple poles. The fundamental solution $x(t)$ is expressed as a sum of the oscillatory part $f_1(t)$, and the monotone part $f_2(t)$. In this regard, it also follows that $f_1(t)$ represents a decaying oscillation, whose amplitude decays exponentially, whereas $f_2(t)$ exhibits an algebraic decay in the form of a negative-exponent power function [32].

5 Forced fractional oscillations

The damping of the fractional oscillator can be compensated by an external periodic influence, whose effect on the system can be expressed by a separate term; a periodic function of the time, in the differential equation of motion. We are interested in the response of the system to the periodic external force. The behavior of oscillatory systems under periodic external forces is one of the most important topics in any theory of oscillations, including fractional dynamics. Forced oscillations are accompanied by the phenomenon of resonance found everywhere in physics. The fractional oscillator under an external force, with the initial conditions $x(0) = 0$ and $\dot{x}(0) = 0$, reads:

$$x(t) = -\frac{\omega_0^{\alpha}}{\Gamma(\alpha)} \int_0^t (t-\tau)^{\alpha-1} x(\tau)\,d\tau + \frac{1}{\Gamma(\alpha)} \int_0^t (t-\tau)^{\alpha-1} F(\tau)\,d\tau, \tag{16}$$

where $1 < \alpha < 2$, and F is the external force. Here, the parameter α serves as a damping constant characterizing the strength of intrinsic friction. The perturbation approach to this problem was developed in [50]. The dynamic response of the driven fractional oscillator was investigated in [17, 43, 44]:

$$x(t) = \int_0^t F(\tau)\,(t-\tau)^{\alpha-1} E_{\alpha,\alpha}(-\omega_0^{\alpha}(t-\tau)^{\alpha})\,d\tau. \tag{17}$$

If $F(t)$ has a periodic character $F(t) = F_0\,e^{j\omega t}$, then the solution of Equation (16) includes the steady-state oscillation and the normal mode damping in this system. The first term is more interesting and found by the substitution of $x_0 e^{j\omega t}$ for $x(t)$ in Equation(16). As $t \to \infty$, it is easy to make sure that

$$x_0 = \frac{F_0}{[\omega_0^\alpha + \omega^\alpha \exp(j\pi\alpha/2)]}. \tag{18}$$

This implies the forced solution $\rho \exp(j\omega t + \theta)$ with

$$\rho^2 = \frac{F_0}{[\omega_0^{2\alpha} + \omega^{2\alpha} + 2\omega_0^\alpha \omega^\alpha \cos(\pi\alpha/2)]},$$

$$\tan\theta = -\frac{\omega^\alpha \sin(\pi\alpha/2)}{\omega_0^\alpha + \omega^\alpha \cos(\pi\alpha/2)}.$$

Note that the maximum of $\rho^2\omega_0^{2\alpha}/F_0$ is not attained for $\omega = \omega_0$. Differentiating $\rho^2\omega_0^{2\alpha}/F_0$, with respect to $z = \omega/\omega_0$, we obtain the maximum $z_{max} = \sqrt[\alpha]{\cos[\pi(2-\alpha)/2]}$.

It is interesting that the fractional oscillator can be considered as a dispersion medium model [55], just as the ensemble of classical harmonic oscillators with exponential damping. The latter case is a basic topic for consideration in the classical theory of dispersion, for example, necessary for the dispersion analysis of propagating electromagnetic waves into air [68], and more. As applied to the fractional oscillator, the real and imaginary parts of the permittivity take the form

$$\mathrm{Re}\,\epsilon(\omega) = 1 + \frac{4\pi e^2[\omega_0^\alpha + \omega^\alpha \cos(\pi\alpha/2)]}{m\,[\omega_0^{2\alpha} + \omega^{2\alpha} + 2\omega_0^\alpha\omega^\alpha \cos(\pi\alpha/2)]}, \tag{19}$$

$$\mathrm{Im}\,\epsilon(\omega) = -\frac{4\pi e^2\omega^\alpha \sin(\pi\alpha/2)}{m\,[\omega_0^{2\alpha} + \omega^{2\alpha} + 2\omega_0^\alpha\omega^\alpha \cos(\pi\alpha/2)]}, \tag{20}$$

where e is the electron charge. For $\alpha = 2$, we arrive at the Sellmeier formula. From relations (19) and (20) it follows that there is a frequency band, where the absorption is small, and the refraction coefficient increases with frequency (normal dispersion). Moreover, in the frequency range, where the absorption is large, the anomalous dispersion happens to be the case for the refraction coefficient decreasing with frequency. Note that the presence of both normal and anomalous dispersions is typical for such an ensemble of ordinary harmonic oscillators; this is well known. However, a new conclusion is that the normal and anomalous dispersions are also typical for the medium described as a fractional oscillator.

Using characteristic functionals, the effect of a Gaussian white noise on the time-dependent response of the fractional oscillator has been studied in [16]. In this case, for any order of the fractional derivative $1 < \alpha < 2$ and for any time t, the coordinate $x(t)$ is a random variable with zero as mean, whereas the variance of $x(t)$ tends to a finite limit as $t \to \infty$. Moreover, this limit grows monotonically with α and approaches infinity with $\alpha = 2$ (harmonic oscillator).

The simplest case recorded is when the linear fractional oscillations, which tend to be self-excited, were considered in [58]. It is based on a generalization of

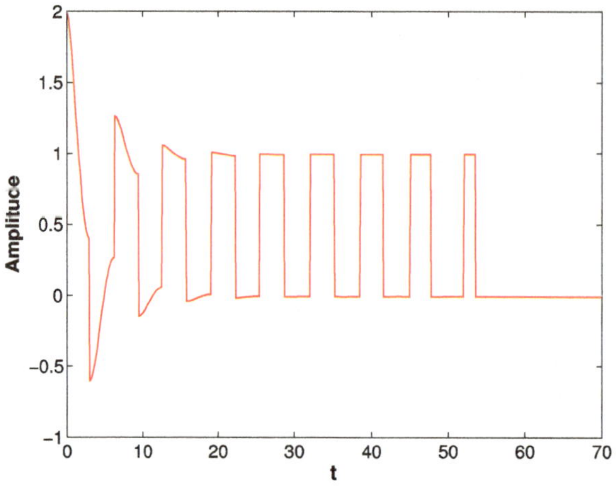

Figure 2: A finite number of self-oscillations in the dynamical system following Equations (21) and (22) for $\alpha = 1.75$.

the Adronov–Vitt–Khaikin model, where the limit cycle arises from solutions of two linear differential equations describing damped oscillations. Its fractional form is written as

$$ {}^{C}\!D^{\alpha}x + \omega_0^{\alpha}x = \omega_0^{\alpha}g, \quad \dot{x} > 0, \tag{21} $$

$$ {}^{C}\!D^{\alpha}x + \omega_0^{\alpha}x = 0, \quad \dot{x} \le 0, \tag{22} $$

where g is the constant, and $1 < \alpha < 2$ as before. It is reasonable to compare the behavior, demonstrated by the dynamical system, with the conventional Adronov–Vitt–Khaikin case. Each solution of two linear differential equations of the Adronov–Vitt–Khaikin model has an infinite number of damped oscillations with respect to zero. As a result, the nonlinear term excites a self-oscillation, and the phase portrait shows a limit cycle. The system, described by Equations (21) and (22), behaves in another way. A finite number of zeros of the fractional equation cannot lead to an ordinary limit cycle. This dynamical system generates only a short-living limit cycle. The numerical simulation of the self-oscillations is represented in Figure 2.

6 Coupled fractional oscillators

Many real physical systems are not usually isolated. In particular, oscillators can interact with other oscillators. Fractional oscillators are no exception. Let us consider two identical fractional oscillators mutually coupled [57]. For $1 < \alpha \le 2$, they are described by the following equations:

$$ {}^{C}\!D^{\alpha}x_1 + \omega_0^2 x_1 + \kappa^2(x_1 - x_2) = 0, \tag{23} $$

$$ {}^{C}\!D^{\alpha}x_2 + \omega_0^2 x_2 + \kappa^2(x_2 - x_1) = 0, \tag{24} $$

where the two oscillators are labeled by 1 and 2; κ^2 is the measure of the coupling, and ω_0, the circular frequency. Introducing new variables $u_1 = x_1 + x_2$ and $u_2 = x_1 - x_2$, the system of Equations (23) and (24) transforms into the equations of two independent fractional oscillators with the frequencies ω_0 and $\sqrt{\omega_0^2 + 2\kappa^2}$. If the oscillators rest initially, then their solutions take the form

$$x_1(t) = \frac{B_0}{2}\left[\omega_0 t^{\alpha/2} E_{\alpha,1+\alpha/2}(-\omega_0^2 t^\alpha) + t^{\alpha/2} E_{\alpha,1+\alpha/2}(-(\omega_0^2 + 2\kappa^2)t^\alpha)\right], \tag{25}$$

$$x_2(t) = \frac{B_0}{2}\left[\omega_0 t^{\alpha/2} E_{\alpha,1+\alpha/2}(-\omega_0^2 t^\alpha) - t^{\alpha/2} E_{\alpha,1+\alpha/2}(-(\omega_0^2 + 2\kappa^2)t^\alpha)\right]. \tag{26}$$

When the coupling is weak ($\kappa \ll 1$), the fractional oscillators exchange energies with one to other. The effect depends on the value of α, characterizing a strength of dissipation and on the magnitude of the parameter κ. Nevertheless, the dissipation finally decrease the fractional oscillations in amplitude after a sufficient period of time to zero.

The decomposition of coupled fractional oscillators in a superposition of normal modes, which are represented by simple fractional oscillators, is a major result in the physics of coupled fractional oscillators. This decomposition is not restricted only to two coupled oscillators. A similar decomposition is possible for any number of coupled fractional oscillators, and the conclusions are similar.

7 Multiple fractional system with forced oscillations

Forced oscillations may be observed in a multiple fractional systems [57]. For two degrees of freedom, the equations of motion in the coordinates x and y yield:

$$\begin{cases} A\,{}^{C}D^\alpha x + H\,{}^{C}D^\alpha y + ax + hy = X\cos pt, \\ H\,{}^{C}D^\alpha x + B\,{}^{C}D^\alpha y + hx + by = Y\cos pt, \end{cases} \tag{27}$$

with $1 < \alpha \le 2$ and the frequency p. Here A, B, H, a, b, h are constants. The analysis of Equations (27) is greatly simplified if one uses the representation in complex numbers, that is, $x = \beta e^{jpt}$ and $y = \gamma e^{jpt}$. This method allows one to transform Equations (27) into algebraic expressions solvable by the ordinary matrix method. From this it follows that

$$\beta = \frac{1}{\Delta}\begin{vmatrix} X & h + Hp^\alpha e^{j\pi\alpha/2} \\ Y & b + Bp^\alpha e^{j\pi\alpha/2} \end{vmatrix}, \quad \gamma = \frac{1}{\Delta}\begin{vmatrix} a + Ap^\alpha e^{j\pi\alpha/2} & X \\ h + Hp^\alpha e^{j\pi\alpha/2} & Y \end{vmatrix}, \tag{28}$$

where

$$\Delta = \begin{vmatrix} a + Ap^\alpha e^{j\pi\alpha/2} & h + Hp^\alpha e^{j\pi\alpha/2} \\ h + Hp^\alpha e^{j\pi\alpha/2} & b + Bp^\alpha e^{j\pi\alpha/2} \end{vmatrix}$$

is the determinant. The steady-state solutions of Equations (27) are derived from (28) by taking the "real" parts of β and γ. Then, it is not difficult to prove the remarkable theorem of reciprocity (for fractional oscillators), first proved for the theory of aerials by Helmholtz and afterward greatly extended by Rayleigh. If $X \neq 0$ and $Y = 0$, then

$$\gamma = -\frac{X(h + H\,p^{\alpha}e^{j\pi\alpha/2})}{\Delta}. \tag{29}$$

On the other hand, we have $X = 0$, $Y \neq 0$. This gives

$$\beta = -\frac{Y(h + H\,p^{\alpha}e^{j\pi\alpha/2})}{\Delta}.$$

Comparing with (29), we see that

$$\gamma : X = \beta : Y. \tag{30}$$

The result only concerns the theorem of reciprocity. Its interpretation is most easy to understand when the "forces" X, Y are of the same character; putting $X = Y$, one obtains $\beta = \gamma$. The coupled fractional oscillators forced by harmonic oscillations definitely support the theorem, as their equations of motion are linear.

8 Fractional diffusive waves

The ordered structure of oscillators, connected together in a certain way, is a very simple model, which can naturally describe waves. If one of them is displaced from its equilibrium position, it will move its neighbor and, hence, a wave will run along the whole of the ordered structure. This general idea is easily extended from the fractional oscillator model to fractional waves [57]. If the oscillators interact with each other, the equation of the nth fractional oscillator takes the form

$$\frac{d^{\alpha}\phi_n}{dt^{\alpha}} + \omega_0^2\phi_n = M(\phi_{n-1} - 2\phi_n + \phi_{n+1}), \tag{31}$$

where M is constant, and $1 < \alpha \leq 2$ as usual. The second spatial derivative $\partial^2\phi/\partial x^2$ can be approximated by the second-order central difference, as a combination of ϕ_{n-1}, ϕ_n, and ϕ_{n+1}. Passing from the discrete representation to a continuous limit in the unbounded one-dimensional case, the system is written in partial derivatives as

$$\frac{\partial^{\alpha}\phi}{\partial t^{\alpha}} - v^2\frac{\partial^2\phi}{\partial x^2} + \omega_0^2\phi = 0, \tag{32}$$

where $v^2 = Ma^2$ is a positive constant of dimension L^2/T^{α}, and a is the mesh size. This approach is easily generalized for the multidimensional wave equation. If $\omega_0 = 0$, then

Equation (32) is simplified to $\partial^\alpha \phi/\partial t^\alpha = v^2\,\partial^2\phi/\partial x^2$. The comprehensive study of basic boundary-value problems of this equation were made by Mainardi [36]. He considered the Cauchy problem (on the whole space axis) and the Signalling problem (on the semi-infinite space axis), showing that the problems can be solved by the technique of the Laplace transform. The Green functions exhibit scaling properties, as they are expressed in terms of the auxiliary functions in the variable $\xi = |x|/(vt^{\alpha/2})$. This makes plotting their distance (at fixed instant) and versus their time (at fixed position) easier. The functions relate to the transcendental functions of the Wright type. More details can be found in [34, 38]. For $1 < \alpha < 2$, the behavior of the Green functions turns out to be intermediate between diffusion and wave propagation, so that it is justifiable to call them fractional diffusive waves.

9 Conclusions

Nature teems with oscillatory phenomena. Friction causes their damping. Fractional calculus extends our representation of oscillatory phenomena. The temporal evolution of fractional oscillator models occupies a place between nonexponential relaxation and conventional harmonic oscillations. Just as there are many applications of fractional calculus (see, for example, [25, 33, 61, 63] and references therein), describing various physical processes and systems, the fractional oscillator model does not fall behind. This is a very successful physical concept. The goal of this chapter was to review various forms of the model. But the model penetrated so deeply into the description of various fractional systems, it is quite difficult to outline the limits covering the subject entirely. The concept continues to develop, and some problems still require solutions. We confined ourselves to a linear representation. Nevertheless, there is also a wide range of nonlinear oscillatory models in fractional derivatives [21, 29, 52, 65, 69], and quantum dynamics with memory [30, 31]. In particular, the fractional oscillator solutions turn out to be helpful in solving the free fractional Schrödinger equation [19, 25]. We have shown that the linear fractional oscillator can be considered as a macroscopic system consisting of the ensemble of ordinary harmonic oscillators, governed by their own internal clock. Although dynamics of each classical oscillator is described by the ordinary Hamiltonian equations, the averaging procedure modifies the ensemble description in the fractional oscillator equation. With the continuously growing number of applications of fractional differential equations, the impact of fractional-order oscillators, with their analytical tools in practical and theoretical applications, will be of considerable interest in the future. There are many different types of fractional derivatives named after famous mathematicians Erdelyi, Hadamard, Kober, Letnikov, Liouville, Riemann, Riesz, Sonin, Weyl, and others (see, for example, [28, 51]). In general, the fractional derivatives of noninteger orders is a special class of integro-differential operators with nonstandard properties.

To describe the processes with memory and systems with spatial and temporal nonlocality in applied mathematics, physics, and economics, fractional derivatives became very popular. Numerous and, even unexpected, recent applications of such operators stimulate the development of this field [2, 3, 9]. Real fractional derivatives are expressed in terms of an infinite series of derivatives of integer orders. This indicates their nonlocality in space or/and time. Such operators can be a part of new types of fractional oscillator equations awaiting their explorers.

Bibliography

[1] O. P. Agrawal, Formulation of Euler–Lagrange equations for fractional variational problems, *Journal of Mathematical Analysis and Applications*, **272** (2002), 368–379.

[2] T. M. Atanackovic, M. Budincevic, and S. Pilipovic, On a fractional distributed-order oscillator, *Journal of Physics. A, Mathematical and General*, **38** (2005), 6703–6713.

[3] T. M. Atanackovic, L. Oparnica, and S. Pilipovic, On a nonlinear distributed order fractional differential equation, *Journal of Mathematical Analysis and Applications*, **328** (2007), 590–608.

[4] R. L. Bagley and P. J. Torvik, A generalized derivative model for an elastomer damper, *The Shock and Vibration Bulletin*, **49** (1979), 135–143.

[5] E. Bazhlekova, Subordination principle for a class of fractional order differential equations, *Mathematics*, **2** (2015), 412–427.

[6] G. W. Bohannan and B. Knauber, A physical experimental study of the fractional harmonic oscillator, in *International Symposium on Circuits and Systems (ISCAS'15)*, Lisbon, Portugal, 24–27 May 2015, pp. 2341–2344, 2015.

[7] M. Caputo, A model for the fatigue in elastic materials with frequency independent Q, *The Journal of the Acoustical Society of America*, **66** (1979), 176–179.

[8] W. S. Chung and M. Jung, Fractional damped oscillators and fractional forced oscillators, *Journal of the Korean Physical Society*, **64** (2014), 186–191.

[9] M. Concezzi, R. Garra, and R. Spigler, Fractional relaxation and fractional oscillation models involving Erdelyi–Kober integrals, *Fractional Calculus and Applied Analysis*, **18** (2015), 1212–1231.

[10] J. Cresson, Fractional embedding of differential operators and Lagrangian systems, *Journal of Mathematical Physics*, **48** (2007), 033504.

[11] J. Cresson and P. Inizan, About fractional Hamiltonian systems, *Physica Scripta*, **T136** (2009), 014007.

[12] E. C. de Oliveira and J. A. Tenreiro Machado, A review of definitions for fractional derivatives and integrals, *Mathematical Problems in Engineering*, **2014** (2014), 238459.

[13] P. A. M. Dirac, The relation between mathematics and physics, *Proceedings of the Royal Society A. Mathematical, Physical and Engineering Sciences*, **59** (1990), 122–129.

[14] D. W. Dreisigmeyer and P. M. Young, Nonconservative Lagrangian mechanics: a generalized function approach, *Journal of Physics. A, Mathematical and General*, **36** (2003), 8297–8310.

[15] D. W. Dreisigmeyer and P. M. Young, Extending Bauer's corollary to fractional derivatives, *Journal of Physics. A, Mathematical and General*, **37** (2004), L117–L121.

[16] A. D. Drozdov, Fractional oscillator driven by a Gaussian noise, *Physica A*, **376** (2007), 237–245.

[17] J.-S. Duan, The periodic solution of fractional oscillation equation with periodic input, *Advances in Mathematical Physics*, **2013** (2013), Article ID 869484.

[18] J.-S. Duan, Z. Wang, and S.-Z. Fu, The zeros of the solutions of the fractional oscillation equation, *Fractional Calculus and Applied Analysis*, **17** (2014), 10–22.
[19] R. Eid, S. Muslih, D. Baleanu, and E. M. Rabei, Fractional dimensional harmonic oscillator, *Romanian Journal of Physics*, **56** (2011), 323–331.
[20] W. Feller, *An Introduction to Probability Theory and Its Applications*, vol. 2, John Wiley, New York, 1966.
[21] V. Gafiychuk, B. Datsko, and V. Meleshko, Analysis of fractional order Bonhoeffer–van der Pol oscillator, *Physica A*, **387** (2008), 418–424.
[22] R. Gorenflo and F. Mainardi, Fractional oscillations and Mittag-Leffler functions, in *Proceedings of International Workshop on the Recent Advances in Applied Mathematics (RAAM '96)*, Kuwait University, Department of Mathematics & Computer Science, pp. 193–208, 1996.
[23] R. Gorenflo and F. Mainardi, Fractional calculus: integral and differential equations of fractional order, in A. Carpinteri and F. Mainardi (eds.) *Fractals and Fractional Calculus in Continuum Mechanics*, pp. 223–276, Springer-Verlag, Wien/New York, 1997.
[24] R. Gorenflo and F. Mainardi, Fractional calculus and stable probability distributions, *Archives of Mechanics*, **50** (1998), 377–388.
[25] R. Herrmann, *Fractional Calculus: an Introduction for Physicists*, World Scientific Publishing Company, Singapore, 2011.
[26] P. Inizan, Homogeneous fractional embeddings, *Journal of Mathematical Physics*, **49** (2008), 082901.
[27] E. Kaslik and S. Sivasundaram, Non-existence of periodic solutions in fractional-order dynamical systems and a remarkable difference between integer and fractional-order derivatives of periodic functions, *Nonlinear Analysis: Real World Applications*, **13** (2012), 1489–1497.
[28] V. Kiryakova, *Generalized Fractional Calculus and Applications*, Longman and J. Wiley, New York, 1994.
[29] N. Korabel, G. M. Zaslavsky, and V. E. Tarasov, Coupled oscillators with power-law interaction and their fractional dynamics analogues, *Communications in Nonlinear Science and Numerical Simulation*, **12** (2007), 1405–1417.
[30] N. Laskin, Fractional quantum mechanics, *Physical Review E*, **62** (2000), 3135.
[31] N. Laskin, Principles of fractional quantum mechanics, in J. Klafter, S. C. Lim, and R. Metzler (eds.) *Fractional Dynamics*, pp. 393–427, World Scientific, Singapore, 2012.
[32] L.-L. Liu and J.-S. Duan, A detailed analysis for the fundamental solution of fractional vibration equation, *Open Mathematics*, **13** (2015), 826–838.
[33] C. F. Lorenzo and T. T. Hartley, *The Fractional Trigonometry: With Applications to Fractional Differential Equations and Science*, John Wiley & Sons, New Jersey, 2017.
[34] Yu. Luchko, Fractional wave equation and damped waves, *Journal of Mathematical Physics*, **54** (2013), 031505.
[35] F. Mainardi, On the initial value problem for the fractional diffusion-wave equation, in S. Rionero and T. Ruggeri (eds.) *Waves and Stability in Continuous Media*, pp. 246–251, World Scientific, Singapore, 1994.
[36] F. Mainardi, The time fractional diffusion-wave equation, *Radiophysics and Quantum Electronics*, **38** (1995), 20–36.
[37] F. Mainardi, Fractional relaxation-oscillation and fractional diffusion-wave phenomena, *Chaos, Solitons and Fractals*, **7** (1996), 1461–1477.
[38] F. Mainardi, *Fractional Calculus and Waves in Linear Viscoelasticity*, Imperial College Press, London, 2010.
[39] A. M. Mathai and H. J. Haubold, *Special Functions for Applied Scientists*, Springer, New York, 2008.

[40] M. M. Meerschaert and H.-P. Scheffler, Limit theorems for continuous-time random walks with infinite mean waiting times, *Journal of Applied Probability*, **41** (2004), 623–638.
[41] S. Muslih and D. Baleanu, Formulation of Hamiltonian equations for fractional variational problems, *Czechoslovak Journal of Physics*, **55** (2005), 633–642.
[42] M. Naber, Linear fractionally damped oscillator, *International Journal of Differential Equations*, **2010** (2010), 197020.
[43] B. N. Narahari Achar, J. W. Hanneken, and T. Clarke, Response characteristics of a fractional oscillator, *Physica A*, **309** (2002), 275–288.
[44] B. N. Narahari Achar, J. W. Hanneken, and T. Clarke, Damping characteristics of a fractional oscillator, *Physica A*, **339** (2004), 311–319.
[45] B. N. Narahari Achar, J. W. Hanneken, T. Enck, and T. Clarke, Dynamics of the fractional oscillator, *Physica A*, **297** (2001), 361–367.
[46] I. Podlubny, *Fractional Differential Equations*, Academic Press, San Diego, 1999.
[47] F. Riewe, Nonconservative Lagrangian and Hamiltonian mechanics, *Physical Review E*, **53** (1996), 1890–1899.
[48] F. Riewe, Mechanics with fractional derivatives, *Physical Review E*, **55** (1997), 3581–3592.
[49] Yu. A. Rossikhin and M. V. Shitikova, Application of fractional falculus for dynamic problems of solid mechanics: novel trends and recent results, *Applied Mechanics Reviews*, **63** (2010), 010801.
[50] Ya. E. Ryabov and A. Puzenko, Damped oscillations in view of the fractional oscillator equation, *Physical Review. B, Condensed Matter and Materials Physics*, **66** (2002), 184201.
[51] S. G. Samko, A. A. Kilbas, and O. I. Marichev, *Fractional Integrals and Derivatives Theory and Applications*, Gordon and Breach, New York, 1993.
[52] M. Seredyńska and A. Hanyga, Nonlinear Hamiltonian equations with fractional damping, *Journal of Mathematical Physics*, **41** (2000), 2135–2156.
[53] A. A. Stanislavsky, Fractional dynamics from the ordinary Langevin equation, *Physical Review E*, **67** (2003), 021111.
[54] A. A. Stanislavsky, Probabilistic interpretation of the integral of fractional order, *Theoretical and Mathematical Physics*, **138** (2004), 418–431.
[55] A. A. Stanislavsky, Fractional oscillator, *Physical Review E*, **70** (2004), 051103.
[56] A. A. Stanislavsky, Twist of fractional oscillations, *Physica A*, **354** (2005), 101–110.
[57] A. A. Stanislavsky, Hamiltonian formalism of fractional systems, *The European Physical Journal. B, Condensed Matter Physics*, **49** (2006), 93–101.
[58] A. A. Stanislavsky, The peculiarity of self-excited oscillations in fractional systems, *Acta Physica Polonica. B*, **37** (2006), 319–329.
[59] V. E. Tarasov, Fractional generalization of Liouville equations, *Chaos*, **14** (2004), 123–127.
[60] V. E. Tarasov, Fractional Liouville and BBGKI equations, *Journal of Physics: Conference Series*, **7** (2005), 17–33.
[61] V. E. Tarasov, *Fractional Dynamics: Application of Fractional Calculus to Dynamics of Particles, Fields and Media*, HEP, Springer, Beijing, Berlin, 2010.
[62] V. E. Tarasov, The fractional oscillator as an open system, *Central European Journal of Physics*, **10** (2012), 382–389.
[63] V. E. Tarasov, Review of some promising fractional physical models, *International Journal of Modern Physics B*, **27** (2013), 1330005.
[64] V. E. Tarasov and G. M. Zaslavsky, Dynamics with low-level fractionality, *Physica A*, **368** (2006), 399–415.
[65] V. E. Tarasov and G. M. Zaslavsky, Fractional dynamics of coupled oscillators with long-range interaction, *Chaos*, **16** (2006), 023110.
[66] A. Tofighi, The intrinsic damping of the fractional oscillator, *Physica A*, **329** (2003), 29–34.

[67] A. Tofighi, An especial fractional oscillator, *Journal of Statistical Mechanics: Theory and Experiment*, **2013** (2013), 175273.
[68] A. A. Voronin and A. M. Zheltikov, The generalized Sellmeier equation for air, *Scientific Reports*, **7** (2017), 46111.
[69] G. M. Zaslavsky, A. A. Stanislavsky, and M. Edelman, Chaotic and pseudochaotic attractors of perturbed fractional oscillator, *Chaos*, **16** (2006), 013102.

Francesco Mainardi

Fractional viscoelasticity

Abstract: In this chapter we consider the basic properties of fractional viscoelasticity, restricting our analysis to linear theory in one-dimensional case. Fractional refers to the nature of the constitutive laws that contain nonlocal operators interpreted in terms of fractional integrals and derivatives. The memory effects in time turn out to be expressed in terms of functions of the Mittag-Leffler type.

Keywords: Linear viscoelasticity, stress–strain relations, fractional derivatives, Mittag-Leffler function, complete monotonicity, Laplace transforms

MSC 2000: 26A33, 35E12, 44A10, 74DXX, 76DXX

1 Introduction

A topic of continuum mechanics, where fractional calculus is suited to be applied, is without doubt the linear theory of viscoelasticity. In fact, an increasing number of authors have used fractional calculus as an empirical method of describing the properties of linear viscoelastic materials.

The purpose of this chapter (essentially based on the 2010 book by Mainardi [30], and on the 2011 article by Mainardi and Spada [34]) is to provide, after a general survey to the linear theory of viscoelasticity, a (more) systematic discussion and a graphical representation of the main properties of the basic models described by stress–strain relationships of fractional order. The properties under discussion concern the standard creep and relaxation tests that have relevance in experiments.

The plan of the paper is as following: In Section 2 we recall the essential notions of linear viscoelasticity in order to present our notations for the analog mechanical models. We limit our attention to the basic mechanical models, characterized by two, three, and four parameters, that we refer to as Kelvin–Voigt, Maxwell, Zener, anti-Zener, and Burgers.

In Section 3 we consider our main topic concerning the creep, relaxation, and viscosity properties of the previous basic models generalized by replacing in their differential constitutive equations the derivatives of integer orders 1 and 2 with derivatives of fractional orders ν and $1 + \nu$, respectively, with $0 < \nu \leq 1$. We provide analytical

Acknowledgement: The work of FM has been carried out in the framework of the activities of the National Group of Mathematical Physics (INdAM-GNFM).

Francesco Mainardi, Department of Physics and Astronomy, University of Bologna, Via Irnerio 46, 40126 Bologna, Italy, e-mail: francesco.mainardi@bo.infn.it

https://doi.org/10.1515/9783110571707-007

expressions and the plots of the creep compliance, the relaxation modulus, and the effective viscosity for all the considered fractional models.

Some conclusions close the chapter.

2 Essentials of linear viscoelasticity

In this section we present the fundamentals of linear viscoelasticity restricting our attention to the one–axial case, and assuming that the viscoelastic body is quiescent for all time prior to some starting instant that we assume as $t = 0$.

For convenience, both stress $\sigma(t)$ and strain $\epsilon(t)$ are intended to be normalized, that is, scaled with respect to a suitable reference state $\{\sigma_0, \epsilon_0\}$.

2.1 Generalities

According to the linear theory, the viscoelastic body can be considered as a linear system with the stress (or strain) as the excitation function (input), and strain (or stress) as the response function (output). In this respect, the response functions to an excitation, expressed by the Heaviside step function $\Theta(t)$, and are known to play a fundamental role both from a mathematical and physical point of view. We denote by $J(t)$ the strain response to the unit step of stress, according to the *creep test*, and by $G(t)$ the stress response to a unit step of strain, according to the *relaxation test*.

The functions $J(t)$ and $G(t)$ are usually referred to as the *creep compliance* and *relaxation modulus*, respectively, or, simply, the *material functions* of the viscoelastic body. In view of the causality requirement, both functions are vanishing for $t < 0$.

The limiting values of the material functions as $t \to 0^+$ and $t \to +\infty$ are related to the instantaneous (or glass) and equilibrium behaviors of the viscoelastic body, respectively. As a consequence, it is usual to denote by $J_g := J(0^+)$ the *glass compliance*; $J_e := J(+\infty)$, the *equilibrium compliance*; $G_g := G(0^+)$, the *glass modulus*; $G_e := G(+\infty)$, the *equilibrium modulus*. As a matter of fact, both the material functions are nonnegative. Furthermore, for $0 < t < +\infty$, $J(t)$ is a *nondecreasing function*, and $G(t)$ is a *nonincreasing function*.

The monotonicity properties of $J(t)$ and $G(t)$ are related, respectively, to the physical phenomena of strain *creep* and stress *relaxation*. Moreover, we require that $G(t)$ be a completely monotonic (CM) function, and $J(t)$ be a function of Bernstein type (that is, nonnegative with a CM derivative), in order to ensure the existence of nonnegative spectra to be consistent with the physical realizability of the corresponding models, outlined formerly in 1975 by Molinari [36], and more recently in several papers by Hanyga (see, for example, [20], and by the author in his book [30]. We point out that the existence of positive spectra for the material functions was outlined by Gross in his excellent 1953 monography [19], but without the relationship with the CM properties, unknown to the author. We also note that in some cases the material functions can contain terms represented by *generalized functions* (distributions), in the sense of Gel'fand–Shilov [12], or *pseudo-functions* in the sense of Doetsch [11].

Under the hypotheses of causal histories, we get the stress–strain relationships:

$$\begin{cases} \epsilon(t) = \int_{0^-}^{t} J(t-\tau)\, d\sigma(\tau) = \sigma(0^+) J(t) + \int_0^t J(t-\tau)\, \frac{d}{d\tau}\sigma(\tau)\, d\tau, \\ \sigma(t) = \int_{0^-}^{t} G(t-\tau)\, d\epsilon(\tau) = \epsilon(0^+)\, G(t) + \int_0^t G(t-\tau)\, \frac{d}{d\tau}\epsilon(\tau)\, d\tau, \end{cases} \tag{2.1}$$

where the passage to the RHS is justified if differentiability is assumed for the stress–strain histories (see also the excellent book by Pipkin [39]). Being of convolution type, Equations (2.1) can be conveniently treated by the technique of Laplace transforms, so they read in the Laplace domain, in an obvious notation:

$$\widetilde{\epsilon}(s) = s\widetilde{J}(s)\,\widetilde{\sigma}(s), \quad \widetilde{\sigma}(s) = s\,\widetilde{G}(s)\,\widetilde{\epsilon}(s), \tag{2.2}$$

from which we derive the *reciprocity relation*

$$s\widetilde{J}(s) = \frac{1}{s\,\widetilde{G}(s)}. \tag{2.3}$$

Because of the limiting theorems for the Laplace transform, we deduce that $J_g = 1/G_g$, $J_e = 1/G_e$, with the convention that 0 and $+\infty$ are reciprocal to each other. The above remarkable relations allow us to classify the viscoelastic bodies according to their instantaneous and equilibrium responses in four types, as stated by Caputo and Mainardi in their 1971 review paper [7] (see Table 2.1).

We note that the viscoelastic bodies of type I exhibit both instantaneous and equilibrium elasticity, so their behavior appears close to the purely elastic one for sufficiently short and long times. The bodies of type II and IV exhibit a complete stress relaxation (at constant strain), since $G_e = 0$, and an infinite strain creep (at constant stress), since $J_e = \infty$. So, they do not present equilibrium elasticity. Finally, the bodies of type III and IV do not present instantaneous elasticity, since $J_g = 0$ ($G_g = \infty$). Other properties will be pointed out later on.

Table 2.1: The four types of viscoelasticity.

Type	J_g	J_e	G_g	G_e
I	> 0	$< \infty$	$< \infty$	> 0
II	> 0	$= \infty$	$< \infty$	$= 0$
III	$= 0$	$< \infty$	$= \infty$	> 0
IV	$= 0$	$= \infty$	$= \infty$	$= 0$

2.2 The mechanical models

To get some feeling for linear viscoelastic behavior, it is useful to consider the simpler behavior of analog *mechanical models*. They are constructed from linear springs and dashpots, disposed singly and in branches of two (in series, or in parallel). As analog of stress and strain, we use the total extending force and the total extension. We note that when two elements are combined in series [in parallel], their compliances [moduli] are additive. This can be stated as a combination rule: *creep compliances add in series, while relaxation moduli add in parallel.*

The mechanical models play an important role in the literature, which is justified by the historical development. In fact, the early theories were established with the aid of these models, which are still helpful to visualize properties and laws of the general theory, using the combination rule.

Now, it is worthwhile to consider the simple models of Figures 1, 2, and 3, providing their governing stress–strain relations along with the related material functions.

2.2.1 The Hooke and the Newton models

The spring (Figure 1(a)) is the elastic (or storage) element, as for it, the force is proportional to the extension; it represents a perfect elastic body obeying the Hooke law (ideal solid). This model is thus referred to as the *Hooke* model. If we denote by m the pertinent elastic modulus, we have

$$\text{Hooke model:} \quad \sigma(t) = m\,\epsilon(t) \quad \text{and} \quad \begin{cases} J(t) = 1/m, \\ G(t) = m. \end{cases} \tag{2.4}$$

In this case, we have no creep and no relaxation; so the creep compliance and the relaxation modulus are constant functions: $J(t) \equiv J_g \equiv J_e = 1/m$; $G(t) \equiv G_g \equiv G_e = m$.

The dashpot (Figure 1(b)) is the viscous (or dissipative) element, the force being proportional to rate of extension; it represents a perfectly viscous body obeying the Newton law (perfect liquid). This model is thus referred to as the *Newton* model. If we denote by b_1 the pertinent viscosity coefficient, we have

$$\text{Newton model:} \quad \sigma(t) = b_1 \frac{d\epsilon}{dt} \quad \text{and} \quad \begin{cases} J(t) = t/b_1, \\ G(t) = b_1\,\delta(t). \end{cases} \tag{2.5}$$

In this case, we have a linear creep $J(t) = J_+ t$ and instantaneous relaxation $G(t) = G_-\,\delta(t)$, with $G_- = 1/J_+ = b_1$.

We note that the *Hooke* and *Newton* models represent the limiting cases of viscoelastic bodies of type *I* and *IV*, respectively.

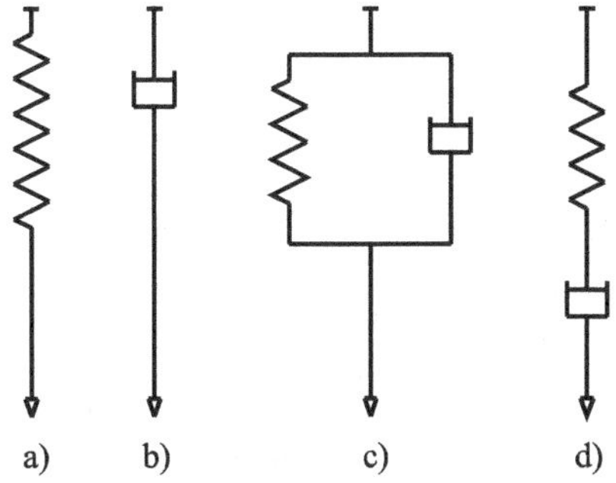

Figure 1: The representations of the basic mechanical models: (a) spring for Hooke; (b) dashpot for Newton; (c) spring and dashpot in parallel for Voigt; (d) spring and dashpot in series for Maxwell.

2.2.2 The Kelvin–Voigt and the Maxwell models

A branch constituted by a spring in parallel with a dashpot is known as the *Kelvin–Voigt* model (Figure 1(c)). We have

$$\text{Kelvin–Voigt model:} \quad \sigma(t) = m\,\epsilon(t) + b_1\,\frac{d\epsilon}{dt}, \tag{2.6a}$$

and

$$\begin{cases} J(t) = J_1[1 - \mathrm{e}^{-t/\tau_\epsilon}], & J_1 = \frac{1}{m}, \quad \tau_\epsilon = \frac{b_1}{m}, \\ G(t) = G_e + G_-\,\delta(t), & G_e = m, \quad G_- = b_1, \end{cases} \tag{2.6b}$$

where τ_ϵ is referred to as the *retardation time.*

A branch constituted by a spring in series with a dashpot is known as the *Maxwell* model (Figure 1(d)). We have

$$\text{Maxwell model:} \quad \sigma(t) + a_1\,\frac{d\sigma}{dt} = b_1\,\frac{d\epsilon}{dt}, \tag{2.7a}$$

and

$$\begin{cases} J(t) = J_g + J_+\,t, & J_g = \frac{a_1}{b_1}, \quad J_+ = \frac{1}{b_1}, \\ G(t) = G_1\,\mathrm{e}^{-t/\tau_\sigma}, & G_1 = \frac{b_1}{a_1}, \quad \tau_\sigma = a_1, \end{cases} \tag{2.7b}$$

where τ_σ is referred to as the *relaxation time.*

The *Voigt* and *Maxwell* models are thus the simplest viscoelastic bodies of type *III* and *II*, respectively. The *Voigt* model exhibits an exponential (reversible) strain creep, but no stress relaxation; it is also referred to as the retardation element. The *Maxwell* model exhibits an exponential (reversible) stress relaxation and a linear (nonreversible) strain creep; it is also referred to as the relaxation element.

2.2.3 The Zener and anti-Zener models

Based on the combination rule, we can continue the previous procedure in order to construct the simplest models of type *I* and *IV* that require three parameters.

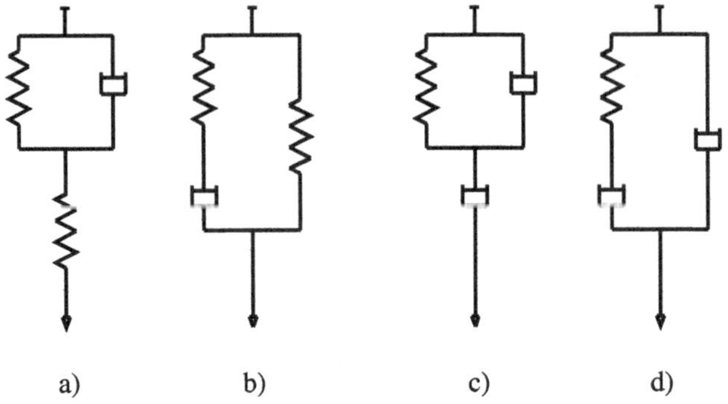

Figure 2: The mechanical representations of the Zener model: see (a), (b), and of the anti-Zener model, see (c), (d), where: (a) spring in series with Voigt; (b) spring in parallel with Maxwell; (c) dashpot in series with Voigt; (d) dashpot in parallel with Maxwell.

The simplest viscoelastic body of type *I* is obtained by adding a spring, either in series to a Voigt model, or in parallel to a Maxwell model (Figure 2(a) and Figure 2(b), respectively). So doing, according to the combination rule, we add a positive constant both to the Voigt-like creep compliance and to the Maxwell-like relaxation modulus, so that we obtain $J_g > 0$ and $G_e > 0$. Such a model was introduced by Zener [42] with the denomination of *Standard Linear Solid* (S.L.S.). We have

$$\text{Zener model:} \quad \left[1 + a_1 \frac{d}{dt}\right]\sigma(t) = \left[m + b_1 \frac{d}{dt}\right]\epsilon(t) \tag{2.8a}$$

and

$$\begin{cases} J(t) = J_g + J_1[1 - \mathrm{e}^{-t/\tau_\epsilon}], & J_g = \frac{a_1}{b_1}, \quad J_1 = \frac{1}{m} - \frac{a_1}{b_1}, \quad \tau_\epsilon = \frac{b_1}{m}, \\ G(t) = G_e + G_1\,\mathrm{e}^{-t/\tau_\sigma}, & G_e = m, \quad G_1 = \frac{b_1}{a_1} - m, \quad \tau_\sigma = a_1. \end{cases} \tag{2.8b}$$

We point out the condition $0 < m < b_1/a_1$ in order that J_1, G_1 be positive and, hence, $0 < J_g < J_e < \infty$ and $0 < G_e < G_g < \infty$. As a consequence, we note that, for the *S.L.S.* model, the retardation time must be greater than the relaxation time, that is, $0 < \tau_\sigma < \tau_\epsilon < \infty$.

Also the simplest viscoelastic body of type *IV* requires three parameters, that is, a_1, b_1, b_2; it is obtained adding a dashpot either in series to a Voigt model or in parallel to the Maxwell model (Figure 2(c) and Figure 2(d), respectively). According to the combination rule, we add a linear term to the Voigt-like creep compliance and a delta impulsive term to the Maxwell-like relaxation modulus so that we obtain $J_e = \infty$ and $G_g = \infty$. We may refer to this model to as the *anti-Zener* model. We have

$$\text{anti-Zener model:} \quad \left[1 + a_1 \frac{d}{dt}\right]\sigma(t) = \left[b_1 \frac{d}{dt} + b_2 \frac{d^2}{dt^2}\right]\epsilon(t), \tag{2.9a}$$

and

$$\begin{cases} J(t) = J_+ t + J_1[1 - \mathrm{e}^{-t/\tau_\epsilon}], & J_+ = \frac{1}{b_1}, \quad J_1 = \frac{a_1}{b_1} - \frac{b_2}{b_1^2}, \quad \tau_\epsilon = \frac{b_2}{b_1}, \\ G(t) = G_-\,\delta(t) + G_1\,\mathrm{e}^{-t/\tau_\sigma}, & G_- = \frac{b_2}{a_1}, \quad G_1 = \frac{b_1}{a_1} - \frac{b_2}{a_1^2}, \quad \tau_\sigma = a_1. \end{cases} \tag{2.9b}$$

We point out the condition $0 < b_2/b_1 < a_1$ in order that J_1, G_1 be positive. As a consequence, we note that, for the *anti-Zener* model, the relaxation time must be greater than the retardation time, that is, $0 < \tau_\epsilon < \tau_\sigma < \infty$, contrary to the Zener (*S.L.S.*) model.

In Figure 2, we exhibit the mechanical representations of the Zener model ((2.8a) and (2.8b)) (see (a) and (b)), and of the anti-Zener model ((2.9a) and (2.9b)) (see (c) and (d)).

Remark. We note that the constitutive equation of the anti-Zener model is formally obtained from that of the Zener model by replacing the strain $\epsilon(t)$ by the stain-rate $\dot{\epsilon}(t)$. However, the Zener model, introduced by Zener in 1948 [42] for anelastic metals, was formerly introduced by Jeffreys with respect to bodily imperfection of elasticity in tidal friction. From the first 1924 edition of his (Jeffreys) treatise on the Earth [21, 23], Sir Harold Jeffreys usually referred to the rheology of the Kelvin–Voigt model (that was suggested to him by Sir J. Larmor) as *firmoviscosity,* and to the rheology of the Maxwell model to as *elastoviscosity*. We observe that in the literature of rheology of viscoelastic fluids (including polymeric liquids), our anti-Zener model is (surprisingly for us) known as *Jeffreys fluid* (see, for example, the review paper by Bird and Wiest [2]). Presumably this is due to the replacement of the strain with the strain-rate (suitable for fluids) in the stress–strain relationship introduced by Jeffreys. As a matter of fact, in Earth rheology, the Jeffreys model is known to be the creep model introduced by him in 1958 (see [22]), as generalization of the Lomnitz logarithmic creep law, and well described in the subsequent editions of Jeffreys' treatise on the Earth. The 1958 model by Jeffreys has been discussed and extended by Mainardi and Spada in [35]. In view of above considerations, we are tempted to call our anti-Zener model *Standard Linear Fluid,* in analogy with the terminology *Standard Linear Solid,* commonly adopted for the Zener model.

2.2.4 The Burgers model

In Rheology literature, it is customary to consider the so-called *Burgers* model, which is obtained by adding a dashpot or a spring to the representations of the Zener or of the anti-Zener model, respectively. Assuming the creep representation, the dashpot or the spring is added in series, so the Burgers model results in a series combination of a Maxwell element with a Voigt element. Assuming the relaxation representation, the dashpot or the spring is added in parallel, so the Burgers model results in two Maxwell elements disposed in parallel. We refer the reader to Figure 3 for the two mechanical representations of the Burgers model.

According to our general classification, the Burgers model is, thus, a four-element model of type II, defined by the four parameters $\{a_1, a_2, b_1, b_2\}$.

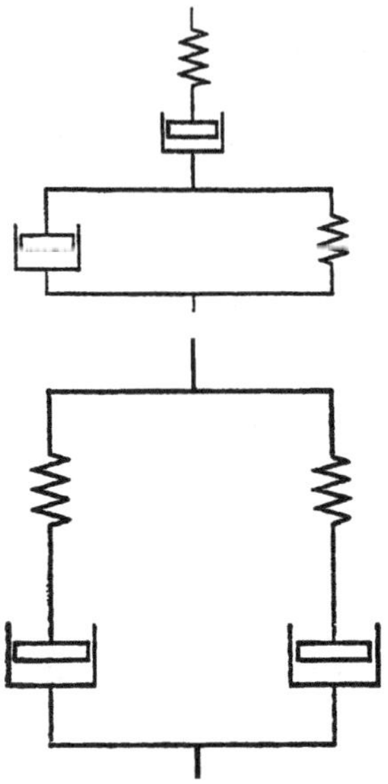

Figure 3: The mechanical representations of the Burgers model: the creep representation (top) and the relaxation representation (bottom).

We have

$$\text{Burgers model:} \quad \left[1 + a_1 \frac{d}{dt} + a_2 \frac{d^2}{dt^2}\right] \sigma(t) = \left[b_1 \frac{d}{dt} + b_2 \frac{d^2}{dt^2}\right] \epsilon(t), \tag{2.10a}$$

so

$$\begin{cases} J(t) = J_g + J_+ \, t + J_1 (1 - \mathrm{e}^{-t/\tau_\epsilon}), \\ G(t) = G_1 \, \mathrm{e}^{-t/\tau_{\sigma,1}} + G_2 \, \mathrm{e}^{-t/\tau_{\sigma,2}}. \end{cases} \tag{2.10b}$$

We leave to the reader to express as an exercise the physical quantities J_g, J_+, τ_ϵ and $G_1, \tau_{\sigma,1}, G_2, \tau_{\sigma,2}$ in terms of the four parameters $\{a_1, a_2, b_1, b_2\}$ in the operator equation (2.10a).

2.2.5 The operator equation for the mechanical models

Based on the combination rule, we can construct models whose material functions are of the following type:

$$\begin{cases} J(t) = J_g + \sum_n J_n [1 - \mathrm{e}^{-t/\tau_{\epsilon,n}}] + J_+ \, t, \\ G(t) = G_e + \sum_n G_n \, \mathrm{e}^{-t/\tau_{\sigma,n}} + G_- \, \delta(t), \end{cases} \tag{2.11}$$

where all the coefficient are nonnegative, and interrelated because of the *reciprocity relation* (2.3) in the Laplace domain. We note that the four types of viscoelasticity of

Table 2.1 are obtained from Equations (2.11) by taking into account that

$$\begin{cases} J_e < \infty \iff J_+ = 0, \quad J_e = \infty \iff J_+ \neq 0, \\ G_g < \infty \iff G_- = 0, \quad G_g = \infty \iff G_- \neq 0. \end{cases} \tag{2.12}$$

Appealing to the theory of Laplace transforms, we write

$$\begin{cases} s\widetilde{J}(s) = J_g + \sum_n \frac{J_n}{1+s\,\tau_{\epsilon,n}} + \frac{J_+}{s}, \\ s\widetilde{G}(s) = (G_e + \beta) - \sum_n \frac{G_n}{1+s\,\tau_{\sigma,n}} + G_-\, s, \end{cases} \tag{2.13}$$

where we have put $\beta = \sum_n G_n$.

Furthermore, as a consequence of (2.13), $s\widetilde{J}(s)$ and $s\widetilde{G}(s)$ turn out to be *rational* functions in **C**, with simple poles and zeros on the negative real axis and, possibly, with a simple pole or with a simple zero at $s = 0$, respectively.

In these cases the integral constitutive Equations (2.1) can be written in differential form. Following Bland [3] with our notations, for these models, we obtain

$$\left[1 + \sum_{k=1}^{p} a_k \frac{d^k}{dt^k}\right] \sigma(t) = \left[m + \sum_{k=1}^{q} b_k \frac{d^k}{dt^k}\right] \epsilon(t), \tag{2.14}$$

where q and p are integers with $q = p$ or $q = p + 1$, and m, a_k, b_k are nonnegative constants, subjected to proper restrictions in order to meet the physical requirements of realizability. The general Equation (2.14) is referred to as the *operator equation* of the mechanical models.

In the Laplace domain, we thus get

$$s\widetilde{J}(s) = \frac{1}{s\widetilde{G}(s)} = \frac{P(s)}{Q(s)}, \quad \text{where} \quad \begin{cases} P(s) = 1 + \sum_{k=1}^{p} a_k\, s^k, \\ Q(s) = m + \sum_{k=1}^{q} b_k\, s^k, \end{cases} \tag{2.15}$$

with $m \geq 0$ and $q = p$ or $q = p+1$. The polynomials at the numerator and denominator turn out to be *Hurwitz polynomials* (since they have no zeros for $\mathbb{R}\,\{s\} > 0$) whose zeros are alternating on the negative real axis ($s \leq 0$). The least zero in absolute magnitude is a zero of $Q(s)$. The four types of viscoelasticity then correspond to whether the least zero is ($J_+ \neq 0$) or is not ($J_+ = 0$) equal to zero, and to whether the greatest zero in absolute magnitude is a zero of $P(s)$ ($J_g \neq 0$) or a zero of $Q(s)$ ($J_g = 0$).

In Table 2.2 we summarize the four cases, which are expected to occur in the *operator equation* (2.14), corresponding to the four types of viscoelasticity.

We recognize that, for $p = 1$, Equation (2.14) includes the operator equations for the classical models with two parameters: Voigt and Maxwell, illustrated in Figure 1, with three parameters: Zener and anti-Zener, illustrated in Figure 2. In fact, we recover the Voigt model (type III) for $m > 0$ and $p = 0, q = 1$, the Maxwell model (type II) for $m = 0$ and $p = q = 1$, the Zener model (type I) for $m > 0$ and $p = q = 1$, and the

Table 2.2: The four cases of the operator equation.

Type	Order	m	J_g	G_e	J_+	G_-
I	$q = p$	> 0	a_p/b_p	m	0	0
II	$q = p$	$= 0$	a_p/b_p	0	$1/b_1$	0
III	$q = p + 1$	> 0	0	m	0	b_q/a_p
IV	$q = p + 1$	$= 0$	0	0	$1/b_1$	b_q/a_p

anti-Zener model (type IV) for $m = 0$ and $p = 1, q = 2$. Finally, with four parameters we can construct two models; the former with $m = 0$ and $p = q = 2$, the latter with $m > 0$ and $p = 1, q = 2$, referred to by Bland [3] as four-element models of the first kind and of the second kind, respectively. According to our convention, they are of type II and III, respectively. We have restricted our attention to the former model, the Burgers model (type II), illustrated in Figure 3, because it has found numerous applications, specially in geosciences (see, for example, the books by Klausner [25] and by Carcione [8]).

2.3 Complex modulus, effective modulus, and effective viscosity

In Earth rheology and seismology, it is customary to write the one-dimensional stress–strain relation in the Laplace domain in terms of a *complex shear modulus* $\widehat{\mu}(s)$ as

$$\widetilde{\sigma}(s) = 2\widehat{\mu}(s)\,\widetilde{\epsilon}(s), \tag{2.16}$$

which is expected to generalize the relation for a perfect elastic solid (*Hooke model*)

$$\sigma(t) = 2\mu_0\,\epsilon(t), \tag{2.17}$$

where μ_0 denotes the *shear modulus*.

Adopting this notation, we note comparing (2.16) with (2.2) that the functions $\widetilde{J}(s)$ and $\widetilde{G}(s)$ can be expressed in terms of the complex shear modulus $\widehat{\mu}(s)$ as

$$\widetilde{J}(s) = \frac{1}{2s\,\widehat{\mu}(s)}, \quad \widetilde{G}(s) = \frac{2\widehat{\mu}(s)}{s}. \tag{2.18}$$

As a consequence, we are led to introduce an *effective modulus* defined as

$$\mu(t) := \frac{1}{2}\left[\frac{d}{dt}\,G(t) + G_g\right]. \tag{2.19}$$

Recalling that for perfect viscous fluid (*Newton model*), we have

$$\sigma(t) = 2\eta_0\,\frac{d}{dt}\,\epsilon(t), \tag{2.20}$$

where η_0 denotes the *viscosity coefficient*, likewise we are led to define, following Müller [37] and Krönig and Müller [27], the *effective viscosity* as

$$\eta(t) := \frac{1}{2\dot{J}(t)}, \tag{2.21}$$

where the dot denotes the derivative w.r.t. time t.

We easily recognize that, for the Hooke model (2.4), we recover $\mu(t) \equiv \mu_0 = m/2$, and for the Newton model (2.5), $\eta(t) \equiv \eta_0 = b_1/2$. Furthermore, we recognize that for a spring (Hooke model), the corresponding complex modulus is a constant, that is,

$$\widehat{\mu}_H(s) = \mu_0, \tag{2.22}$$

whereas for a dashpot (Newton model),

$$\widehat{\mu}_N(s) = \eta_0 s = \mu_0 s \tau_0, \tag{2.23}$$

where $\tau_0 = \eta_0/\mu_0$ denotes a characteristic time related to viscosity.

To avoid possible misunderstanding, we explicitly note that the complex modulus is not the Laplace transform of the effective modulus, but its Laplace transform multiplied by s.

The appropriate form of the complex modulus $\widehat{\mu}(s)$ can be obtained, recalling the combination rule for which creep compliances add in series, whereas relaxation moduli add in parallel, as stated at the beginning of Section 2.2. Accordingly, for a serial combination of two viscoelastic models with individual complex moduli $\widehat{\mu}_1(s)$ and $\widehat{\mu}_2(s)$, we have

$$\frac{1}{\widehat{\mu}(s)} = \frac{1}{\widehat{\mu}_1(s)} + \frac{1}{\widehat{\mu}_2(s)}, \tag{2.24}$$

whereas for a combination in parallel,

$$\widehat{\mu}(s) = \widehat{\mu}_1(s) + \widehat{\mu}_2(s). \tag{2.25}$$

We close this section with a discussion about the definition of solid–like and fluid-like behavior for viscoelastic materials. The matter is of course subjected to personal opinions.

Generally, one may define a fluid if it can creep indefinitely under constant stress ($J_e = \infty$), namely, when it relaxes at zero under constant deformation ($G_e = 0$). According to this view, viscoelastic models of type II and IV are fluid–like, whereas models of type I and III are solid-like. However, in his interesting book [39], Pipkin calls a solid if the integral of $G(t)$ from zero to infinity diverges. This includes those cases in which the equilibrium modulus G_e is not zero. It also includes the cases in which $G_e = 0$, but the approach to the limit is not fast enough for integrability, for example, $G(t) \approx t^{-\alpha}$ with $0 < \alpha \leq 1$ as $t \to \infty$.

3 Fractional viscoelasticity

The straightforward way to introduce fractional derivatives in linear viscoelasticity is to replace in the constitutive Equation (2.5) of the Newton model, the first derivative, with a fractional derivative of order $\nu \in (0,1)$, given that $\epsilon(0^+) = 0$ may be intended both in the Riemann–Liouville or Caputo sense.

For the theory and applications of fractional calculus we recommend e. g. the treatise by Kilbas, Srivastava and Trujillo [24]. For the basic results on fractional integrability and differentiability we refer the interested reader to the survey paper by Li and Zhao [28].

For the use of Laplace transform to fractional viscoelasticity we refer the reader to the works by Caputo and by Caputo & Mainardi, see [4, 5] and [6, 7]. We refer also to the more recent paper by Mainardi & Gorenflo [33].

Some people call the fractional model of the Newtonian dashpot (fractional dashpot) with the suggestive name *pot*; see Koeller [26]. We prefer to refer such model as *Scott–Blair* model, to honor the scientist who already in the middle of the past century proposed such a constitutive equation to explain a material property that is intermediate between the elastic modulus (Hooke solid) and the coefficient of viscosity (Newton fluid) (see, for example, [41]). G.W. Scott Blair was surely a pioneer of fractional calculus even if he did not provide a mathematical theory accepted by mathematicians of his time, as pointed out by the author in [31]. More details on Scott Blair are found in the paper by Rogosin and Mainardi [40].

It is known that the creep and relaxation power laws of the Scott Blair model can be interpreted in terms of a continuous spectrum of retardation and relaxation times, respectively, see, for example, p. 58 of the author's book [30]. Starting from these continuous spectra, in [38] Papoulia et al. interpreted the fractional dashpot by an infinite combination of Kelvin–Voigt or Maxwell elements in series or in parallel, respectively. We also note that Liu and Xu [29] computed the relaxation and creep functions for higher-order fractional rheological models involving more parameters than in our analysis.

3.1 Fractional derivatives in mechanical models

The use of *fractional calculus* in linear viscoelasticity leads to generalizations of the classical mechanical models: the basic Newton element is substituted by the more general Scott Blair element (of order ν). In fact, we can construct the class of these generalized models from Hooke and Scott Blair elements, disposed singly and in branches of two (in series or in parallel). Then, extending the procedures of the classical mechanical models (based on springs and dashpots), we will get the *fractional operator equation* (that is an operator equation with fractional derivatives) in the form which properly generalizes (2.14), that is,

$$\left[1+\sum_{k=1}^{p} a_k \frac{d^{\nu_k}}{dt^{\nu_k}}\right]\sigma(t) = \left[m+\sum_{k=1}^{q} b_k \frac{d^{\nu_k}}{dt^{\nu_k}}\right]\epsilon(t), \quad \nu_k = k+\nu-1, \tag{3.1}$$

where the fractional derivatives may be intended both in the Riemann–Liouville or Caputo sense. Indeed their equivalence in fractional viscoelasticity was proven by Mainardi [32] and by Bagley [1], independently. As a generalization of (2.11), we state as a conjecture

$$\begin{cases} J(t) = J_g + \sum_n J_n\{1 - \mathrm{E}_\nu[-(t/\tau_{\epsilon,n})^\nu]\} + J_+ \frac{t^\nu}{\Gamma(1+\nu)}, \\ G(t) = G_e + \sum_n G_n \mathrm{E}_\nu[-(t/\tau_{\sigma,n})^\nu] + G_- \frac{t^{-\nu}}{\Gamma(1-\nu)}, \end{cases} \tag{3.2}$$

where all the coefficient are nonnegative. Here, Γ denotes the well-known gamma function, and E_ν denotes the Mittag-Leffler function of order ν, briefly discussed hereafter along with its generalized form in two parameters $E_{\nu,\mu}$. For more details on this special function, we refer the reader to the articles by Gorenflo and Mainardi [18], by Gorenflo et al. [17] and the recent monograph by Gorenflo et al. [16]. We also note that for the consistency among Equations (3.1) and (3.2), we would require a general proof; however, these equations turn out to be correct for the particular models dealt in the next subsections. Of course, for the fractional operator Equation (3.1), the four cases summarized in Table 2.2 are expected to occur in analogy with the operator Equation (2.14). The definitions in the complex plane of the Mittag-Leffler functions in one and two parameters are provided by their Taylor powers series around $z = 0$, that is,

$$E_\nu(z) := \sum_{n=0}^{\infty} \frac{z^n}{\Gamma(\nu n+1)}, \quad E_{\nu,\mu}(z) : \sum_{n=0}^{\infty} \frac{z^n}{\Gamma(\nu n+\mu)}, \quad \nu,\mu > 0, \tag{3.3}$$

relating them by the following expressions (see, for example, [30] and Appendix E):

$$E_\nu(z) = E_{\nu,1}(z) = 1 + z\,E_{\nu,1+\nu}(z), \quad \frac{d}{dz} E_\nu(z^\nu) = z^{\nu-1} E_{\nu,\nu}(z^\nu). \tag{3.4}$$

In our case, the Mittag-Leffler functions appearing in (3.2) are all of order $\nu \in (0.1]$, an argument real and negative, namely, they are of the type

$$E_\nu[-(t/\tau)^\nu] = \sum_{n=0}^{\infty} (-1)^n \frac{(t/\tau)^{\nu n}}{\Gamma(\nu n+1)}, \quad 0<\nu<1, \quad \tau>0, \tag{3.5}$$

which reduce to $\exp(-t/\tau)$ for $\nu = 1$.

Let us now outline some noteworthy properties of the Mittag-Leffler function (3.5), assuming for brevity that $\tau = 1$. Since the asymptotic behaviors for small and large times are as following:

$$\mathrm{E}_\nu(-t^\nu) \sim \begin{cases} 1 - \frac{t^\nu}{\Gamma(1+\nu)}, & t \to 0^+, \\ \frac{t^{-\nu}}{\Gamma(1-\nu)}, & t \to +\infty, \end{cases} \tag{3.6}$$

we recognize that, for $t \geq 0$, the Mittag-Leffler function decays for short times like a stretched exponential and for long periods with a negative power law. Furthermore, it turns out to be *completely monotonic* in $0 < t < \infty$ (that is, its derivatives of successive order exhibit alternating signs, such as e^{-t}); so it can be expressed in terms of a continuous distribution of elementary relaxation processes:

$$E_\nu(-t^\nu) = \int_0^\infty e^{-rt} K_\nu(r)\, dr, \quad K_\nu(r) = \frac{1}{\pi r} \frac{\sin(\nu\pi)}{r^\nu + 2\cos(\nu\pi) + r^{-\nu}} \geq 0. \tag{3.7}$$

For completeness, we also recall its relations with the corresponding Mittag-Leffler function of two parameters, as obtained from (3.4):

$$E_\nu(-t^\nu) = E_{\nu,1}(-t^\nu) = 1 - t^\nu E_{\nu,1+\nu}(-t^\nu), \quad \frac{d}{dt} E_\nu(-t^\nu) = -t^{\nu-1} E_{\nu,\nu}(-t^\nu). \tag{3.8}$$

We recognize that in fractional viscoelasticity, governed by the operator Equation (3.1), the corresponding material functions are obtained by using the combination rule valid for the classical mechanical models. Their determination is made easy if we take into account the following *correspondence principle* between the classical and fractional mechanical models, outlined in 1971 by Caputo and Mainardi [7]. Taking $0 < \nu \leq 1$ and denoting by $\tau > 0$ a characteristic time related to viscosity, such a correspondence principle can be formally stated by the following three equations, where the Laplace transform pairs are outlined as well:

$$\delta(t/\tau) \div \tau \Rightarrow \frac{(t/\tau)^{-\nu}}{\Gamma(1-\nu)} \div \frac{1}{s}(s\tau)^\nu, \tag{3.9}$$

$$t/\tau \div \frac{1}{s}\frac{1}{(s\tau)} \Rightarrow \frac{(t/\tau)^\nu}{\Gamma(1+\nu)} \div \frac{1}{s}\frac{1}{(s\tau)^\nu}, \tag{3.10}$$

$$e^{-t/\tau} \div \frac{\tau}{1+s\tau} \Rightarrow E_\nu[-(t/\tau)^\nu] \div \frac{1}{s}\frac{(s\tau)^\nu}{1+(s\tau)^\nu}. \tag{3.11}$$

In the following, we will provide the creep compliance $J(t)$, the relaxation modulus $G(t)$, and the effective viscosity $\eta(t)$ for a set of fractional models, which properly generalize with fractional derivatives the basic mechanical models discussed in Section 2, that is, Kelvin–Voigt, Maxwell, Zener, anti-Zener, and Burgers, by using their connections with Hooke and Scott Blair elements. Henceforth, for brevity, we refer to all the basic models with their first letters, that is, *H*, *SB*, *KV*, *M*, *Z*, *AZ*, and *B*. Our analysis will be carried out by using the Laplace transform and the complex shear modulus of the elementary Hooke and Scott Blair models. For this purpose, we recall the constitutive equations for the *H* and *SB* models:

$$\text{Hooke model:} \quad \sigma(t) = m\,\epsilon(t), \quad \text{Scott Blair model:} \quad \sigma(t) = b_1 \frac{d^\nu}{dt^\nu}\epsilon(t), \tag{3.12}$$

and, setting

$$\mu := \frac{m}{2}, \quad \tau^{\nu} := \frac{b_1}{2\mu}, \tag{3.13}$$

we get the complex shear moduli for these elements

$$\widehat{\mu}_H(s) = \mu, \quad \widehat{\mu}_{SB}(s) = \mu(s\tau)^{\nu}, \tag{3.14}$$

where $\nu \in (0, 1]$, and $\tau > 0$ is a characteristic time of the *SB* element. So, whereas the *H* element is characterized by a unique parameter, its elastic modulus μ, and the *SB* element is characterized by a triplet of parameters, that is, $\{\mu, \tau, \nu\}$.

3.2 Fractional Kelvin–Voigt model

The constitutive equation for the fractional Kelvin–Voigt model (referred to as *KV* body) is obtained from (2.6a) in the form

$$\text{fractional Kelvin–Voigt model:} \quad \sigma(t) = m\,\epsilon(t) + b_1 \frac{d^{\nu}\epsilon}{dt^{\nu}}. \tag{3.15}$$

The mechanical analogue of the fractional *KV* body is represented by a Hooke (*H*) element in parallel with a Scott Blair (*SB*) element. The parallel combination rule (2.25) provides the complex modulus

$$\widehat{\mu}_{KV}(s) = \mu[1 + (s\tau)^{\nu}], \quad \mu := \frac{m}{2}, \quad \tau^{\nu} := \frac{b_1}{2\mu}, \tag{3.16}$$

where our time constant τ reduces for $\nu = 1$ to the retardation time τ_ϵ of the classical *KV* body (see (2.6b)). Hence, substitution of (3.16) into Equations (2.18) gives

$$\widetilde{J}_{KV}(s) = \frac{1}{2\mu s}\left[1 - \frac{(s\tau)^{\nu}}{1 + (s\tau)^{\nu}}\right], \quad \widetilde{G}_{KV}(s) = \frac{2\mu}{s}[1 + (s\tau)^{\nu}]. \tag{3.17}$$

By inverting the above Laplace transforms according to Equations (3.11) and (3.9), we get, for $t \geq 0$,

$$J_{KV}(t) = \frac{1}{2\mu}[1 - E_{\nu}(-(t/\tau)^{\nu})], \quad G_{KV}(t) = 2\mu\left[1 + \frac{(t/\tau)^{-\nu}}{\Gamma(1-\nu)}\right]. \tag{3.18}$$

For plotting purposes, it is convenient to introduce a nondimensional time $\xi := t/\tau$ and define normalized, nondimensional material functions. With $J'_{KV}(\xi) = \mu J_{KV}(t)|_{t=\tau\xi}$ and $G'_{KV}(\xi) = (1/\mu)G_{KV}(t)|_{t=\tau\xi}$, these can be written in the nondimensional form

$$J'_{KV}(\xi) = \frac{1}{2}[1 - E_{\nu}(-\xi^{\nu})], \quad G'_{KV}(\xi) = 2\left[1 + \frac{\xi^{-\nu}}{\Gamma(1-\nu)}\right], \quad \xi = \frac{t}{\tau}. \tag{3.19}$$

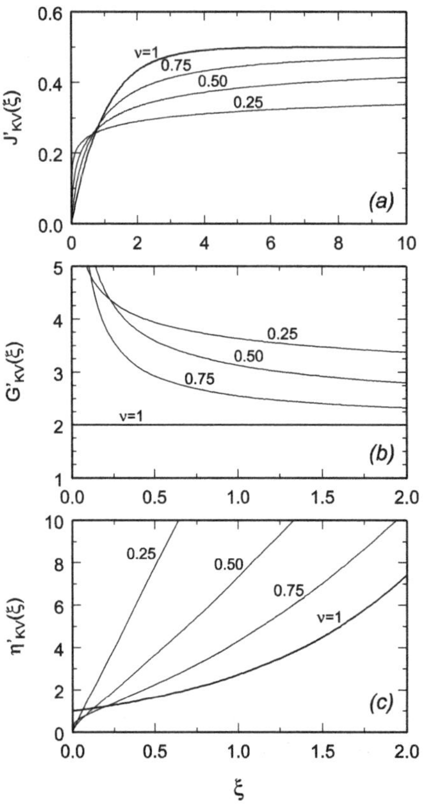

Figure 4: (a) creep compliance; (b) relaxation modulus; (c) effectiveviscosity; for the fractional KV body, for some values of the fractional index ν in the range $(0 < \nu \le 1)$, as a function of normalized time ξ. The thick lines $(\nu = 1)$ represent material functions and effective viscosity for the traditional Kelvin–Voingt body.

Finally, by a straightforward application of Equation (2.21), recalling the derivative rule in (3.8) for the Mittag-Leffler function, the effective viscosity of the KV body turns out to be

$$\eta'_{KV}(\xi) = \frac{\xi^{1-\nu}}{E_{\nu,\nu}(-\xi^{\nu})}. \tag{3.20}$$

Figure 4 shows plots of $J'_{KV}(\xi)$, $G'_{KV}(\xi)$, and $\eta'_{KV}(\xi)$ as a function of nondimensional time ξ. For $\nu \to 1$, the response of the fractional KV body reduces to that of a classical KV body. Taking into account that $E_1(-\xi) = \mathrm{e}^{-\xi}$, the creep compliance reduces, in this limiting case, to $J'_{KV}(\xi) = (1 - \mathrm{e}^{-\xi})/2$. Recalling the Dirac delta representation $\delta(t) = t^{-1}/\Gamma(0)$, the relaxation modulus reduces, as $\nu \to 1$, to $G'_{KV}(t) = 2[1 + \tau\delta(t)] = 2[1 + \delta(t/\tau)]$. Finally, for the effective viscosity, we obtain the classical exponential law $\eta'_{KV}(\xi) = \mathrm{e}^{\xi}$. We note the effective viscosity exceeds the classical value since the early stage of creep.

3.3 Fractional Maxwell model

The constitutive equation for the fractional Maxwell model referred to as fractional M body) is obtained from (2.7a) in the form

$$\text{fractional Maxwell model:} \quad \sigma(t) + a_1 \frac{d^\nu \sigma}{dt^\nu} = b_1 \frac{d^\nu \epsilon}{dt^\nu}. \tag{3.21}$$

The mechanical analogue of the fractional M body is composed by a Hooke (H) element connected in series with a Scott Blair element. From the series combination rule (2.24) we obtain the complex modulus as

$$\widehat{\mu}_M(s) = \mu \frac{(s\tau)^\nu}{1+(s\tau)^\nu}, \quad \mu = \frac{b_1}{2a_1}, \quad \tau^\nu := \frac{b_1}{2\mu}, \tag{3.22}$$

where now our time constant τ reduces for $\nu = 1$ to the relaxation time τ_σ of the classical M body, see (2.7b). Hence, substitution of (3.22) into Equations (2.18) gives the Laplace transforms of the material functions

$$\widetilde{J}_M(s) = \frac{1}{2\mu s}\left[1 + \frac{1}{(s\tau)^\nu}\right], \quad \widetilde{G}_M(s) = \frac{2\mu}{s}\frac{(s\tau)^\nu}{1+(s\tau)^\nu}. \tag{3.23}$$

By inverting the Laplace transforms according to Equations (3.10) and (3.11), we get, for $t \geq 0$,

$$J_M(t) = \frac{1}{2\mu}\left[1 + \frac{(t/\tau)^\nu}{\Gamma(1+\nu)}\right], \quad G_M(t) = 2\mu\, E_\nu(-(t/\tau)^\nu). \tag{3.24}$$

For plotting purposes, it is convenient to write nondimensional forms of $J_M(t)$ and $G_M(t)$. These can be obtained, following the example of the fractional KV body, by introducing a nondimensional time with $\xi = t/\tau$, and defining normalized, nondimensional material functions $J'_M(\xi) = \mu J_M(t)|_{t=\tau\xi}$ and $G'_M(\xi) = (1/\mu)\, G_M(t)|_{t=\tau\xi}$, which provides

$$J'_M(\xi) = \frac{1}{2}\left[1 + \frac{\xi^\nu}{\Gamma(1+\nu)}\right], \quad G'_M(\xi) = 2\, E_\nu(-\xi^\nu), \quad \xi = \frac{t}{\tau}. \tag{3.25}$$

Following this normalization scheme, the effective viscosity (in nondimensional form) can be readily obtained from (2.21) and (3.25) as

$$\eta'_M(\xi) = \frac{\Gamma(1+\nu)}{\nu}\xi^{1-\nu}. \tag{3.26}$$

Plots of $J'_M(\xi)$, $G'_M(\xi)$, and $\eta'_M(\xi)$ are shown in Figure 5. Thick lines show material functions and effective viscosity in the limit of $\nu \mapsto 1$, that is, when the response of the fractional M body degenerates into that of a classical M body. In particular, from Equation (3.25), we obtain $J'_M(\xi) = (1+\xi)/2$ and $G'_M(\xi) = 2e^{-\xi}$, and the effective viscosity is constant ($\eta'_M(\xi) = 1$), which denotes the lack of transient effects. For $0 < \nu < 1$, the effective viscosity always increases with time, and, except during the very early stages of creep ($\xi \approx 1$), it exceeds the value corresponding to the classical M body ($\nu = 1$).

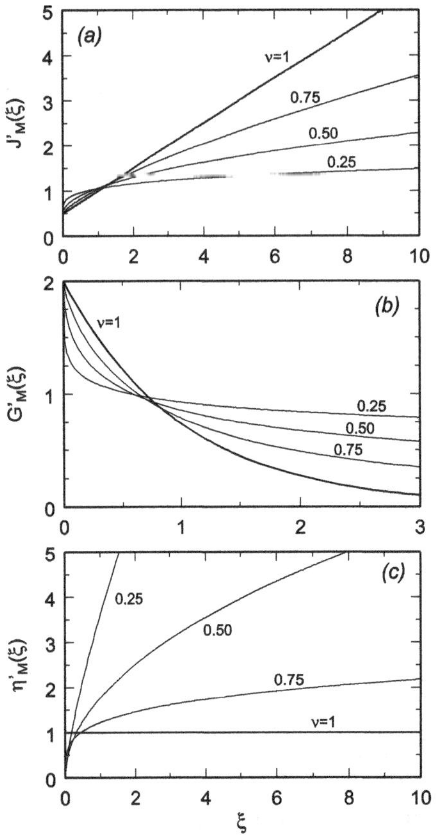

Figure 5: (a) creep compliance; (b) relaxation modulus; (c) effectiveviscosity; for the fractional M body, for some rational values of the index ν in the range $(0 < \nu \leq 1)$, as a function of normalized time ξ. The thick lines, corresponding to $\nu = 1$, show the classical Maxwell body.

3.4 Fractional Zener model

The constitutive equation for the fractional Zener model (referred to as fractional Z body) is obtained from Equation (2.8a) in the form

$$\text{fractional Zener model:} \quad \sigma(t) + a_1 \frac{d^\nu \sigma}{dt^\nu} = m\,\epsilon(t) + b_1 \frac{d^\nu \epsilon}{dt^\nu}. \tag{3.27}$$

The mechanical analogue of the fractional Z body is represented by a Hooke (H) element in series with a Kelvin–Voigt (KV) element. Here, we indicate with μ_1 the shear modulus of the H body, whereas with $\{\mu_2, \tau_2, \nu\}$, the triplet of parameters, the KV body is denoted.

The material functions for the fractional Z body in the time domain can be derived following the same procedure outlined above for the fractional KV and M. However, the algebraic complexity increases because of the increased number of independent rheological parameters involved. The use of the combination rule (2.24), which holds for connections in series, provides the complex modulus

$$\frac{1}{\widetilde{\mu}_Z(s)} = \frac{1}{\mu_1} + \frac{1}{\mu_2}\frac{1}{1+(s\tau_2)^\nu}, \quad \mu_1 = \frac{b_1}{2a_1}, \quad \mu_2 = \frac{m}{2}, \quad \tau_2^\nu = \frac{b_1}{2\mu_2}, \tag{3.28}$$

where now our time constant τ_2 reduces for $\nu = 1$ to the retardation time τ_ϵ of the classical Z body; see (2.8b). Hence, substitution of (3.28) into Equations (2.18) provides the Laplace transform of the creep compliance and relaxation modulus. For the creep compliance, we have

$$\widetilde{J}_Z(s) = \frac{1}{2s}\left[\left(\frac{1}{\mu_1} + \frac{1}{\mu_2}\right) - \frac{1}{\mu_2}\frac{(s\tau_2)^\nu}{1+(s\tau_2)^\nu}\right], \tag{3.29}$$

which can be easily inverted to obtain

$$J_Z(t) = \frac{1}{2}\left[\frac{1}{\mu_1} + \frac{1}{\mu_2}(1 - E_\nu(-(t/\tau_2)^\nu))\right], \quad t \geq 0. \tag{3.30}$$

For the relaxation modulus, the following expression can be obtained:

$$\widetilde{G}_Z(s) = 2\mu^*\left(\frac{\tau_2}{\tau_a}\right)^\nu \frac{1}{s}\frac{s^\nu + 1/\tau_2^\nu}{s^\nu + 1/\tau_a^\nu}, \tag{3.31}$$

where

$$\mu^* = \frac{\mu_1\mu_2}{\mu_1+\mu_2}, \tag{3.32}$$

and we have introduced an additional characteristic time

$$\tau_a^\nu = \frac{1}{1+r_\mu}\tau_2^\nu, \quad r_\mu = \frac{\mu_1}{\mu_2}. \tag{3.33}$$

Applying a partial fraction expansion to Equation (3.31) and using Equations (3.8), the relaxation modulus in the time domain can be cast in the form

$$G_Z(t) = 2\mu^*\left(\frac{\tau_2}{\tau_a}\right)^\nu [E_\nu(-(t/\tau_a)^\nu) + (t/\tau_2)^\nu E_{\nu,\nu+1}(-(t/\tau_a)^\nu)]. \tag{3.34}$$

Hence, after some rearrangement, we finally obtain

$$G_Z(t) = 2\mu^*[1 + r_\mu E_\nu(-(t/\tau_a)^\nu)]. \tag{3.35}$$

We now recognize that for $\nu = 1$ the constant τ_a reduces to the relaxation time τ_σ $(0 < \tau_\sigma < \tau_\epsilon < \infty)$ for the classical Z model; see (2.8b).

Writing $J_Z'(\xi) = \mu^* J_Z(t)|_{t=\tau_2\xi}$ and $G_Z'(\xi) = (1/\mu^*)G_Z(t)|_{t=\tau_2\xi}$, the material functions for the Zener model can be written in the nondimensional form

$$J_Z'(\xi) = \frac{1}{2}\left[1 - \frac{h_1}{h_2}E_\nu(-\xi^\nu)\right], \quad G_Z'(\xi) = 2[1 + h_1 E_\nu(-h_2\xi^\nu)], \quad \xi = \frac{t}{\tau_2}, \tag{3.36}$$

with $h_1 = r_\mu$, $h_2 = 1 + r_\mu$.

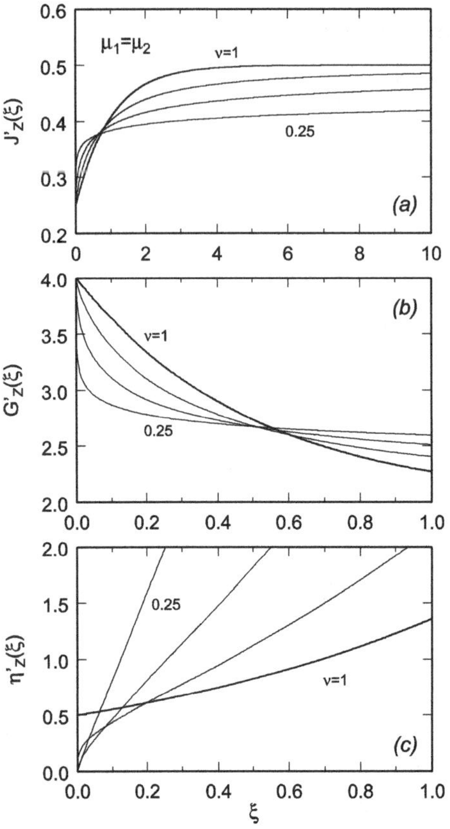

Figure 6: (a) Creep compliance; (b) relaxation modulus; (c) effective viscosity; for the fractional Z body, for some values of the fractional index ν in the range $(0 < \nu \leq 1)$, as a function of normalized time ξ. Here, we adopt the ratio $r_\mu = \mu_2/\mu_1 = 1$. The thick lines $(\nu = 1)$ represent material functions and effective viscosity for the traditional Zener body.

As a consequence, the effective viscosity of the Z body turns out to be

$$\eta'_Z(\xi) = \frac{h_1}{h_2} \frac{\xi^{1-\nu}}{E_{\nu,\nu}(-\xi^\nu)}. \tag{3.37}$$

Figure 6 shows plots of $J'_Z(\xi)$, $G'_Z(\xi)$, and $\eta'_Z(\xi)$ as a function of ξ for various values of ν in the range $0 < \nu \leq 1$. In the limiting case of $\nu = 1$ (thick curves), which corresponds to the classical Zener model, the material functions and the effective viscosity can be written in terms of exponential functions, in agreement with Equations (2.8b). In particular, since $r_\mu = 1$, we obtain $J'_Z(\xi) = {}^1\!/_2(1 - {}^1\!/_2\, e^{-\xi})$ and $G'_Z(\xi) = 2(1 + e^{-2\xi})$. Furthermore, the effective viscosity grows exponentially, since $\eta'_Z(\xi) = {}^1\!/_2\, e^{\xi}$. We note that, for sufficiently long times ($\xi > 0.2$ in Figure 6), the value of $\eta'_Z(\xi)$ for a fractional Zener model $(0 < \nu < 1)$ always exceeds the classical Zener viscosity.

3.5 Fractional anti-Zener model

The constitutive equation for fractional anti-Zener model (referred to as fractional *AZ* body) is obtained from (2.9a) in the form

$$\text{fractional anti-Zener model:}\quad \left[1 + a_1 \frac{d^\nu}{dt^\nu}\right]\sigma(t) = \left[b_1 \frac{d^\nu}{dt^\nu} + b_2 \frac{d^{1+\nu}}{dt^{1+\nu}}\right]\epsilon(t). \tag{3.38}$$

The fractional anti-Zener model results from the combination in series of an *SB* element (with material parameters $\{\mu, \tau_1, \nu\}$) and a fractional *KV* element ($\{\mu, \tau_2, \nu\}$).

Using the series combination rule (2.24), we obtain

$$\begin{aligned} \frac{1}{\widehat{\mu}_{AZ}(s)} &= \frac{1}{\mu}\frac{1}{(s\tau_1)^\nu} + \frac{1}{\mu}\left[1 - \frac{(s\tau_2)^\nu}{1+(s\tau_2)^\nu}\right], \\ \frac{1}{\mu} &= 2\left(\frac{a_1}{b_1} - \frac{b_2}{b_1^2}\right), \quad \tau_1{}^\nu = \frac{1}{2\mu\, b_1}, \quad \tau_2{}^\nu = \frac{b_2}{b_1}, \end{aligned} \tag{3.39}$$

where now our time constant τ_2 reduces for $\nu = 1$ to the retardation time τ_ϵ of the classical *AZ* body, see (2.9b). Using (2.18), the above equation easily provides the Laplace transform of the creep compliance

$$\widetilde{J}_{AZ}(s) = \frac{1}{2\mu s}\left[\frac{1}{(s\tau_1)^\nu} + 1 - \frac{(s\tau_2)^\nu}{1+(s\tau_2)^\nu}\right], \tag{3.40}$$

whose inverse is

$$J_{AZ}(t) = \frac{1}{2\mu}\left[\frac{(t/\tau_1)^\nu}{\Gamma(1+\nu)} + (1 - E_\nu(-(t/\tau_2)^\nu))\right], \quad t \geq 0. \tag{3.41}$$

The derivation of the relaxation modulus is more complicated, but straightforward. Starting from the general relationship (2.18) and using (3.40), by a partial fraction expansion, we obtain

$$\widetilde{G}_{AZ}(s) = \frac{2\mu}{s}\left[(s\tau_0)^\nu + \left(\frac{\tau_1}{\bar{\tau}}\right)^{2\nu} \frac{(s\bar{\tau})^\nu}{1+(s\bar{\tau})^\nu}\right], \tag{3.42}$$

where we have defined the new time constants

$$\bar{\tau} = (\tau_1^\nu + \tau_2^\nu)^{\frac{1}{\nu}}, \quad \tau_0 = \frac{\tau_1\tau_2}{\bar{\tau}}. \tag{3.43}$$

Hence, using (3.9) and (3.11), the inverse Laplace transform of (3.43) is

$$G_{AZ}(t) = 2\mu\left[\frac{(t/\tau_0)^{-\nu}}{\Gamma(1-\nu)} + \left(\frac{\tau_1}{\bar{\tau}}\right)^{2\nu} E_\nu(-(t/\bar{\tau})^\nu)\right], \quad t \geq 0. \tag{3.44}$$

We now recognize that, for $\nu = 1$, the constant $\bar{\tau}$ reduces to the relaxation time τ_σ ($0 < \tau_\epsilon < \tau_\sigma < \infty$) for the classical *AZ* model; see (2.9b).

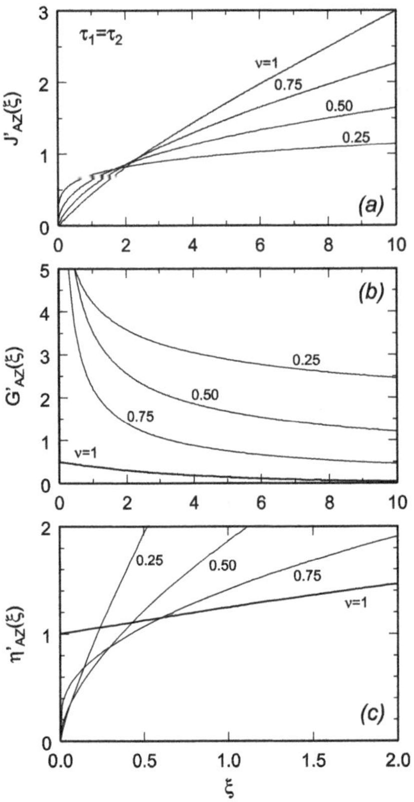

Figure 7: (a) Creep compliance; (b) relaxation modulus; (c) effective viscosity; (c) for the fractional AZ body, for some values of the fractional index ν in the range $(0 < \nu \leq 1)$, as a function of normalized time ξ. The particular case $\tau_1 = \tau_2$ is considered. The thick lines pertaining to $\nu = 1$ show results for the traditional AZ body.

From (3.41) and (3.44), it is easy to derive nondimensional expressions for the material functions in the time domain. The choice of the scaling for the time variable is of course arbitrary. Here we define

$$\xi = \frac{t}{\tau^*}, \quad \tau^* = \frac{\tau_1 \tau_2}{\tau_1 + \tau_2}. \tag{3.45}$$

With this choice, defining $J'_{AZ}(\xi) = \mu J_{AZ}(t)|_{t=\tau^*\xi}$ and manipulating (3.41), we obtain

$$J'_{AZ}(\xi) = \frac{1}{2}\left[\frac{(c_2\xi)^\nu}{\Gamma(1+\nu)} + (1 - E_\nu(-(c_1\xi)^\nu))\right], \tag{3.46}$$

where

$$c_1 = \frac{r_\tau}{1+r_\tau}, \quad c_2 = \frac{1}{1+r_\tau}, \quad r_\tau \equiv \frac{\tau_1}{\tau_2}. \tag{3.47}$$

Likewise, from (3.44), the relaxation modulus has the nondimensional form

$$G'_{AZ}(\xi) = 2\left[\frac{(e_1\xi)^{-\nu}}{\Gamma(1-\nu)} + e_2^{2\nu} E_\nu(-(e_3\xi)^\nu)\right], \tag{3.48}$$

where $G'_{AZ}(\xi) = (1/\mu)G_{AZ}(t)|_{t=\tau^*\xi}$, and we have introduced the constants

$$e_1 = \frac{(1+r_\tau^\nu)^{\frac{1}{\nu}}}{1+r_\tau}, \quad e_2 = \frac{r_\tau}{(1+r_\tau^\nu)^{\frac{1}{\nu}}}, \quad e_3 = \frac{r_\tau}{(1+r_\tau)(1+r_\tau^\nu)^{\frac{1}{\nu}}}. \tag{3.49}$$

Using the normalization scheme above, the expression for the effective viscosity of the *AZ* body can be obtained from Equations (3.45) and (3.46) as

$$\eta'_{AZ}(\xi) = \frac{1}{\frac{\nu}{\Gamma(1+\nu)}c_2(c_2\xi)^{\nu-1} + c_1(c_1\xi)^{\nu-1}E_{\nu,\nu}(-(c_1\xi)^\nu)}. \tag{3.50}$$

Figure 7 shows functions $J'_{AZ}(\xi)$, $G'_{AZ}(\xi)$, and $\eta'_{AZ}(\xi)$ as a function of ξ in the particular case $\tau_1 = \tau_2$ ($r_\tau = 1$, according to Equation (3.46)). The behavior for $\nu = 1$ (thick curves), which corresponds to the classical anti-Zener model, can be expressed in terms of elementary functions. Since $r_\tau = 1$, $c_1 = c_2 = 1/2$, see Equation (3.47). Hence, from (3.46), in this particular case, the normalized creep function is $J'_{AZ}(\xi) = 1/2(1+\xi/2+\mathrm{e}^{-\xi})$, which, in the range of ξ values considered in Figure 7, appears to be dominated by the linear term. In the same limit, since from Equations (3.49) we have $e_1 = 1$, $e_2 = 1/2$, and $e_3 = 1/4$, omitting an additive $\delta(\xi)$ term, the normalized relaxation modulus decays exponentially with ξ, according to $G'_{AZ}(\xi) = \mathrm{e}^{-\xi/4}/2$. The effective viscosity, according to (3.50), turns out to be $\eta'_{AZ}(\xi) = 2/(1+\mathrm{e}^{-\xi/2})$, which approximately is increasing linearly in the range of ξ values considered.

3.6 Fractional Burgers model

The constitutive equation for the fractional Burgers model (referred to as fractional *B* body) is obtained from (2.10a) as

$$\text{fractional Burgers model:} \quad \left[1 + a_1\frac{d^\nu}{dt^\nu} + a_2\frac{d^{1+\nu}}{dt^{1+\nu}}\right]\sigma(t) = \left[b_1\frac{d^\nu}{dt^\nu} + b_2\frac{d^{1+\nu}}{dt^{1+\nu}}\right]\epsilon(t). \tag{3.51}$$

The mechanical analogue of the fractional Burgers (*B*) model can be represented as the combination in series of a fractional *KV* element (with material parameters $\{\mu_1, \tau_1, \nu\}$) and a fractional *M* fractional element ($\{\mu_2, \tau_2, \nu\}$).

Using the expressions (3.16) and (3.22) for complex moduli of fractional *KV* and *M* bodies into the series combination rule (2.24) provides the complex modulus

$$\frac{1}{\widehat{\mu}_B(s)} = \frac{1}{\mu_1}\left[1 + \frac{1}{(s\tau_1)^\nu}\right] + \frac{1}{\mu_2}\left[1 - \frac{(s\tau_2)^\nu}{1+(s\tau_2)^\nu}\right], \tag{3.52}$$

where the four constants μ_i and τ_i are related in some way to the four coefficients a_i, b_i of the constitutive equation with $i = 1, 2$.

Using (2.18), from (3.52) it is possible to obtain immediately the Laplace transformed creep compliance

$$\widetilde{J}_B(s) = \frac{1}{2s}\left[\left(\frac{1}{\mu_1}+\frac{1}{\mu_2}\right)+\frac{1}{\mu_1}\frac{1}{(s\tau_1)^\nu}-\frac{1}{\mu_2}\frac{(s\tau_2)^\nu}{1+(s\tau_2)^\nu}\right], \tag{3.53}$$

which can be easily inverted with the aid of (3.10) and (3.11), obtaining

$$J_B(t) = \frac{1}{2}\left[\left(\frac{1}{\mu_1}+\frac{1}{\mu_2}\right)+\frac{1}{\mu_1}\frac{(t/\tau_1)^\nu}{\Gamma(1+\nu)}-\frac{1}{\mu_2}E_\nu(-(t/\tau_2)^\nu)\right], \quad t \geq 0. \tag{3.54}$$

The computation of the Laplace-transformed relaxation modulus for the fractional B body requires more cumbersome algebra compared to J_B. Using (3.52) into (2.18) provides, after some algebra,

$$\widetilde{G}_B(s) = 2\mu^* \tau_0^\nu s^{\nu-1} \frac{z+\gamma_0}{z^2\delta_2 + z\delta_1 + \delta_0}. \tag{3.55}$$

Above, for convenience, we have introduced the variable

$$z \equiv (s\tau_0)^\nu, \tag{3.56}$$

where τ_0 is defined as in the fractional AZ model (3.43), and several constants as follows:

$$\mu^* = \frac{\mu_1\mu_2}{\mu_1+\mu_2}, \tag{3.57}$$

$$\begin{aligned}\gamma_0 &= \frac{r_\tau^\nu}{1+r_\tau^\nu}, \quad \delta_2 = \frac{1}{1+r_\mu},\\ \delta_1 &= \frac{1}{1+r_\mu}+\frac{r_\mu}{1+r_\mu}\cdot\frac{r_\tau^\nu}{1+r_\tau^\nu}, \quad \delta_0 = \frac{1}{1+r_\mu}\cdot\frac{r_\tau^\nu}{(1+r_\tau^\nu)^2},\end{aligned} \tag{3.58}$$

where $r_\tau = \tau_1/\tau_2$ and $r_\mu = \mu_1/\mu_2$.

Denoting by z_1 and z_2 the roots of algebraic equation

$$z^2\delta_2 + z\delta_1 + \delta_0 = 0, \tag{3.59}$$

we obtain

$$\widetilde{G}_B(s) = 2\mu^* \tau_0^\nu s^{\nu-1} \sum_{i=1}^{2} \frac{G_i}{z-z_i}, \tag{3.60}$$

where the evaluation of constants $G_i = G_i(r_\mu, r_\tau)$ is straightforward (the details are left to the reader). Since from (3.58) it can be easily recognized that the discriminant of Equation (3.59) is positive, the roots are real. Hence, observing that $\delta_i > 0$ $(i = 0, 1, 2)$, Descartes' rule of signs ensures that $z_i < 0$ $(i = 1, 2)$. Setting $z_i = -1/\rho_i^\nu$ with $\rho_i > 0$ and recalling (3.56), we obtain a convenient expression of the relaxation modulus in the Laplace domain:

$$\widetilde{G}_B(s) = 2\mu^* \sum_{i=1}^{2} \frac{G_i}{s} \cdot \frac{(s\tau_0\rho_i)^\nu}{1 + (s\tau_0\rho_i)^\nu}, \tag{3.61}$$

which can be easily inverted with the aid of (3.11), obtaining the relaxation modulus in the time domain as a combination of two Mittag-Leffler functions with different arguments:

$$G_B(t) = 2\mu^* \sum_{i=1}^{2} G_i E_\nu\left(-\left(\frac{t}{\tau_0\rho_i}\right)^\nu\right). \tag{3.62}$$

Useful nondimensional forms for the material functions in the time domain can be easily obtained. For the creep compliance (3.54), we have

$$J_B'(\xi) = \frac{1}{2}\left[1 + c_2' \frac{(c_2\xi)^\nu}{\Gamma(1+\nu)} - c_1' E_\nu(-(c_1\xi)^\nu)\right], \quad \xi = \frac{t}{\tau^*}, \quad \tau^* = \frac{\tau_1\tau_2}{\tau_1 + \tau_2}, \tag{3.63}$$

where $J_B'(\xi) = \mu^* J_B(t)|_{t=\tau^*\xi}$, and

$$c_2' = \frac{1}{1 + r_\mu}, \quad c_1' = \frac{r_\mu}{1 + r_\mu}, \quad r_\mu = \frac{\mu_1}{\mu_2}. \tag{3.64}$$

For the nondimensional form for the relaxation modulus, we obtain

$$G_B'(\xi) = 2\sum_{i=1}^{2} G_i E_\nu\left(-\left(\frac{e_1}{\rho_i}\xi\right)^\nu\right), \tag{3.65}$$

where $G_B'(\xi) = (1/\mu^*) G_B(t)|_{t=\tau^*\xi}$, $\xi = t/\tau^*$.

By the above normalization scheme, it is possible to find the expression for the effective viscosity of the B body by using Equations (2.21) and (3.63) as

$$\eta_B'(\xi) = \frac{1}{\frac{\nu}{\Gamma(1+\nu)} c_2 c_2' (c_2\xi)^{\nu-1} + c_1 c_1' (c_1\xi)^{\nu-1} E_{\nu,\nu}(-(c_1\xi)^\nu)}. \tag{3.66}$$

As an illustration of the behavior of the material functions for the B body, in Figure 8, we consider $\tau_1 = 2\tau_2$ and $\mu_1 = \mu_2$, and $J_B'(\xi)$, $G_B'(\xi)$, and $\eta_B'(\xi)$ as functions of the variable ξ. Whereas the results for the creep compliance (a) and the effective viscosity (c) are qualitatively similar to those obtained for the AZ model in Figure 7, those pertaining to the relaxation modulus (c) differ significantly, being now removed, the singularity for $\xi \mapsto 0$ displayed in Figure 7(c) for $\nu \neq 1$. The curves obtained for $\nu = 1$ (thick), which correspond to the classical B model, can be expressed in terms of elementary functions. Since $r_\tau = 2$ and $r_\mu = 1$, in this case, we have $c_1 = 2/3$, $c_2 = 1/3$,

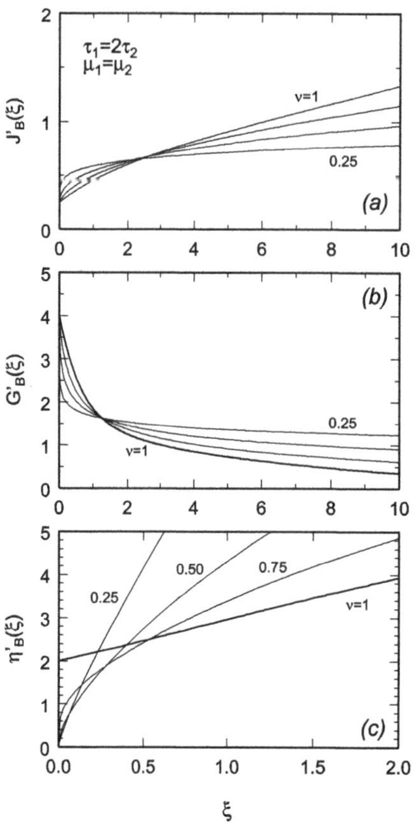

Figure 8: (a) Creep compliance: (b) relaxation modulus; (c) effective viscosity; for the fractional B body, for values of the fractional power $\nu = 1, 0.75, 0.50,$ and 0.25 as a function of normalized time ξ. The thick lines ($\nu = 1$), show results for the traditional B body.

$c_1' = c_2' = 1/2$. Hence, with these numerical values for the rheological parameters, the creep function varies with time as

$$J_{AZ}'(\xi) = \frac{1}{2}\left(1 + \frac{\xi}{6} - \frac{1}{2}\,e^{-2\xi/3}\right). \tag{3.67}$$

In the same limit, since $E_1(-x) = e^{-x}$, the relaxation modulus $G_B'(\xi)$ reduces to the sum of two exponentially decaying functions, and its initial value is only determined by the numerical value of the elastic parameters μ_1 and μ_2. The effective viscosity, according to (3.66), for $\nu = 1$ asymptotically approaches a constant value as $\xi \to \infty$, as in the case of the *AZ* body; for all the other values of the fractional order, $\eta_B'(\xi)$ increases indefinitely, as in the case of the *M* body.

4 Conclusions

Starting from the mechanical analog of the basic fractional models of linear viscoelasticity, we have verified the consistency of the *correspondence principle*. This principle allows one to formally get the material functions and the effective viscosity of the fractional models, starting from the known expressions of the corresponding classical models by using the corresponding rules (3.9), (3.10), and (3.11).

In particular, in the present chapter we were able to plot all these functions versus a suitable time scale in order to visually show the effect of the order $\nu \in (0, 1]$ entering in our basic fractional models. This can help the researchers to guess which fractional model can better fit the experimental results.

In Earth rheology, the concept of effective viscosity is often introduced to describe the behavior of composite materials that exhibit both linear and a nonlinear (that is, non-Newtonian) stress–strain components of deformation) (see, for example, Giunchi and Spada [13]). However, according to Müller [37] and to the results outlined in this paper, it is clear that this concept has a role also within the context of linear viscoelasticity, both for describing the transient creep of classical models and for characterizing the mechanical behavior of fractional models.

Finally, let us conclude this chapter by recalling a few novel models related in some way to fractional viscoelasticity, involving several special functions and non-local operators, authored by PhD A. Giusti and his collaborators (see, for example, [9, 10, 14, 15]).

Bibliography

[1] R. L. Bagley, On the equivalence of the Riemann–Liouville and the Caputo fractional order derivatives in modeling of linear viscoelastic materials, *Fractional Calculus & Applied Analysis. An International Journal for Theory and Applications*, **10**(2) (2007), 123–126.

[2] R. B. Bird and J. M. Wiest, Constitutive equations for polymeric liquids, *Annual Review of Fluid Mechanics*, **27** (1995), 169–193.

[3] D. R. Bland, *The Theory of Linear Viscoelasticity*, Pergamon, Oxford, 1960.

[4] M. Caputo, Linear models of dissipation whose *Q* is almost frequency independent, Part II, *Geophysical Journal of the Royal Astronomical Society*, **13** (1967), 529–539. [Reprinted in *Fract. Calc. Appl. Anal.* **11** (2008), 4–14.]

[5] M. Caputo, *Elasticità e Dissipazione*, Zanichelli, Bologna, 1969 [in Italian].

[6] M. Caputo and F. Mainardi, A new dissipation model based on memory mechanism, *Pure and Applied Geophysics*, **91** (1971a), 134–147. [Reprinted in *Fract. Calc. Appl. Anal.*, **10**(3) (2007), 309–324.]

[7] M. Caputo and F. Mainardi, Linear models of dissipation in anelastic solids, *Rivista del Nuovo Cimento (Ser. II)*, **1** (1971b), 161–198.

[8] J. M. Carcione, *Wave Fields in Real Media*, 3rd Ed., Elsevier, Amsterdam, 2015.

[9] I. Colombaro, R. Garra, A. Giusti, and F. Mainardi, Scott-Blair models with time-varying viscosity, *Applied Mathematics Letters*, **86** (2018), 57–63.

[10] I. Colombaro, A. Giusti, and F. Mainardi, A class of linear viscoelastic models based on Bessel functions, *Meccanica*, **52**(4–5) (2016), 825–832.

[11] G. Doetsch, *Introduction to the Theory and Application of the Laplace Transformation*, Springer Verlag, Berlin, 1974.

[12] I. M. Gel'fand and G. E. Shilov, *Generalized Functions*, vol. 1, Academic Press, New York, 1964.

[13] C. Giunchi and G. Spada, Postglacial rebound in a non-Newtonian spherical Earth, *Geophysical Research Letters*, **27**(14) (2000), 2065–2068.
[14] A. Giusti, On infinite order differential operators in fractional viscoelasticity, *Fractional Calculus & Applied Analysis. An International Journal for Theory and Applications*, **20**(4) (2017), 854–867.
[15] A. Giusti and I. Colombaro, Prabhakar-like fractional viscoelasticity, *Communications in Nonlinear Science and Numerical Simulation*, **56** (2018), 138–143.
[16] R. Gorenflo, A. A. Kilbas, F. Mainardi, and S. V. Rogosin, *Mittag-Leffler Functions. Related Topics and Applications*, Springer, Berlin, 2014, 2nd edition in preparation.
[17] R. Gorenflo, J. Loutchko, and Yu. Luchko, Computation of the Mittag-Leffler function $E_{\alpha,\beta}(z)$ and its derivatives, *Fractional Calculus & Applied Analysis. An International Journal for Theory and Applications*, **5**(4) (2002), 491–518. Corrections in *Fract. Calc. Appl. Anal.* **6**(1) (2003), 111–112.
[18] R. Gorenflo and F. Mainardi, Fractional calculus: integral and differential equations of fractional order, in A. Carpinteri and F. Mainardi (eds.) *Fractals and Fractional Calculus in Continuum Mechanics*, pp. 223–276, Springer Verlag, Wien, 1997. [E-print: http://arxiv.org/abs/0805.3823.]
[19] B. Gross, *Mathematical Structure of the Theories of Viscoelasticity*, Hermann & Cie, Paris, 1953.
[20] A. Hanyga, Viscous dissipation and completely monotone stress relaxation functions, *Rheologica Acta*, **44** (2005), 614–621.
[21] H. Jeffreys, *The Earth*, 1st Ed., § 14.423, pp. 224–225, Cambridge University Press, Cambridge, 1924.
[22] H. Jeffreys, A modification of Lomnitz's law of creep in rocks, *Geophysical Journal of the Royal Astronomical Society*, **1** (1958), 92–95.
[23] H. Jeffreys, *The Earth*, 5th Ed., § 8.13, pp. 321–322, Cambridge University Press, Cambridge, 1970.
[24] A. A. Kilbas, H. M. Srivastava, and J. J. Trujillo, *Theory and Applications of Fractional Differential Equations*, Elsevier, Amsterdam, 2006.
[25] Y. Klausner, *Fundamentals of Continuum Mechanics of Soils*, Springer Verlag, Berlin, 1991.
[26] R. C. Koeller, Application of fractional calculus to the theory of viscoelasticity, *Journal of Applied Mechanics*, **51**(2) (1984), 229–307.
[27] M. Krönig and G. Müller, Rheological models and interpretation of postglacial uplift, *Geophysical Journal International*, **98** (1989), 243–253.
[28] C. P. Li and Z. G. Zhao, Introduction to fractional integrability and differentiability, *The European Physical Journal Special Topics*, **193** (2011), 5–26.
[29] J. G. Liu and M. Y. Xu, Higher-order fractional constitutive equations of viscoelastic materials involving three different parameters and their relaxation and creep functions, *Mechanics of Time-Dependant Materials*, **10** (2006), 263–279.
[30] F. Mainardi, *Fractional Calculus and Waves in Linear Viscoelasticity*, Imperial College Press–World Scientific, London, 2010, 2nd edition in preparation.
[31] F. Mainardi, An historical perspective on fractional calculus in linear viscoelasticity, *Fractional Calculus & Applied Analysis. An International Journal for Theory and Applications*, **15**(4) (2012), 712–717. [E-print http://arxiv.org/abs/1007.2959.]
[32] F. Mainardi, Physical and mathematical aspects of fractional calculus in linear viscoelasticity, in A. Le Méhauté, J. A. Tenreiro Machado, J. C. Trigeassou, and J. Sabatier (eds.) *Proceedings the 1-st IFAC Workshop on Fractional Differentiation and Its Applications (FDA'04) [ENSEIRB]*, Bordeaux, France, July 19–21, 2004, pp. 62–67.
[33] F. Mainardi and R. Gorenflo, Time-fractional derivatives in relaxation processes: a tutorial survey, *Fractional Calculus & Applied Analysis. An International Journal for Theory and Applications*, **10** (2007), 269–308. [E-print: http://arxiv.org/abs/0801.4914.]

[34] F. Mainardi and G. Spada, Creep, relaxation and viscosity properties for basic fractional models in rheology, *The European Physical Journal Special Topics*, **193** (2011), 133–160. [E-print http://arxiv.org/abs/1110.3400.]
[35] F. Mainardi and G. Spada, On the viscoelastic characterization of the Jeffreys–Lomnitz law of creep, *Rheologica Acta*, **51** (2012), 783–791. [E-print: http://arxiv.org/abs/1112.5543.]
[36] A. Molinari, Viscoélasticité linéaire and functions complètement monotones, *Journal de Mécanique* **12** (1975), 541–553.
[37] G. Müller, Generalized Maxwell bodies and estimates of mantle viscosity, *Geophysical Journal of the Royal Astronomical Society*, **87** (1986), 1113–1145. Erratum, *Geophys. J. R. Astr. Soc.* **91** (1987), 1135.
[38] K. D. Papoulia, V. P. Panoskaltsis, N. V. Kurup, and I. Korovajchuk, Rheological representation of fractional order viscoelastic material models, *Rheologica Acta*, **49** (2010), 381–400.
[39] A. C. Pipkin, *Lectures on Viscoelastic Theory*, 2nd Ed., Springer Verlag, New York, 1986.
[40] S. Rogosin and F. Mainardi, George William Scott Blair—the pioneer of fractional calculus in rheology, *Communications in Applied and Industrial Mathematics*, **6**(1) (2014), 20 pages; DOI: 10.1685/journal.caim.481. [A larger version (22 pages) is posted as E-print at arXiv:1404.3295.]
[41] G. M. Scott-Blair, *Survey of General and Applied Rheology*, Pitman, London, 1949.
[42] C. Zener, *Elasticity and Anelasticity of Metals*, University of Chicago Press, Chicago, 1948.

Raoul R. Nigmatullin and Dumitru Baleanu

Relationships between 1D and space fractals and fractional integrals and their applications in physics

Abstract: In this paper, the *exact* relationships between the averaging procedure of a smooth function over 1D-fractal sets and the fractional integral of the RL-type are found. The numerical verifications are realized for confirmation of the analytical results and the physical meaning of these obtained formulas is discussed. Besides, the generalizations of the results for a combination of fractal circuits having a discrete set of fractal dimensions were obtained. We suppose that these new results help to deeper understand the intimate links between fractals and fractional integrals of different types, especially in applications of the fractional operators in complex systems. These results can be used in different branches of the interdisciplinary physics, where the different equations describing the complex physical phenomena, and the fractional derivatives and integrals with complex-conjugated power-law exponents are used. We consider also possibilities of applications of these results in classical mechanics. Besides these exact results, in Section 3, we consider the difficulties that can arise in attempting to generalize them for 2D and 3D fractals. We suggest one approximate approach (tested numerically) that can solve these arising difficulties.

Keywords: Fractals, fractional integrals, complex systems

MSC 2000: 26A33, 28A80, 37F05

1 Introduction

Now the acronym FDA (Fractional Derivative and its Applications) received a very wide propagation. The reference "hot spot" was formed in the end of the 80s of the last century, when many researches working in different application fields understood that this new tool, suggested by the mathematicians working in the fractional calculus area, could give rise to new features and generalizations in addition to those of the previous phenomena associated with fractal geometry studied. For beginners one can recommend some monographs [1, 16–19] and reviews [7, 8] included in the old and recent historical survey, where the foundations of this "hot spot" are ex-

Raoul R. Nigmatullin, Kazan National Research Technical University (KNRTU-KAI), Radioelectronic and Information-Measurements, Technics Department, Karl Marx str. 10, 420011, Kazan, Tatarstan, Russian Federation, e-mail: renigmat@gmail.com
Dumitru Baleanu, Department of Mathematics, Cankaya University, 06530 Balgat, Ankara, Turkey; and Institute of Space Sciences, Magurele–Bucharest, Romania, e-mail: dumitru@cankya.edu.tr

https://doi.org/10.1515/9783110571707-008

plained. The interest to find accurate relationships between fractals and fractional calculus is renewed again. Some original approaches (but without proper physical interpretation) were outlined in papers [4, 5, 20]. One of the basic problems affecting the fractional calculus community, during the era of such approaches, was that it did *not* yet accurately find the justified and exact relationships between the smoothed functions of fractal objects and fractional operators. This problem was solved partly for the time-dependent functions averaged over Cantor sets in monograph [6] and paper [15], where the influence of unknown log-periodic function (leading finally to the understanding of the meaning of the fractional integral with the complex-conjugated power-law exponents) was taken into account. Some important and original results from [15] will be reproduced below. Possible generalizations helping to understand the role of a spatial fractional integral as a mathematical operator replacing the operation of averaging of the smoothed functions over fractal objects were considered in monograph [6] as well. However, in order to receive as a generalization the desired expressions for the gradient, divergence and curl expressed by means of the fractional operator in the limits of mesoscale (when the current scale η lies in the interval ($\lambda < \eta < \Lambda$) determining the limits of a possible self-similarity), it was necessary to apply the additional averaging procedure over possible places of location of the fractal object considered. This procedure provides the correct convergence of the microscopic function $f(z)$ at small ($\eta \simeq \lambda$) and large ($\eta \simeq \Lambda$) scales. Nevertheless, the basic reason that serves as a specific mathematical *obstacle* in accurate establishing of the desired relationship between the fractal object and the corresponding fractional integral is the *absence* of the 2D- and 3D-Laplace transformations. Therefore, the *basic* problem that is considered in this paper can be formulated as:

What accurate form of the fractional operator is generated in the results of the averaging procedure of a smoothed function over the given fractal set if we want to realize this procedure without *any* approximations?

We should stress also the results obtained in the recent paper [2], where new attempt to relate a fractal object (branching flow stream of a liquid passes through porous medium) with the fractional integral is presented. However, in this model the definition of the averaged effective velocity does not allow receiving typical log-periodic oscillations that naturally appears in any fractal object with discrete structure.

In this paper, we want to discuss some new ideas that can help to understand deeper the desired relationship between the accurate averaging of the smoothed functions over fractal sets and fractional integrals in time and space. We consider these new results as a natural generalization of the previous and approximate results achieved in [6, 15].

2 New exact relationships connecting 1D fractals and fractional integrals

2.1 The exact relationship between temporal fractional integral and the smoothed function averaged over the Cantor set with M bars

In paper [6], the relationship between fractional integral with complex power-law exponent and Cantor set has been established. But this relationship was *approximate* and obtained in one-mode approximation, and it would be desirable to establish the *exact* relationship between 1D fractals and fractional integrals in time-domain. Attentive analysis shows that one important point in the previous results leading to the desired exact relationship was *missed*. In order to show it, let us reproduce some mathematical expressions that will be helpful for further manipulations and understanding the problem posed. As it has been shown in [6], the Laplace image of the kernel of the Cantor set is described by the expression

$$\lim_{N\to\infty} K^{(N)}(z) \equiv K_\nu(z) = \frac{\pi_\nu(\ln(z))}{z^\nu}, \tag{1}$$

where $z = pT(1-\xi)$, p defines the Laplace parameter, T is a period of location of the Cantor set, and ξ is some scaling factor. The kernel $K^{(N)}(z)$ can be presented as

$$K^{(N)}(z) = \prod_{n=-(N-1)}^{N-1} g(z\xi^n) = \prod_{n=0}^{N-1} g(z\xi^n) \prod_{n=1}^{N-1} g(z\xi^{-n}). \tag{2}$$

Here, $g(z)$ describes the structure of the given fractal with asymptotics given below. In particular, the Laplace image of the function $g(z)$ for the Cantor set having M bars has the form

$$g(z) = \frac{1}{M}\frac{1-\exp(-\frac{zM}{M-1})}{1-\exp(-\frac{z}{M-1})}. \tag{3}$$

In order to satisfy to the functional equation of the type (3) in the limit ($N \gg 1$) for the kernel

$$K(\xi z) = \frac{1}{g} K(z), \tag{4}$$

the function $g(z)$ should have the following decompositions for small and large values of z:

- for $\mathrm{Re}(z) \ll 1$,

$$g(z) = 1 + c_1 z + c_2 z^2 + \cdots; \tag{5}$$

– for $\mathrm{Re}(z) \gg 1$,

$$g(z) = \bar{g} + \frac{A_1}{z} + \frac{A_2}{z^2} + \cdots. \tag{6}$$

It is easy to note that for the function $g(z)$ from (3) having M bars in each self-similar stage, these requirements are satisfied ($\bar{g} = 1/M$).

The solution of the functional Equation (4) has the form (1) with power-law exponent ν equaled [4, 20]

$$\nu = \frac{\ln(\bar{g})}{\ln(\xi)} = \frac{\ln(1/M)}{\ln(\xi)}, \quad 0 < \nu < 1. \tag{7}$$

The log-periodic function $\pi_\nu(\ln z \pm \ln \xi) = \pi_\nu(\ln z)$ with period $\ln(\xi)$ that figures in (1) can be decomposed to the infinite Fourier series [6, 15]

$$\pi_\nu\left(\frac{\ln(z)}{\ln(\xi)}\right) = \sum_{n=-\infty}^{\infty} C_n \exp\left(2\pi ni\frac{\ln(z)}{\ln(\xi)}\right) \equiv C_0 + \sum_{n=1}^{\infty}(C_n z^{i\Omega_n} + C_n^* z^{-\Omega_n}). \tag{8}$$

Here $\Omega_n = 2\pi n/\ln\xi$ is a set of frequencies providing a periodicity with $\ln\xi$ of product (4). Taking into account this decomposition, one can present the limiting solution of (1) in the form

$$K_\nu(z) = C_0 z^{-\nu} + \sum_{n=1}^{\infty}(C_n z^{-\nu+i\Omega_n} + C_n^* z^{-\nu-i\Omega_n}). \tag{9}$$

Here the real exponent ν is defined by expression (7). The presentation of kernel (2) in the form (9) allows reproducing the previous cases (when the sum in (9) is negligible and/or it can be presented in one-mode approximation [6]), and finding the desired *exact* relationship. Indeed, taking into account the well-known relationship [2]

$$(p)^{-a} =: \frac{1}{\Gamma(a)}\int_0^t \tau^{a-1}\exp(-p\tau)d\tau, \quad \mathrm{Re}(a) \geq 0, \tag{10}$$

it is easy to find the desired original for the kernel (9)

$$K_\nu(t) = C_0\frac{t^{\nu-1}}{\Gamma(\nu)} + \sum_{n=1}^{\infty}\left(C_n\frac{t^{\nu-1+i\Omega_n}}{\Gamma(\nu+i\Omega_n)} + C_n^*\frac{t^{\nu-1-i\Omega_n}}{\Gamma(\nu-i\Omega_n)}\right). \tag{11}$$

Here we should take into account that the temporal variable t should be dimensionless. Based on the determination of the variable $z = pT(1-\xi)$ (see expression (1)), it is easy to restore the dimension of the constant $t \to t/T(1-\xi)$. Based on the definition of the fractional integral in the form of the Riemann–Liouville-type (RL-type), one can write the desired relationship as

$$\int_0^t K(t-\tau)\cdot f(\tau)d\tau = C_0 J^\nu f(t) + \sum_{n=1}^{\infty}[C_n J^{\nu+i\Omega_n} f(t) + C_n^* J^{\nu-i\Omega_n} f(t)]. \tag{12}$$

Here we define the RL-integral of the complex order as

$$J_0^{\nu\pm i\Omega}f(t) = \frac{1}{\Gamma(\nu \pm i\Omega)} \int_0^t (t-\tau)^{\nu-1\pm i\Omega} f(\tau)d\tau. \tag{13}$$

From the exact relationship (12), it follows that the averaging procedure of a smooth function over the generalized Cantor set (having M bars) is accompanied *always* by an infinite set of fractional integrals having complex-conjugated power-law exponents. Only in the partial case, when the contribution of log-periodic function becomes negligible, relationship (13) restores the desired RL fractional integral with real power-law exponent [6, 15]. When the total sum in (11) can be replaced approximately by one term

$$\begin{aligned} &\sum_{n=1}^{\infty}\left(C_n \frac{t^{\nu-1+i\Omega_n}}{\Gamma(\nu+i\Omega_n)} + (C_n)^* \frac{t^{\nu-1-i\Omega_n}}{\Gamma(\nu-i\Omega_n)} \right) \\ &\quad \cong C \frac{t^{\nu-1+i\langle\Omega\rangle}}{\Gamma(\nu+i\langle\Omega\rangle)} + C^* \frac{t^{\nu-1-i\langle\Omega\rangle}}{\Gamma(\nu-i\langle\Omega\rangle)}, \end{aligned} \tag{14}$$

we restore the basic result of paper [15] in the so-called one-mode approximation. In relationship (14), $\langle\Omega\rangle$ defines the leading mode, which replaces approximately other modes that figure in the left-hand side of expression (14).

2.2 Numerical verification

For verification purposes, one can take time-domain expression for the kernel defined by (11), and the corresponding one-mode approximation expression from (14).

Numerical calculations were realized with the help of the following procedure:

1. Calculation of the right side of expression (14) using one-mode approximation constants for some given M (number of the Cantor columns) and the scaling parameter ξ from [15] (the first 3 rows of Table 1).
2. Calculation of the fitting constants C_n ($n = 1, \ldots, N_m$), where N_m defines the finite number of modes in the corresponding sum (11) by the linear least-square method (LLSM)).

The values of fitting parameters for the given values of M and ξ are collected in last 3 rows of Table 1. One can see that the fitting parameters C_0, A_1, and Ω_1 are very close to the corresponding initial ones obtained in one-mode approximation.

Results of further numerical calculations are shown in Figures 1 and 2. There one can see that contribution of each next term in the sum (11) decreases drastically in comparison with the previous one. This observation serves as a good proof of the validity of one-mode approximation approach proposed in [6].

Analyzing these results obtained in this section, we found that higher terms $n = 1, 2, \ldots, N_m$ give relatively insignificant contribution in the total result and the value

Table 1: The basic initial (the first 3 rows) and the fitting parameters (rest rows) obtained in the result of numerical verification of expression (11).

M	ξ	v	C_0	A	$\langle\Omega\rangle$	$\langle n\rangle$
2	0.125	0.3333	0.63	0.0082	3.01161	0.9967
5	0.05	0.5372	0.6117	0.0217	2.09144	0.9972
15	0.0167	0.6614	0.606	0.0353	1.5331	0.9991
M	ξ	v	C_0	C_1	Ω_1	N_m
2	0.125	0.3333	0.62328	0.00805	3.02157	20
5	0.05	0.5372	0.61139	0.02157	2.09738	20
15	0.0167	0.6614	0.60640	0.03523	1.5346	20

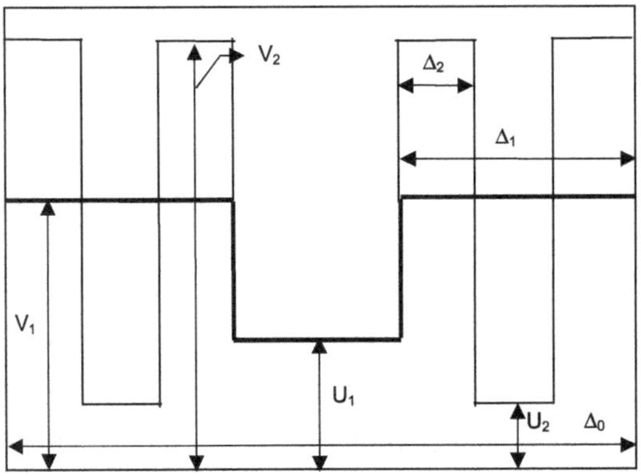

Figure 1: Here two successive stages of the modified binary set are shown. The set of velocities outside of the Cantor set are random. The widths of Cantor bars are denoted as $\{\Delta_i\ (i = 0, 1, 2)\}$, the set of velocities $\{V_i\}$ defines the movement inside the set.

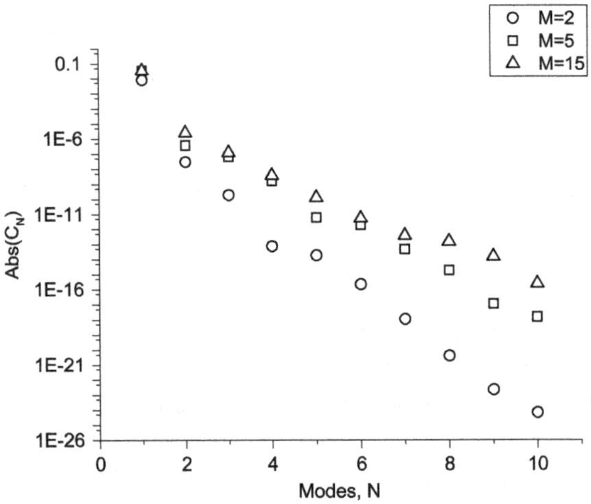

Figure 2: The fitted C_n coefficients presented in log-scale (y-axis) correspond to the successive terms of the sum in (11) (x-axis) for $M = 2$ (open circles), $M = 5$ (open squares) and $M = 15$ (open triangles).

of the fitting error has a plateau with increasing of N_m. One can notice that fitting error mostly depends on $\langle n\rangle$ parameter, characterizing the contribution of the main harmonics.

2.3 Physical interpretation of this result and possible generalizations

To understand deeper the results from a physical point of view, it makes sense to start from the simplest physical examples, which can clarify the procedure considered above. For this purpose, we consider at first the following mechanical problem: how to calculate the total and averaged path $\langle L(t)\rangle$ for some interval of the time t for a set of material points (particles) if every particle is moving in one direction with the constant velocity V coinciding with the points of Cantor set, and is idle outside the set? The expression for the path $L_N(t)$ on the Nth stage of Cantor set construction has the form

$$L_N(t) = V\int_0^t K_{T,\nu}^{(N)}(\tau)d\tau. \tag{15}$$

Here the value $K_{T,\nu}^{(N)}$ determines the Cantor set located on the temporal interval T on the N-th stage of its self-similarity, having M bars and dimension ν defined by expression (7).

The value of the normalization interval T for the given case can be found from the condition that during the interval T, the body passes the distance L^*. Hence, $T = L^*/V$. Then, integrating expression (11), we obtain

$$\begin{aligned} L_\nu(t) &= \frac{L^*}{\Gamma(1+\nu)}\left(\frac{t}{T}\right)^\nu \\ &\times\left[1+\sum_{n=1}^{\infty}\left(c_n\frac{\Gamma(1+\nu)}{\Gamma(\nu+1+i\Omega_n)}\left(\frac{t}{T}\right)^{i\Omega_n}+c_n^*\frac{\Gamma(1+\nu)}{\Gamma(\nu+1-i\Omega_n)}\left(\frac{t}{T}\right)^{-i\Omega_n}\right)\right]. \end{aligned} \tag{16}$$

The sum entering in (16) determines the log-periodic corrections appearing in the result of discretization of the Cantor set. In Figure 3 we demonstrate the distribution of the relative errors that fit the log-periodic corrections in expression (16). Then, realizing the averaging procedure for the total assembly of particles described in a book [6], we obtain

$$\langle L(t)\rangle = L^*\frac{[B(\nu)]}{\Gamma(1+\nu)}\left(\frac{Vt}{L^*}\right)^\nu. \tag{17}$$

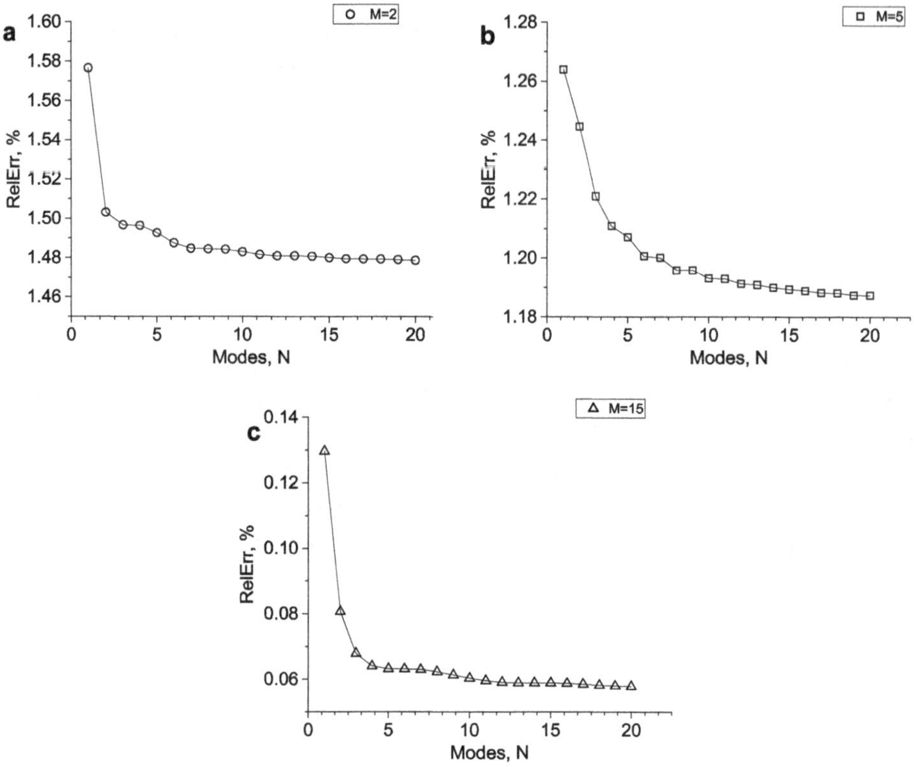

Figure 3: Relative fitting error in % (y-axis) versus number of modes (x-axis) for (a) $M = 2$, (b) $M = 5$ and (c) $M = 15$.

Here $[B(\nu)]$ determines the averaged value of the sum located in the square brackets (16). The binary Cantor set can be evaluated analytically, which gives

$$B(\nu) = \langle \pi_{T,\nu}(z) \rangle = \int_{-1/2}^{1/2} \pi_{T,\nu}(z + x \ln \xi) dx = \frac{2^{-(1+\nu/2)}}{\ln 2}. \tag{18}$$

Actually, formula (17) expresses the dependence of the averaged path as the function of time for a set of particles if every particle is moving only inside the Cantor set with the fractal dimension ν, and we are interested only in the averaged path of the total assembly. The fractal dimension ν shows what part of states of the assembly are involved in this movement. We also want to note that exact intervals, where one type of movement is transformed into the state of idleness, are averaged and after the performing of the averaging procedure we have the information only about a part of all the states involved in the distribution over Cantor sets. The averaged procedure of a smoothed function over Cantor set described above represents a specific Cantor's "filter", which enables to filtrate one type of movement and delete (because of the normalization of the states) another one. The averaged log-periodic states are described by a constant (18). It represents the result of this fractal “filtration” procedure. To stress the filtration properties of the Cantor set, we consider more complicated examples. Let us assume that in the intervals between Cantor stripes an assembly of bodies is moved with various and random velocities defined as $\{U_i\}$. The index $i = 1, 2, \dots, N, \dots$ refers to the stages of Cantor construction. Two stages of the modified Cantor set are shown in Figure 1.

It is interesting to set up the following question: *Is there a condition for the distribution of velocities $\{U_i\}$ when the influence of a movement outside the set becomes negligible?*

In this case the recurrence relationship for the density is expressed by the following formula:

$$\begin{aligned} K_{\Delta_N}^{(N)}(t) &= V_N\big[K_{\Delta_N}^{(N-1)}(t) + K_{\Delta_N}^{(N-1)}(t - (\Delta_{N-1} - \Delta_N))\big] \\ &\quad + U_N\big[K_{\Delta_N}^{(N-1)}(t - (\Delta_{N-1} - \Delta_N)) - K_{\Delta_N}^{(N-1)}(t - \Delta_N)\big]. \end{aligned} \tag{19}$$

For Laplace-image with the help of retardation theorem

$$F(t - a) =: \exp(-pa)F(p), \tag{20}$$

one can obtain the following expression:

$$\begin{aligned} K_{\Delta_N}^{(N)}(p) &= V_N \frac{1 - \exp(-p\Delta_N)}{p} \prod_{n=0}^{N-1} (1 + \exp(-p(\Delta_n - \Delta_{n+1}))) \\ &\quad + \sum_{k=0}^{N-1} U_{k+1} \frac{\exp(-p\Delta_{k+1}) - \exp(-p(\Delta_k - \Delta_{k+1}))}{p} \\ &\quad \times \prod_{n=0}^{k-1} (1 + \exp(-p(\Delta_n - \Delta_{n+1}))). \end{aligned} \tag{21}$$

The total area on the Nth stage is given by the following expression:

$$2^N V_N \Delta_N + \sum_{k=0}^{N-1} 2^k U_{k+1}(\Delta_k - 2\Delta_{k+1}) = S_N. \tag{22}$$

We require that the total area on every stage remains the *same* (it is equivalent to the condition of conservation of the total number of states):

$$S_N = S_{N-1} = \cdots = S_0. \tag{23}$$

Let us assume that $U_i = \bar{U} + \delta U_i$, where

$$\bar{U} = \frac{1}{N} \sum_{i=1}^{N} U_i, \tag{24}$$

is the arithmetic mean of random velocity. The set δU_i $(i = 1, 2, \ldots, N, \ldots)$ defines the random deviations of U_i. We assume that these deviations $\{\delta U_i\}$ satisfy the condition

$$\sum_{k=0}^{N-1} 2^k \delta U_{k+1}(\Delta_k - 2\Delta_{k+1}) = 0. \tag{25}$$

Based on this condition, we can rewrite (21) as

$$\begin{aligned} K_{\Delta_N}^{(N)}(p) = {} & \frac{(S_0 - \bar{U}T)}{(2\xi)^N T}\left[\frac{1-\exp(-pT\xi^N)}{p}\right]\prod_{n=0}^{N-1}(1+\exp(-pT(1-\xi)\xi^n)) \\ & + \bar{U}\frac{1-\exp(-pT)}{p}. \end{aligned} \tag{26}$$

If we propagate the binary Cantor set on the whole temporal interval (“in” and “out” of the given T), then we obtain

$$K_{\Delta_N}^{(N\gg1)}(z) \cong \frac{S_0 - \bar{U}T}{T}\prod_{n=-(N-1)}^{N-1}\left[\frac{1+\exp(-z\xi^n)}{2}\right] + \bar{U}\cdot T\frac{1-\exp(-pT)}{pT}. \tag{27}$$

The first part of (27) satisfies the scaling Equation (4) with $\bar{g} = 2$, and the second part is proportional to t. If we, again, apply the averaging procedure, then we obtain

$$\langle L(t)\rangle = L^*\left(1 - \frac{\bar{U}}{V}\right)\frac{B(\nu)}{\Gamma(1+\nu)}\left(\frac{Vt}{L^*}\right)^{\nu} + \bar{U}t. \tag{28}$$

The last expression confirms again the filtration properties of the Cantor set. It divides all motion on two parts: (a) the first motion of the particles located inside the Cantor set, (b) the second part describes the motion that takes place *outside* the fractal set.

2.4 New relationships connecting a fractal process in time with the smoothed function by means of the fractional integral

It is natural to consider another self-similar fractal process, which can be presented in the form of an additive summation [6, 15]. These sums appeared naturally when the self-similar RLC elements connected in parallel or in-series are considered [6, 15]:

$$S(z) = s_0 \sum_{n=-N}^{N} b^n f(z\xi^n). \tag{29}$$

Here z is a dimensionless Laplace parameter, b and ξ are some constant scaling factors. Each term in (29) can be associated, for example, with the scaled resistor ($R_n = R_0 b^n$), capacitance ($C_n = R/z\xi^n$, $z = j\omega RC$), or inductance ($L_n = Rz\xi^n$, $z = j\omega L_0/R$), which can form a self-similar fractal circuit [6], or with additive contribution of a bar belonging to some Cantor set. In this case, an additive contribution of each bar is expressed by the function

$$f(z\xi^n) = \frac{1 - \exp(-z\xi^n)}{z\xi^n}. \tag{30}$$

It is easy to note that sum (29) satisfies the following equation:

$$\begin{aligned} S(z\xi) &= s_0 \sum_{n=-N}^{N} b^n f(z\xi^{n+1}) = s_0 \sum_{n=-N+1}^{N+1} b^{n-1} f(z\xi^n) \\ &= \frac{1}{b} S(z) + b^N f(z\xi^{N+1}) - b^{-N-1} f(z\xi^{-N}). \end{aligned} \tag{31}$$

We *suppose* that the contributions of the last two terms on the ends of the finite interval are negligible:

$$b^N f(z\xi^{N+1}) \overset{N\gg 1}{\cong} 0, \quad b^{-N-1} f(z\xi^{-N}) \overset{N\gg 1}{\cong} 0. \tag{32}$$

Below, we will discuss these suppositions in detail. Equation (31) at conditions (32) satisfies the following functional equation that formally coincides with Equation (4) considered above:

$$S(z\xi) = \frac{1}{b} S(z). \tag{33}$$

So, based on the results obtained in the previous section, one can write

$$S(z) = z^{-\nu} \operatorname{Pr}(\ln z), \quad \nu = \ln(b)/\ln(\xi), \quad \operatorname{Pr}(\ln z \pm \ln \xi) = \operatorname{Pr}(\ln z). \tag{34}$$

In spite of the formal coincidence of Equation (33) with (4), in order to satisfy conditions (32), the function $f(z)$ in (31) should have *another* asymptotic decomposition:

- for $\operatorname{Re}(z) \ll 1$,

$$f(z) = c_1 z + c_2 z^2 + \cdots, \tag{35}$$

- for $\operatorname{Re}(z) \gg 1$,

$$f(z) = \frac{A_1}{z} + \frac{A_2}{z^2} + \cdots, \tag{36}$$

  and the condition

$$1 < b < \xi. \tag{37}$$

As it follows from (34), the last condition provides in addition the obvious inequality

$$0 < \nu = \frac{\ln(b)}{\ln \xi} < 1. \tag{38}$$

So, the results obtained for the sum (31) coincides with the results obtained earlier for the product (2), and the exact relationship in time-domain (11) with subsequent convolution (12) of the kernel (11) with a smooth function remained *invariant*. Is it possible to consider a more complicated self-similar process than is presented by an additive combination of two sums of the same type

$$S(z) = s_1 \sum_{n=-N}^{N} b_1^n f_1(z\xi^n) + s_2 \sum_{n=-N}^{N} b_2^n f_2(z\xi^n) \equiv S_1(z) + S_2(z)? \tag{39}$$

If we assume, by analogy with requirement (18), that the contribution of each sum on the end of the intervals are negligible

$$b_{1,2}^N f_{1,2}(z\xi^{N+1}) \overset{N\gg 1}{\cong} 0, \quad b_{1,2}^{-N-1} f_{1,2}(z\xi^{-N}) \overset{N\gg 1}{\cong} 0 \tag{40}$$

(here the number of elements N and the scaling factor ξ for both sums are kept the same values), then from (39) we have the system of equations

$$\begin{aligned} S(z\xi) &= \frac{1}{b_1} S_1(z) + \frac{1}{b_2} S_2(z), \\ S(z\xi^2) &= \frac{1}{b_1^2} S_1(z) + \frac{1}{b_2^2} S_2(z). \end{aligned} \tag{41}$$

Excluding from (39) and the first row of (41) the unknown sums $S_{1,2}(z)$ (defined above by expressions (39)) and inserting them into the second row of (41), we obtain the generalized scaling equation of the second order

$$S(z\xi^2) = \left(\frac{1}{b_1} + \frac{1}{b_2}\right) S(z\xi) - \frac{1}{b_1 b_2} S(z). \tag{42}$$

Solutions of the scaling equations of the second order was considered in paper [14], and, therefore, we can write

$$S(z) = z^{\nu_1} \Pr_1(\ln z) + z^{\nu_2} \Pr_2(\ln z), \tag{43}$$

where the power-law exponents are found from the equation

$$\xi^{2\nu} - \left(\frac{1}{b_1} + \frac{1}{b_2}\right)\xi^{\nu} + \frac{1}{b_1 b_2} = 0, \quad \nu_{1,2} = \frac{\ln(1/b_{1,2})}{\ln \xi}. \tag{44}$$

In solution (33), we have now two log-periodic functions $\Pr_{1,2}(\ln z \pm \ln \xi) = \Pr_{1,2}(\ln z)$ that can be decomposed to the Fourier series relative to the variable $\ln(z)$. Therefore, one can present the solution (33) in the form

$$\begin{aligned} S(z) &= C_0^{(1)} z^{-\nu_1} + \sum_{n=1}^{\infty} [C_n^{(1)} z^{-\nu_1 + i\Omega_n} + (C_n^{(1)})^* z^{-\nu_1 - i\Omega_n}] \\ &+ C_0^{(2)} z^{-\nu_2} + \sum_{n=1}^{\infty} [C_n^{(2)} z^{-\nu_2 + i\Omega_n} + (C_n^{(2)})^* z^{-\nu_2 - i\Omega_n}], \end{aligned} \tag{45}$$

where we introduced new definitions of the power-law exponents and discrete frequencies reflecting the discrete structure of the fractal process considered.

Moreover, these results can be further generalized. Let us suppose that we have an additive combination of three self-similar processes

$$\begin{aligned} S(z) &= s_1 \sum_{n=-N}^{N} b_1^n f_1(z\xi^n) + s_2 \sum_{n=-N}^{N} b_2^n f_2(z\xi^n) + s_3 \sum_{n=-N}^{N} b_3^n f_3(z\xi^n) \\ &\equiv S_1(z) + S_2(z) + S_3(z). \end{aligned} \tag{46}$$

The functions $f_k(z)$ $(k = 1, 2, 3)$ are chosen in such a way that their contributions on the ends of the interval remain negligible:

$$b_k^N f_k(z\xi^{N+1}) \overset{N \gg 1}{\cong} 0, \quad b_k^{-N-1} f_k(z\xi^{-N}) \overset{N \gg 1}{\cong} 0, \quad k = 1, 2, 3. \tag{47}$$

We *suppose* again that for each function $f_k(z)$ $(k = 1, 2, 3)$, the asymptotic behavior similar to expressions (35) and (37) is conserved. Therefore, suppositions (47) are satisfied if the constants b_k and ξ follow the requirements

$$1 < b_k < \xi, \quad k = 1, 2, 3. \tag{48}$$

Therefore, we obtain the following system of equations for finding unknown $S_k(z)$:

$$\begin{aligned} S(z) &= S_1(z) + S_2(z) + S_3(z), \\ S(z\xi) &= \frac{1}{b_1} S_1(z) + \frac{1}{b_2} S_2(z) + \frac{1}{b_3} S_3(z), \\ S(z\xi^2) &= \frac{1}{b_1^2} S_1(z) + \frac{1}{b_2^2} S_2(z) + \frac{1}{b_3^2} S_3(z). \end{aligned} \tag{49}$$

Excluding from system (49) the unknown sums $S_k(z)$ and inserting them into the relationship

$$S(z\xi^3) = \frac{1}{b_1^3} S_1(z) + \frac{1}{b_2^3} S_2(z) + \frac{1}{b_3^3} S_3(z), \tag{50}$$

after simple algebraic manipulations, we obtain the following functional equation with respect to the function $S(z)$:

$$\begin{aligned} &S(z\xi^3) = a_2 S(z\xi^2) + a_1 S(z\xi) + a_0 S(z), \\ &a_2 = \frac{1}{b_1} + \frac{1}{b_2} + \frac{1}{b_3}, \quad a_1 = -\left(\frac{1}{b_1 b_2} + \frac{1}{b_1 b_3} + \frac{1}{b_3 b_2}\right), \quad a_0 = \frac{1}{b_1 b_2 b_3}. \end{aligned} \tag{51}$$

The solution of the functional Equation (51) can be written as

$$S(z) = z^{\nu_1} \underset{1}{\Pr}(\ln z) + z^{\nu_2} \underset{2}{\Pr}(\ln z) + z^{\nu_3} \underset{3}{\Pr}(\ln z). \tag{52}$$

The power-law exponents entering into (52) are found from the cubic equation

$$\xi^{3\nu} - a_2 \xi^{2\nu} - a_1 \xi^{\nu} - a_0 = 0 \tag{53}$$

and can expressed as

$$\nu_k = \frac{\ln(1/b_k)}{\ln \xi} = -\frac{\ln(b_k)}{\ln \xi}, \quad 0 < |\nu_k| < 1, \quad k = 1, 2, 3. \tag{54}$$

Therefore, expression (45) conserves its form, and for this case, solution (52) can be rewritten as

$$S(z) = \sum_{k=1}^{3} C_0^{(k)} z^{-\nu_k} + \sum_{k=1}^{3} \sum_{n=1}^{\infty} \left[C_n^{(k)} z^{-|\nu_k| + i\Omega_n} + (C_n^{(k)})^* z^{-|\nu_k| - i\Omega_n}\right]. \tag{55}$$

It is quite obvious that this result admits the further generalization for the case of $k = 1, 2, \ldots, K$ sums:

$$S(z) = \sum_{k=1}^{K} s_k \sum_{n=-N}^{N} b_k^n f_k(z\xi^n) \equiv \sum_{k=1}^{K} S_k(z). \tag{56}$$

Using the mathematical induction method, one can write the final result for this case:

$$\begin{aligned} S(z) &= \sum_{k=1}^{K} z^{\nu_k} \underset{k}{\Pr}(\ln z) \\ &= \sum_{k=1}^{K} C_0^{(k)} z^{-\nu_k} + \sum_{k=1}^{K} \sum_{n=1}^{\infty} \left[C_n^{(k)} z^{-|\nu_k| + i\Omega_n} + (C_n^{(k)})^* z^{-|\nu_k| - i\Omega_n}\right]. \end{aligned} \tag{57}$$

The desired roots are found from the relationships

$$|\nu_k| = \frac{\ln(b_k)}{\ln \xi}, \quad 1 < b_k < \xi, \quad 0 < |\nu_k| < 1. \tag{58}$$

So, the kernel connecting the given fractal process with a smooth function in time-domain accept the following general form:

$$K_\nu(t) = \sum_{k=1}^{K} C_0^{(k)} \frac{t^{\nu_k - 1}}{\Gamma(\nu_k)} + \sum_{k=1}^{K} \sum_{n=1}^{\infty} \left(C_n^{(k)} \frac{t^{\nu_k - 1 + i\Omega_n}}{\Gamma(\nu_k + i\Omega_n)} + (C_n^{(k)})^* \frac{t^{\nu_k - 1 - i\Omega_n}}{\Gamma(\nu_k - i\Omega_n)} \right). \tag{59}$$

To the best of our knowledge, this is the most general result that can be obtained for a set of additive self-similar processes in time. As one may note from the calculations, all dynamical functions describe a self-similar process with the *same* scale ξ. If the scales are different and the discrete boundaries of the processes are different, also $N_k \neq N$, then these results become questionable and require the further research. But nevertheless, one can show the situation taking place for the different scales $\xi_i \neq \xi$ when the result can be reduced to the case considered above. Let us suppose that we have a product combining a random combination of different scales

$$P_n = \prod_{s=1}^{n} \xi_s, \tag{60}$$

but the deviations from some averaged scale (considered as the *dominant* scale) are small. For this case, we have

$$P_n = \prod_{s=1}^{n} \xi_s = \prod_{s=1}^{n} (\langle\xi\rangle + \delta_s) = \langle\xi\rangle^n \prod_{s=1}^{n}\left(1 + \frac{\delta_s}{\langle\xi\rangle}\right) = \langle\xi\rangle^n \exp\left(\sum_{s=1}^{n} \ln\left(1 + \frac{\delta_s}{\langle\xi\rangle}\right)\right). \tag{61}$$

For the condition $\delta_s/\langle\xi\rangle \ll 1$, we have approximately

$$\begin{gathered} P_n \cong \left(\langle\xi\rangle \exp\left[\frac{\langle\delta\rangle}{\langle\xi\rangle} - \frac{\langle\delta^2\rangle}{2\langle\xi\rangle^2} + \frac{\langle\delta^3\rangle}{3\langle\xi\rangle^3}\right]\right)^n, \\ \langle\delta^p\rangle = \frac{1}{n}\sum_{s=1}^{n} \delta_s^p, \quad p = 1, 2, \ldots \end{gathered} \tag{62}$$

Therefore, this result allows considering different scales if they are *not* strongly deviated from each other. All the previous results are valid if we simply replace

$$\xi \to \langle\xi\rangle \exp\left(\frac{\langle\delta\rangle}{\langle\xi\rangle}\right) \exp\left(-\frac{\langle\delta^2\rangle}{2\langle\xi\rangle^2}\right) \cdots \quad \text{with } \langle\xi\rangle = \frac{1}{n}\sum_{s=1}^{n} \xi_s. \tag{63}$$

3 Why these results cannot be simply generalized for 2D and 3D fractals?

In this section we want to show the basic difficulties that are met in attempt to propagate these results for a wide class of fractals.

The basic reason that serves as a specific mathematical obstacle in accurately establishing the desired relationship between the fractal object and the corresponding fractional integral is the *absence* of the 2D- and 3D-Laplace transformations. Application of the Fourier transform, getting rid of the absence of the (2, 3)D-Laplace transformation, leads mathematically to the fact that the behavior of the trigonometric functions on large scales becomes uncertain and the limit of the function

$$\lim_{N \gg 1} \exp(i(\mathbf{k}.\mathbf{r})\xi^N) = ? \tag{64}$$

at $\xi > 1$ does *not* exist in the conventional sense. To provide the convergence of the limit (63) at the conditions ($N \gg 1$, $\xi > 1$), the additional averaging procedure over possible places of location of the fractal object was applied [20]. So, finding the correct mathematical procedure that helps to overcome the ambiguity of expression (64) is the *actual* problem in establishing the desired "bridge" between fractal geometry and fractional calculus.

In this section, we want to demonstrate another approach that can link the procedure of the averaging of a smooth function for a certain class of the given self-similar objects with the fractional integral, and its possible generalizations in space. This alternative approach helps to solve this problem in a more accurate form, and demonstrates new possibilities that can exist between fractals and fractional integrals in space.

3.1 Mathematical part. Properties of the 1D self-similar product

As long well known [6, 15, 16], for many fractal objects located in a space, the most convenient procedure for their description is based on the following property of the Fourier transform:

$$f(r) = \int_{-\infty}^{\infty} F(k)\exp(i\mathbf{k}.\mathbf{r})\,d^3k =: F(k),$$
$$f(r+a) =: \exp(ika)F(k). \tag{65}$$

This property helps to segregate the Fourier image of the averaged smooth function (presented by the function $F(\mathbf{k})$ in 3D-Fourier space) from the structure-factor, which shows the location of the fractal object in space. The current generation of a fractal object can be expressed with the help of a "star" of the **k**-vector $(\mathbf{k}_1, \ldots \mathbf{k}_r)$, where r determines a finite number of new self-similar objects that are created on the current stage of the desired fractal object, and the vector $\mathbf{a} = (\mathbf{a}_1, \mathbf{a}_2, \ldots, \mathbf{a}_r)$ shows the directions of the location of the current generation of the given fractal in space. Based on the property, it is easy to see that from the mathematical point of view, it becomes necessary to consider the product (presenting itself a general definition of the structure-factor) of the following form:

$$P(z_1, z_2 \ldots, z_r) = \prod_{n=-N}^{N} f(z_1, \xi_1^n, \ldots, z_r, \xi_r^n),$$
$$z_q = k\Lambda\cos(\theta_q), \quad q = 1, 2, \ldots, r. \tag{66}$$

In expression (66), the value k defines the modulus of the wave-vector, Λ is the value of the vector referring to the initial (the largest for the case $\xi < 1$ and vice versa) fractal object, θ_q is a set of angles between vectors $\mathbf{k}_q$ and $\boldsymbol{\Lambda}$, respectively. The value ξ defines the scaling parameter. For self-affine fractals having different symmetry, it cannot be the same.

So, the basic problem from the mathematical point of view can be formulated as follows: to consider the properties of the product (66) and relate these properties with some types of the fractional integrals that are considered (for example, in the book [1]) and accepted in the fractional calculus as basic definitions.

As it follows from (66), the consideration of a simple self-similar 1D-object in space is closely related to consideration of the mathematical properties of the product

$$P(z) = \prod_{n=-N_0}^{N} b^n f(z\xi^n). \tag{67}$$

We slightly generalized expression (2), considered in the previous chapter, putting in it the different limits. This product is closely related to the structure-factor for simple fractals considered in [6] that indicates the location of the fractal studied in 3D space. Here b and ξ determine the scaling parameters, and the variable z can take real or complex values (for example, it can coincide with the dimensionless Laplace variable ($z = i\omega + s$) or with the Fourier parameter ($i(\mathbf{ka})$) in space). Here we consider a more general case where the lower limit N_0 does not coincide with the upper limit N. For distinctness, we put ($N_0 < N$). Making a substitution $z \to z\xi$ into (3), we obtain the following *identity*:

$$P(z\xi) = \prod_{n=-N_0}^{N} b^n f(z\xi^{n+1}) = \frac{1}{b}\prod_{n=-N_0+1}^{N+1} b^n f(z\xi^n) = b^{N-|N_0|}\frac{f(z\xi^{N+1})}{f(z\xi^{-N_0})}P(z). \tag{68}$$

If the microscopic function $f(z)$ describing the dynamic process or geometrical location of an elementary fractal on mesoscale is *finite* for large and small values of variable z (as it was supposed in [6, 15] and other papers [10, 11]), then at $N = N_0$ and in the limit $N \gg 1$, one can obtain (for distinctness, we put $\xi > 1$) the following scaling equation:

$$\begin{gathered} P(z\xi) = \frac{A}{c_0}P(z), \\ f(z) \overset{|z|\gg 1}{=} A, \quad f(z) \overset{|z|\ll 1}{=} c_0, \end{gathered} \tag{69}$$

with the well-known solution [6, 15]

$$\begin{gathered} P(z) = \mathrm{PR}_\nu(\ln z)\cdot z^\nu, \quad \nu = \frac{\ln(A/c_0)}{\ln\xi}, \\ \mathrm{PR}_\nu(\ln z \pm \ln\xi) = \mathrm{PR}_\nu(\ln z). \end{gathered} \tag{70}$$

Here expression $\mathrm{PR}_\nu(\ln(z))$ defines the unknown log-periodic function. The main question can be formulated as follows: *how to find the solution for product* (67) *satisfying to the functional Equation* (69) *when the asymptotic behavior of the microscopic* *function $f(z)$ is not finite or does not exist?* Mathematically this condition can be expressed as

$$\lim_{N\gg 1} f(z) = \begin{cases} \text{does not exist,} \\ \infty. \end{cases} \tag{71}$$

The first row in (71) can be associated with condition (65). To find the solution of the functional equation (69) at condition (71), we present identity (68) in the form

$$L(z\xi) = L(z) + \Phi(z\xi^{N+1}) - \Phi(z\xi^{-N_0}) + B, \tag{72}$$

where

$$L(z) = \ln[P(z)], \quad B = (N - N_0)\ln(b), \quad \Phi(z) = \ln[f(z)]. \tag{73}$$

In Equations (71)–(72) and below, N_0 is considered to be a *positive* value. At $N = N_0 B = 0$. For distinctness, we put $\xi > 1$. We should note here that the case $\xi < 1$ is reproduced from expressions (71), (72), (73), and expressions below (containing the parameter $\xi > 1$) by simple replacement $\xi \to 1/\xi$. From identity (72) by replacement $z \to z\xi^{q-1}$, we obtain easily

$$\begin{gathered} L(z\xi^q) = L(z\xi^{q-1}) + \Phi(z\xi^{N+q}) - \Phi(z\xi^{-N_0+q-1}) + B, \\ q = 1, 2, \ldots, k-1, k, \ldots. \end{gathered} \tag{74}$$

Taking into account condition (71), we cannot eliminate the large term $\Phi(z\xi^{N+q})$ from (74). Therefore, in general, the infinite set of Equations (74) contains two types of different variables $L(z\xi^q)$, $\Phi(z\xi^{N+q})$ $(q = 1, 2, \ldots, k, \ldots)$ and cannot be reduced to the system containing only one type of variable. To close the infinite chain of Equations (74) relative to the unknown function $L(z)$ (or equally for the unknown microscopic function $\Phi(z)$), we make the reasonable supposition (it will be justified below numerically and analytically)

$$\Phi(z\xi^{N+k}) \cong \sum_{q=1}^{k-1} w_q \Phi(z\xi^{N+q}). \tag{75}$$

Here $\{w_q\}$ $(q = 1, 2, \ldots, k-1)$ determines a set of constants that approximate the function figuring on the left-hand side. These constants can be found numerically with the help of the linear-least square method (LLSM). From the system of Equations (74), we have

$$\begin{gathered} \Phi(z\xi^{N+q}) = L(z\xi^q) - L(z\xi^{q-1}) + Lm_q - B, \\ q = 1, 2, \ldots, k-1, \end{gathered} \tag{76}$$

$$\Phi(z\xi^{-N_0+q-1}) \overset{1 \ll N_0 \leq N}{\cong} Lm_q. \tag{77}$$

Here we took into account the limit (69), describing approximately the behavior of $f(z)$ at small values of the variable z. For $q = k$, we have from (76)

$$L(z\xi^k) = L(z\xi^{k-1}) + \Phi(z\xi^{N+k}) - Lm_k + B,$$
$$\Phi(z\xi^{-N_0+k-1}) \overset{1\ll N_0\le N}{=} Lm_k. \tag{78}$$

Taking into account the approximate decoupling (75) and relationships (76), we obtain

$$L(z\xi^k) - L(z\xi^{k-1}) + Lm_k - B = \Phi(z\xi^{N+k}) \cong \sum_{q=1}^{k-1} w_q \Phi(z\xi^{N+q})$$
$$= \sum_{q=1}^{k-1} w_q[L(z\xi^q) - L(z\xi^{q-1})] + \sum_{q=1}^{k-1} w_q(Lm_q - B). \tag{79}$$

After some simple algebraic transformations of expression (78), we obtain finally the following closed functional equation relative to the remaining variable $L(z)$:

$$L(z\xi^k) = (1 + w_{k-1})L(z\xi^{k-1}) + \sum_{q=1}^{k-2}(w_q - w_{q+1})L(z\xi^q) - w_1 L(z) + R,$$
$$R = \sum_{q=1}^{k-1} w_q(Lm_q - B) - (Lm_k - B). \tag{80}$$

Therefore, the solutions of the functional Equation (80) help to find new expressions for product (67), when condition (71) is satisfied. The solutions of (80) are closely related with the values of the roots of the polynomial

$$P(\lambda) = \lambda^k - (1 + w_{k-1})\lambda^{k-1} - \sum_{q=1}^{k-2}(w_q - w_{q+1})\lambda^q - w_1 = 0. \tag{81}$$

If the roots of (81) are different, then the general solution of the functional Equation (80) can be presented in the form (see Mathematical Appendix)

$$L(z) = \sum_{s=1}^{k} \mathrm{PR}_s(\ln z) z^{\nu_s} + C(R). \tag{82}$$

Here $C(R)$ is a constant that is found by an arbitrary constant variation method, and the set of the power-law exponents ν_s $(s = 1, 2, \ldots, k)$ is defined as

$$\nu_s = \frac{\ln(\lambda_s)}{\ln(\xi)} \quad (\xi > 1). \tag{83}$$

The set of $\mathrm{PR}_s(\ln(z))$ from (74) determines the *unknown* log-periodic functions. As we saw earlier, these functions admit the decomposition into the infinite Fourier series

$$\mathrm{PR}_s(\ln z) = A_0^{(s)} + \left[\sum_{k=1}^{\infty} Ac_k^{(s)} \cos\left(2\pi k \cdot \frac{\ln(z)}{\ln(\xi)}\right) + As_k^{(s)} \sin\left(2\pi k \cdot \frac{\ln(z)}{\ln(\xi)}\right)\right], \tag{84}$$
$$\mathrm{PR}_s(\ln z \pm \ln \xi) = \mathrm{PR}_s(\ln z),$$

with period $\ln(\xi)$. The decomposition coefficients of the series (84) should be found from initial or some a priori conditions (when the parameter ξ is supposed to be known). When some root ν_s figuring in (83) accepts the *negative* value, then (as it has been shown in [12]), it is necessary to replace the root by its modulus value, and the log-periodic function in (84) should be replaced by some *anti-periodic* function having the following decomposition:

$$\mathrm{PR}_s^{(a)}(\ln z) = \left[\sum_{k=1}^{\infty} Ac_k^{(s)} \cos\left(\pi k \cdot \frac{\ln(z)}{\ln(\xi)}\right) + As_k^{(s)} \sin\left(\pi k \cdot \frac{\ln(z)}{\ln(\xi)}\right)\right], \tag{85}$$

$$\mathrm{PR}_s^{(a)}(\ln z \pm \ln \xi) = -\mathrm{PR}_s^{(a)}(\ln z).$$

As reminded above, the constant $C(R)$ figuring in (82) is determined by the arbitrary constant variation method, and depends totally on the constant value R from (80). If one of the roots of polynomial (81) ν_g is *degenerate*, then the solution for this root is written in the form

$$L_g(z) = \left[\sum_{r=1}^{g} \mathrm{PR}_r(\ln z) \cdot (\ln z)^{r-1}\right] \cdot z^{\nu_g}, \quad \nu_g = \frac{\ln \lambda_q}{\ln \xi}, \quad \xi > 1, \tag{86}$$

where the value g determines the degree of degeneracy. Here, again the unknown log-periodic functions entering into (86) are determined by decompositions (84) or (85). If we take into account relationship (83), then we can obtain the solution for the product $P(z)$. It is useful also to give the solution of (80) when a couple of the roots in (81) is complex-conjugated:

$$\nu = \mathrm{Re}(\nu) \pm i \,\mathrm{Im}\, \nu = \frac{\ln(\mathrm{Re}\,\lambda \pm i\,\mathrm{Im}\,\lambda)}{\ln \xi},$$

$$L(z) = z^{\mathrm{Re}(\nu)}$$

$$\times \left[A_0 + \sum_{k=1}^{\infty}\left(Ac_k \cos\left(\frac{2\pi k}{\ln \xi}\ln z + (\mathrm{Im}\,\nu)\cdot \ln z\right)\right.\right.$$

$$\left.\left. + As_k \sin\left(\frac{2\pi k}{\ln \xi}\ln z + (\mathrm{Im}\,\nu)\cdot \ln z\right)\right)\right].$$

Here we want to demonstrate some general properties of the functional Equation (80) and its corresponding polynomial (81).

Direct test shows that the polynomial (81) has *always* the root $\lambda = 1$, so it can be decomposed as

$$P(\lambda) = (\lambda - 1)(\lambda^{k-1} - w_{k-1}\lambda^{k-2} - \cdots - w_1). \tag{87}$$

If the functions defined by relation (83),

$$f(z\xi^{N+q}) = \exp(\Phi(z\xi^{N+q})), \quad q = 1, 2, \ldots, k-1, \tag{88}$$

have negative values, then from decoupling procedure (75) it follows that

$$f(z\xi^{N+k}) = \prod_{q=1}^{k-1} f(z\xi^{N+q})^{w_q}. \tag{89}$$

Taking the imaginary aspect from both parts of (89), we have

$$i\pi = \sum_{q=1}^{k-1} i\pi w_q \quad \text{or} \quad \sum_{q=1}^{k-1} w_q = 1. \tag{90}$$

Condition (90) leads to the *two-fold* degeneracy of the root $\lambda = 1$, and for this case, instead of decomposition (90), we have

$$\begin{aligned} P(\lambda) &= (\lambda - 1)^2(\lambda^{k-2} + a_{k-2}\lambda^{k-3} + \cdots + a_1), \\ a_1 &= w_1, \quad a_2 = w_1 + w_2, \quad \ldots, \quad a_{k-2} = w_1 + w_2 + \cdots + w_{k-2}. \end{aligned} \tag{91}$$

Before starting to consider some interesting examples, it is instructive to give the solution for some partial cases $k = 2, 3, 4$. These cases admit analytical solutions. As we will see below, these cases can be met frequently in possible applications.
Case $k = 2$. Approximate decoupling (it follows from (75), $w_1 = 1$)

$$f(z\xi^{N+2}) \cong f(z\xi^{N+1}), \quad \text{or} \quad \Phi(z\xi^{N+2}) \cong \Phi(z\xi^{N+1}). \tag{92}$$

The functional equation

$$L(z\xi^2) - 2L(z\xi) + L(z) = R. \tag{93}$$

The desired polynomial and its roots

$$P(\lambda) = (\lambda - 1)^2 = 0. \tag{94}$$

The general solution of the functional equation for this case is

$$\begin{aligned} L(z) &= \mathrm{PR}_1(\ln z) + \mathrm{PR}_2(\ln z) \cdot \ln z + \kappa_2 \ln^2 z, \\ \kappa_2 &= \frac{R}{2\ln^2 \xi}. \end{aligned} \tag{95}$$

Case $k = 3$. For this case, we *suppose* that the following approximate decoupling is satisfied:

$$f(z\xi^{N+3}) \cong f(z\xi^{N+1})^{w_1} f(z\xi^{N+2})^{w_2}. \tag{96}$$

The functional equation

$$L(z\xi^3) - (1 + w_2)L(z\xi^2) - (w_1 - w_2)L(z\xi) + w_1 L(z) = R. \tag{97}$$

The desired polynomial and its roots

$$P(\lambda) = (\lambda - 1)(\lambda^2 - w_2\lambda - w_1) = 0, \quad w_1 + w_2 \neq 1,$$
$$\lambda_{1,2} = \frac{w_2}{2} \pm \sqrt{\left(\frac{w_2}{2}\right)^2 + w_1}. \tag{98}$$

The general solution of the functional Equation (98) ($w_1 + w_2 \neq 1$)

$$L(z) = \mathrm{PR}_0(\ln z) + \mathrm{PR}_1(\ln z) \cdot z^{\nu_1} + \mathrm{PR}_2(\ln z) \cdot z^{\nu_2} + \kappa_1 \ln z,$$
$$\nu_{1,2} = \frac{\ln(\lambda_{1,2})}{\ln \xi}, \quad \kappa_1 = \frac{R}{\ln \xi \cdot (1 - w_1 - w_2)}. \tag{99}$$

The general solution of the functional Equation (98) ($w_1 + w_2 = 1$)

$$P(\lambda) = (\lambda - 1)^2(\lambda + w_1) = 0,$$
$$L(z) = \mathrm{PR}_1(\ln z) + \mathrm{PR}_2(\ln z) \cdot \ln z + \mathrm{PR}_3^{(a)}(\ln z) \cdot z^{\nu_3} + \kappa_2 \ln^2 z, \tag{100}$$
$$\nu_3 = \frac{\ln(|w_1|)}{\ln \xi}, \quad \kappa_2 = \frac{R}{2\ln^2 \xi \cdot (1 + w_1)}.$$

Case $k = 4$. We *suppose* that for this case, the following approximate decoupling is valid:

$$f(z\xi^{N+4}) \cong f(z\xi^{N+1})^{w_1} f(z\xi^{N+2})^{w_2} f(z\xi^{N+3})^{w_3}. \tag{101}$$

The functional equation

$$L(z\xi^4) - (1 + w_3)L(z\xi^3) - \sum_{q=1}^{2}(w_q - w_{q+1})L(z\xi^q) + w_1 L(z) = R. \tag{102}$$

The desired polynomial and its roots

$$P(\lambda) = (\lambda - 1)(\lambda^3 - w_3\lambda^2 - w_2\lambda - w_1) = 0, \quad w_1 + w_2 + w_3 \neq 1. \tag{103}$$

The general solution of the functional Equation (103) ($w_1 + w_2 + w_3 \neq 1$)

$$L(z) = \mathrm{PR}_0(\ln z) + \sum_{s=1}^{3} \mathrm{PR}_s(\ln z) \cdot z^{\nu_s} + \kappa_1 \ln z,$$
$$\nu_{1,2,3} = \frac{\ln(\lambda_{1,2,3})}{\ln \xi}, \quad \kappa_1 = \frac{R}{\ln \xi \cdot (1 - w_1 - w_2 - w_3)}. \tag{104}$$

The general solution of the functional Equation (103) ($w_1 + w_2 + w_3 = 1$)

$$P(\lambda) = (\lambda - 1)^2(\lambda^2 + (w_1 + w_2)\lambda + w_1) = 0,$$
$$L(z) = \mathrm{PR}_1(\ln z) + \mathrm{PR}_2(\ln z) \cdot \ln z + \sum_{s=3}^{4} \mathrm{PR}_s(\ln z) z^{\nu_s} + \kappa_2 \ln^2 z, \tag{105}$$
$$\kappa_2 = \frac{R}{2\ln^2 \xi \cdot (1 + 2w_1 + w_2)}.$$

In expressions given above, the log-periodic functions are defined by decompositions (84) and (85) for periodic and anti-periodic cases, correspondingly. The desired expressions for the product $P(z)$ are obtained from the solutions for $L(z)$ with the use of relationship (73). Solutions for the case $\xi < 1$ are obtained from these expressions by simple replacement $\xi \to 1/\xi$. We want to stress here the *principle* difference of appearance of the power-law exponents in expressions (95), (104), and (105), and a formal absence of the power-law behavior in expression (101). Besides this difference, we obtain the mixed dependence between power-law and logarithmic behavior in expressions (96) and (101).

Before, it was accepted to consider that the power-law exponent is formed from the limiting values of the microscopic function (expression (71)). In new expressions derived in this chapter, we can mark at least two new reasons of appearance of the power-law exponent when the condition (72) is fulfilled. For the nondegenerate case $\sum_{q=1}^{k-1} w_q \neq 1$, the power-law exponent is formed from the value κ_1 figuring in expressions (99), (104); whereas for the degenerate case $\sum_{q=1}^{k-1} w_q = 1$ (expressions (96), (100), and (102)), in formation of the power-law exponent, the constant A_0 from (84) plays the essential role. In these two new cases, marked above, the power-law exponent does *not* coincide with the fractal dimension of the geometrical object considered.

In concluding this section, we want to give some additional arguments and determine some conditions justifying the decoupling supposition (75). Definitely, for each concrete form of the fractal considered, the limits of applicability should be considered independently. The approximate relationship (75) represents the functional equation for the function $\Phi(z)$. If $N \gg 1$, then from this expression we obtain, approximately, the desired solution

$$\begin{aligned}\Phi(z\xi^N) &\cong \left(\sum_{q=1}^{k-1} w_q\right)\Phi(z\xi^N),\\ \Phi(z) &\cong \mathrm{PR}_0(\ln z),\\ \left(\sum_{q=1}^{k-1} w_q\right) &= 1.\end{aligned} \tag{106}$$

Here $\mathrm{PR}_0(\ln z)$ is defined by decomposition (15). So, the decoupling (76) can be realized with high accuracy, if, in turn, the function $\Phi(z)$ can be approximated and finally replaced with high accuracy by log-periodic function, determined by decomposition (84). If condition (106) is realized, then (because of log-periodic properties of (84)) the condition (92), referring to the Case 2, will be realized automatically. However, we stress again, that each specific type of fractal condition (106) should be more accurately tested to find the limits of applicability.

3.2 Numerical test for Cantor set with M bars

As an example for testing of decoupling condition (75), we consider the classical Cantor set containing M bars in each stage of its generation. As it has been shown in the book [6], the structure-factor (defined above by the general expression (90)) for the Cantor set containing M bars, and located along OX axis, is expressed as

$$P_{N_1}(x) = \prod_{n=-N_1}^{N_1} \mathrm{Re}\left[\frac{1-\exp(ixM\xi^n)}{M\cdot(1-\exp(ix\xi^n))}\right] \equiv \prod_{n=-N_1}^{N_1} f_M(x\xi^n),$$
$$x = ka, \quad \xi > 1, \quad \Phi_M(x) = \ln(|f_M(x)|), \quad L_{N_1}(x) = \ln(|P_{N_1}(x)|). \tag{107}$$

For concrete calculations of this expression and testing the supposition (75), we chose the following values of the parameters entering into (107):

$$\xi = 1.5, \quad N_1 = 50, \quad x \in [0,1], \quad M = 2,3,5. \tag{108}$$

The discrete number of points determining the interval of the current variable x was limited by three values of $N = 50, 150$, and 500. Calculations show that it is much convenient to test the value $L_{N_1}(x) = \ln(P_{N_1}(x))$. The typical behavior of this function for $M = 5$, $N = 150$ is shown in Figure 4.

All test calculations are divided into two steps:

S1. The verification of the supposition (75) (or equivalently the accuracy of the decomposition (93)) and calculation of the fitting parameters $\{w_q\}$.

S2. The final fitting of one of the functions (93), (99), (100), (104), (107) to the function $L_{N_1}(x)$ that depends essentially on the number of parameters $\{w_q\}$.

These calculations were realized for the different number of Cantor bars having $M = 2,3,5$ and the number of points $N = 50, 150$, and 500. The results of the verification of

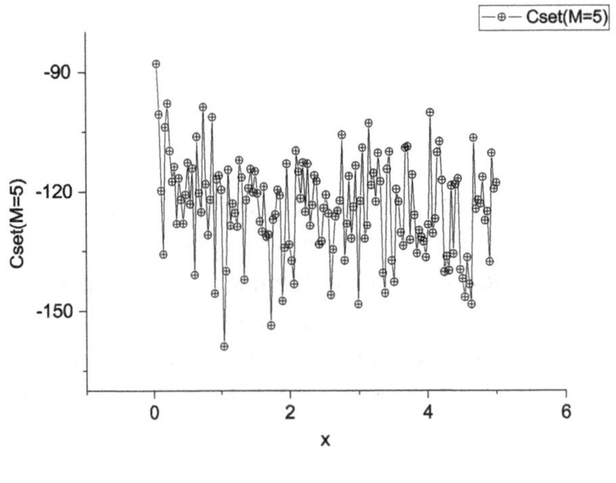

Figure 4: Typical behavior of the function $L_{N_1}(x) = \ln(P_{N_1}(x))$ for $M = 5$ defined by expression (107). The values of parameters are collected in (45). Number of points $N = 150$. Interval of the variable x is [0.05–5.0].

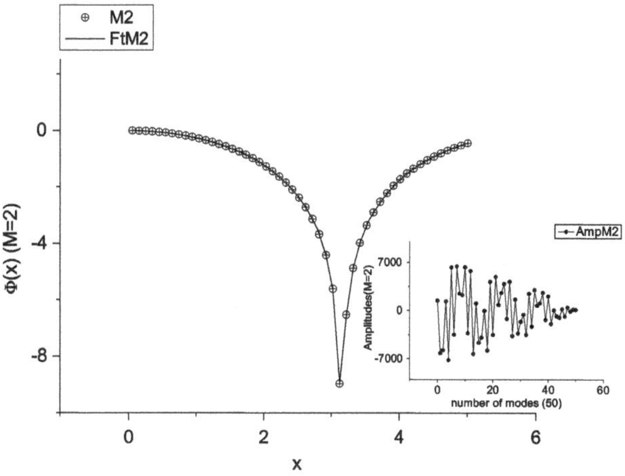

Figure 5(a): The verification of the hypothesis (106) for $M = 2$ ($N = 50$). The points correspond to the function $\Phi_2(x)$ and solid line corresponds to fit of log-periodic function from (20). In the small plot below the distribution of the amplitudes of the corresponding log-periodic function is shown.

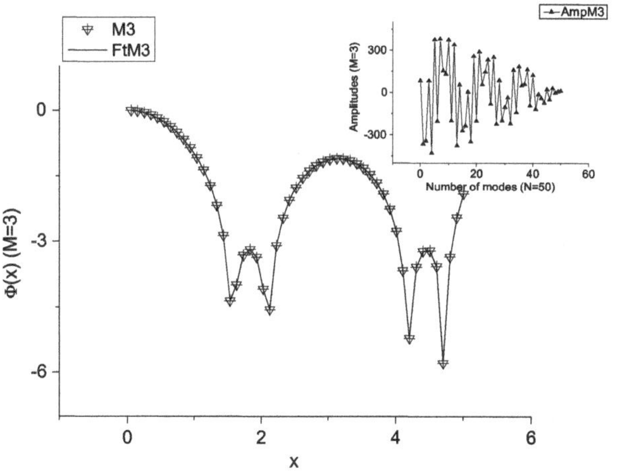

Figure 5(b): The verification of the hypothesis (106) for $M = 3$ ($N = 50$). The points correspond to the function $\Phi_3(x)$ and solid line corresponds to fit of the log-periodic function from (20). In the small plot above the distribution of the amplitudes of the log-periodic function for this case is shown.

hypothesis (92) for the function $\Phi_M(x)$ ($M = 2, 3, 5$) are illustrated by Figures 5(a), 5(b), and 5(c) for $N = 50$.

The value of the cutoff parameter K determining the upper limit of decomposition (84) and the value of the relative error (defined below by (124)) for different M and N are collected in Table 1. The analysis shows that for the relatively large values of $N_1 \gg 1$, supposition (93) is realized with relatively high accuracy (the value of the relative cannot exceed 1 % or can be even less) for all tested $N = 50, 150$, and 500. The value of the constant R entering into expression (95) and calculated from (80) equals zero. So, the fitting function for the fractal object as the Cantor set is reduced (because of expression (107)) to the simplest Case $k = 2$:

$$L(z) = \mathrm{PR}_1(\ln z) + \mathrm{PR}_2(\ln z) \cdot \ln z. \tag{109}$$

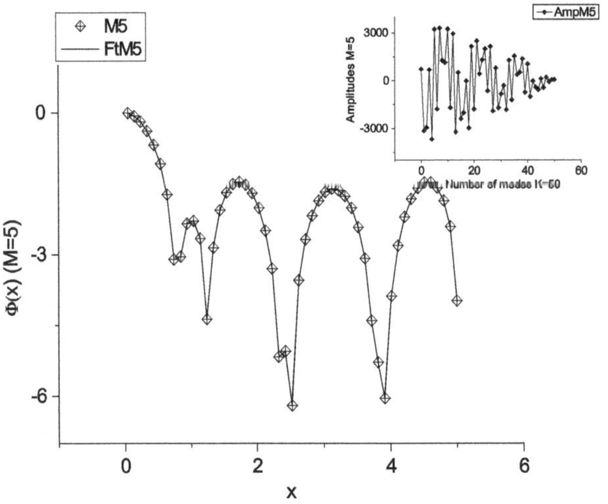

Figure 5(c): The verification of the hypothesis (106) for $M = 5$. The points correspond to the function $\Phi_5(x)$ and solid line corresponds to fit of the log-periodic function from (84). In the small plot above the distribution of the amplitudes of the log-periodic function for this case is shown. The values of the relative error for different $M = 2, 3, 5$ and $N = 50$ is collected in Table 2. For other values of $N = 150, 500$ the fit looks similar and so it is not shown. The values of the relative error for all these cases are collected in Table 2.

Taking into account relationships (84), it is convenient to present function (99) for the fitting purposes in the form

$$\begin{aligned} Lf(x) &= A_0^{(2)} \cdot \ln(x) + \sum_{k=1}^{K} Ac_k^{(2)} Cl_k(\ln x) + \sum_{k=1}^{K} As_k^{(2)} Sl_k(\ln x) \\ &\quad + A_0^{(1)} + \sum_{k=1}^{K} Ac_k^{(1)} C_k(\ln x) + \sum_{k=1}^{K} As_k^{(1)} S_k(\ln x), \\ C_k(\ln x) &= \cos\left(\frac{2\pi k}{\ln \xi} \cdot \ln x\right), \quad S_k(\ln x) = \sin\left(\frac{2\pi k}{\ln \xi} \cdot \ln x\right), \\ Cl_k(\ln x) &= (\ln x) \cdot \cos\left(\frac{2\pi k}{\ln \xi} \cdot \ln x\right), \quad Sl_k(\ln x) = (\ln x) \cdot \sin\left(\frac{2\pi k}{\ln \xi} \cdot \ln x\right). \end{aligned} \tag{110}$$

The function (110) contains $2 + 4K$ fitting parameters. The cutoff parameter K is determined by the value of the relative error

$$\text{RelErr}(\%) = \left(\frac{\text{stdev}(L_{N_1}(x) - Lf(x))}{\text{mean}\,|L_{N_1}(x)|}\right) \cdot 100\,\%. \tag{111}$$

Here $Lf(x)$ implies the corresponding fitting function, the symbols stdev($f(x)$) and mean($f(x)$) determine the conventional value of the standard deviation and expectation value, respectively. The results of the fitting of the function $L_{N_1}(x) = \ln(P_{N_1}(x))$ from (107) to function (110) are illustrated by Figures 6(a), 6(b), 6(c), and 6(d) for $M = 2$ and $M = 5$, correspondingly. The results for $M = 3$ are not shown because they are similar. Analysis of these figures shows that for the accurate fitting, the relatively large number of the fitting parameters entering into (113) is required. The accuracy of the fitting is decreased with increasing of the value of N. The calculated values of $A_0^{(2)}$, RelErr(%) at given $N(K)$ are collected in Table 3.

Table 2: The value of the relative error calculated in testing hypothesis (106).

Number of bars M	Number of the generated points N	Value of the cutoff parameter K	Value of the relative error (%)
2	50	25	$6.161 * 10^{-5}$
3	50	25	$6.662 * 10^{-5}$
5	50	25	$6.671 * 10^{-5}$
2	150	50	0.14491
3	150	50	0.27107
5	150	50	0.09772
2	500	75	1.31705
3	500	75	0.79646
5	500	75	0.40202

Table 3: The value of the relative error calculated in testing hypotheses (109) and (113).

Number of bars M	Number of the generated points N and value of the cutoff parameter K figuring in (110). N (K)	Value of the constant $A_0^{(2)}$ mimicking the fractal dimension	Value of the relative error (%)
2	50 (25)	−5.35978	0.21354
3	50 (25)	−6.36127	0.91534
5	50 (25)	−3.77906	1.24461
2	150 (22)	−4.98805	1.45124
3	150 (22)	−6.92082	0.90255
5	150 (22)	−3.57148	0.77035
2	500 (33)	−3.77267	12.26469
3	500 (33)	−4.80213	6.30386
5	500 (33)	−6.40383	6.72678

This simple test shows that decoupling supposition (75) (which is justified by the fit of expression (107)) to $\Phi_M(x)$) is very *reasonable* and can be applied for establishing the desired relationships between noninteger operators and fractals averaged with smooth functions in space. These results give new understanding of the fractal dimension, and additional evidences that the relationship between the power-law exponent figuring in noninteger operator and power-law exponents appearing in the Cases 2, 3, 4 considered above is *not* simple as it was supposed earlier.

3.3 Possible generalizations

How to generalize the results obtained in the previous section for more complex fractals?

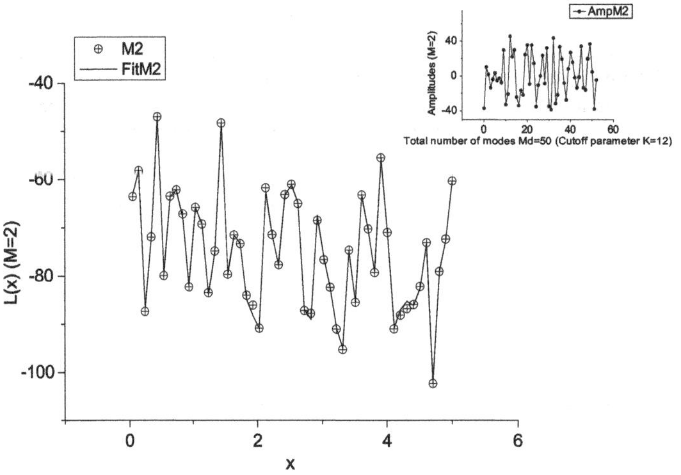

Figure 6(a): Verification of hypothesis (46) for $M - 2$ and $N - 50$. The function $L_{N_1}(x)$ from (107) is presented by crossed points. The fitting function (110) is presented by solid line. The distribution of amplitudes of the log-periodic function is shown above in the small frame.

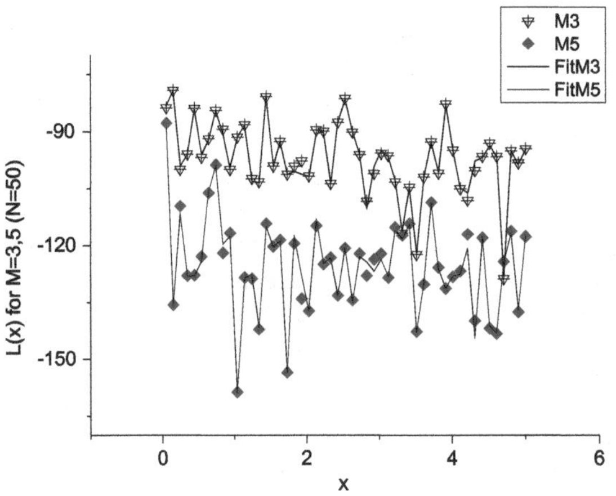

Figure 6(b): Verification of hypothesis (107) for $M = 3, 5$ and $N = 50$. The functions $L_{N_1}(x)$ from (109) are presented by crossed triangles and black rhombs, correspondingly. The fitting functions (110) are presented by solid lines. The distributions of amplitudes for these cases are not shown because they look similar to $M = 2$.

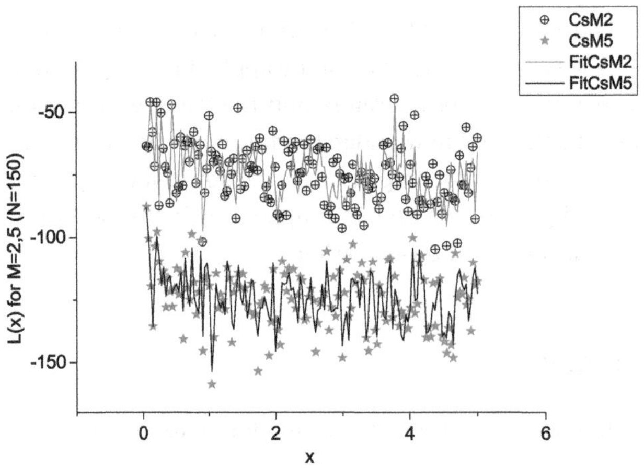

Figure 6(c): Verification of hypothesis (109) for $M = 2, 5$ and $N = 150$. The functions $L_{N_1}(x)$ from (107) are presented by crossed circles and grey stars, correspondingly. The fitting functions (110) are presented by solid lines. The distributions of amplitudes are not shown because they look similar to $M = 2$. Other parameters are collected in Table 1.

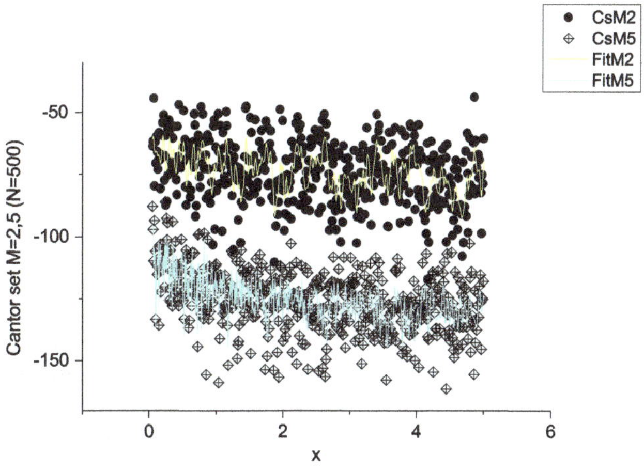

Figure 6(d): Verification of hypothesis (109) for $M = 2, 5$ and $N = 500$. The functions $L_{N_1}(x)$ from (107) are presented by dark circles and crossed rhombs, correspondingly. The fitting functions (110) are presented by yellow and blue lines. The fit of the Cantor set for $M = 3$ and distributions of amplitudes are not shown because they look similar to $M = 2$. Other parameters are collected in Table 3.

For example, if the Cantor set with $M = 2$ is obtained with help of rotation operation realized along OY axis, then the structure-factor for this fractal, having a continuous cylindrical symmetry, is expressed as [6]

$$P_{N_1}(x) = \prod_{n=-N_1}^{N_1} \left(\frac{1}{2}(1 + J_0(x\xi^n))\right), \quad x = \left(\sqrt{k_x^2 + k_y^2}\right)\lambda_{\min} \cdot (1 - \xi^{-1}),$$
$$\xi > 1, \quad \eta \in [\lambda_{\min}, \Lambda_{\max}], \quad L_{N_1}(x) = \ln(P_{N_1}(x)). \tag{112}$$

Here $J_0(x)$ is the ordinary Bessel function of the zeroth order. The scaling parameter ξ (it is equaled to 3/2 for numerical calculations) is counted off from the minimal scale $\lambda_{\min}$. The similar test described above and applied to product (49) shows that the Case $k = 2$ is valid for this case also. Therefore, for the fitting of the function $L_{N_1}(x) = \ln(P_{N_1}(x))$, hypothesis (46) is applicable again. Figure 7 demonstrates the results of the fitting of function (113) to $L_{N_1}(x)$ (from (113)) for $N = 150$ and $K = 10$.

We should note here that contrary to the first example, the Bessel function $J_0(x)$ entering into the product (112) has the finite limits for both cases

$$J_0(z) \overset{z \gg 1}{=} 0, \quad J_0(0) = 1. \tag{113}$$

It means that solution for (112) has the form (98) with power-law exponent $\nu = \ln(1/2)/\ln(\xi)$. To compare the known solution (109) with solution (95), we present the last one in the form

$$P_{N_1}(z) = \mathrm{PR}_\nu(\ln z) \cdot z^{A_0 + \mathrm{RP}_2(\ln z)},$$
$$\mathrm{RP}_2(\ln z) = \sum_{k=1}^{\infty} \left[Ac_k \cos\left(\frac{2\pi}{\ln \xi} \ln z\right) + As_k \sin\left(\frac{2\pi}{\ln \xi} \ln z\right)\right]. \tag{114}$$

From solution (100), it follows that new solution (114) also creates the log-periodic corrections to the power-law exponent, and the constant A_0 should coincide with fractal

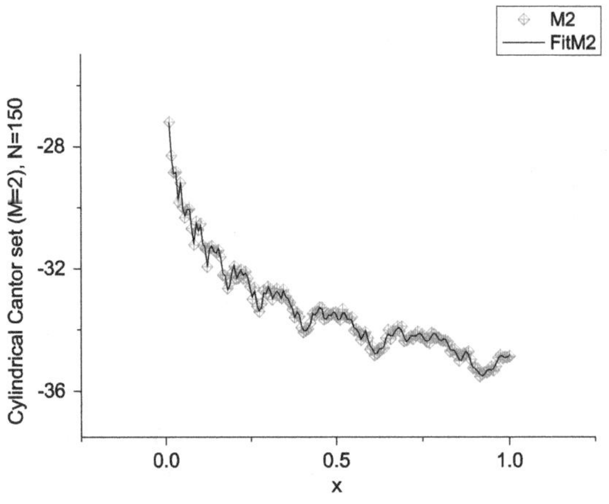

Figure 7: The results of the fitting of the function $L_{N_1}(x)$ from (112) to hypothesis (113). The relative error defined by expression (111) equals 1.233 %. $\xi = 3/2$.

dimension $\nu = \ln(1/2)/\ln(\xi)$. Test calculations realized presumably for $N = 150$ and different values of the cutoff parameter K confirm this relationship. The results of comparison are depicted in Table 2. So, this simple test leads us to one important conclusion: if the microscopic function $f(z)$ from (115) has finite limits for small and large values of z, then the simplest solution (31) allows finding the log-periodic corrections to the fractal dimension coinciding with the constant A_0. This statement is violated when condition (110) is valid.

Finishing this section, let us consider the Cantor set located on the plane XOY and concentrated along two axes. The structure-factor for this case is expressed as [6]

$$P(z_1, z_2) = \prod_{n=-N_1}^{N_1} \left(\frac{\cos(z_1 \xi^n) + \cos(z_2 \xi^n)}{2} \right), \quad z_1 = k_x a, \quad z_2 = k_y b. \tag{115}$$

Direct application of the approach developed above is impossible because the structure of (115) does *not* coincide with the structure of the initial product (116). To apply this approach, it is necessary to *factorize* the product (115) and reduce it to new set of variables. For (115), it can be done easily, and the desired product accepts the form

$$\begin{aligned} P(z_1, z_2) &= \prod_{n=-N_1}^{N_1} \left(\cos\left(\frac{z_1 + z_2}{2} \xi^n \right) \cdot \cos\left(\frac{z_1 - z_2}{2} \xi^n \right) \right) \\ &= \prod_{n=-N_1}^{N_1} \left(\cos\left(\frac{z_1 + z_2}{2} \xi^n \right) \cdot \prod_{n=-N_1}^{N_1} \cos\left(\frac{z_1 - z_2}{2} \xi^n \right) \right). \end{aligned} \tag{116}$$

Now it becomes obvious that this approach can be separately applied for each product figuring in (116).

From this example, one important conclusion follows: if the general structure-factor (product) containing many variables and expressed in the form of expression

(2) can be *factorized* and presented in the form

$$P(z_1,z_2,\ldots,z_r)=\prod_{n=-N_1}^{N_1} f(z_1\xi_1^n)\prod_{n=-N_1}^{N_1} f(z_2\xi_2^n)\cdots\prod_{n=-N_1}^{N_1} f(z_r\xi_r^n), \tag{117}$$

then one can apply the approach developed above to each product figuring in (117). This expression can have wide practical applications because many fractals can be presented in the form (117).

4 Concluding remarks

The new relationships between fractals and fractional integrals presented in this paper help to find new relationships between the procedure of averaging of smooth functions over fractal sets and spatial noninteger integrals. As one can see from the results, these relationship are *not* trivial. If condition (57) is satisfied, then the desired relationship between spatial noninteger integrals that are derived presumably from the structure-factor of type (52) is becoming *questionable*. In this case, there is no direct relationship between the fractal dimension and power-law exponent figuring in the fractional integral. Even in cases where condition (57) is not satisfied, the new approach helps to find the log-periodic corrections (100) for the power-law exponent defining the fractal dimension. Many researches nowadays try to *postulate* simply the desired relationship between the fractal structure and the fractional integral, and the sincere desire to establish this relationship can be violated. These results can be considered as a specific *warning* in attempts to impose simply this "obvious" relationship between fractals and noninteger integrals in space. It is necessary to stress also that the desired original, containing convolution of the spatial integral with smooth function in $\boldsymbol{r}$-space, can be obtained from the corresponding Fourier image in $\boldsymbol{k}$-space by means of expression (65) only *approximately*.

Finally, we want to make two remarks.

R1. The first remark is associated with observation that the linear functional Equation (80) is *not* unique. Other functional equations connecting different $P(z\xi^k)$ are also *possible*. They can be obtained from any decoupling relationship

$$f(z\xi^{N+k})=F(f(z\xi^{N+k-1}),f(z\xi^{N+k-2}),\ldots,f(z\xi^{N+1})), \tag{118}$$

where $F(z_1,z_2,\ldots,z_s)$ determines an arbitrary decoupling function of many variables. Of course, in each specific case, the selection of the decoupling function $F(z_1,z_2,\ldots,z_s)$ should be justified, explained clearly, and tested numerically.

R2. The second remark contains the answer for the following question: are there distributions (which are widely used in the mathematical statistics) that can be derived from the product (53), having self-similar properties? At least one can show three distributions that can be derived from the corresponding functional equations.

(a). *Log-normal distribution.*

Let us suppose that instead of condition (68) and functional Equation (69), we have the following behavior of the microscopic function $f(z)$:

$$\begin{aligned} &P(z\xi) = b^{N-|N_0|}\frac{f(z\xi^{N+1})}{f(z\xi^{-|N_0|})}P(z), \quad \xi > 1, \\ &f(z) \overset{z\gg 1}{=} Az^{-\alpha} + \cdots \quad (\alpha > 0), \\ &f(z) \overset{z\ll 1}{=} c_0 + c_1 z^{\beta} + \cdots \quad (\beta > 0). \end{aligned} \tag{119}$$

Taking the natural logarithm from (119), we obtain the following functional equation:

$$\begin{aligned} &L(z\xi) = L(z) + B - \alpha \ln z, \\ &B = (N - N_0)\ln b + \ln\left(\frac{A}{c_0}\right). \end{aligned} \tag{120}$$

As it was done in (73), in expression (120), we consider N_0 as a positive value ($N_0 > 0$).

The solution of the functional Equation (120) has the form

$$\begin{aligned} &L(z) = \mathrm{PR}_0(\ln z) + a_1 \ln z + a_2 \ln^2 z, \\ &a_1 = \frac{\alpha}{2} + \frac{B}{\ln \xi}, \quad a_2 = -\frac{\alpha}{2\cdot \ln \xi} \quad (\alpha > 0, \xi > 1). \end{aligned} \tag{121}$$

If in expression (119) the constant $c_0 = 0$, then solution (121) keeps the same form, but (in accordance with (120)), it is necessary to make the replacements $c_0 \to c_1$ in (120) and $\alpha \to \alpha + \beta$ in (121), correspondingly. Taking into account the fact that any smooth function taken from the log-periodic function does not change the property of periodicity, one can write the solution for $P(z)$ from (119) in the form

$$P(z) = \mathrm{PR}_0(\ln z)\exp(a_1 \ln z + a_2 \ln^2 z). \tag{122}$$

Here the log-periodic function is defined again by relationship (84). In particular cases, where $\mathrm{PR}_0(\ln z) = A_0$, the last expression is reduced to the conventional log-normal distribution.

(b). *The χ^2-distribution*

If the microscopic function $f(z)$ has asymptotic behavior of type

$$f(z) \overset{z\gg 1}{=} A\exp(-\gamma z) + \cdots \quad (\gamma > 0), \tag{123}$$

then the functional equation for $L(z)$ accepts the form

$$L(z\xi) = L(z) + B - \gamma z. \tag{124}$$

Using again an arbitrary constant variation method, it is easy to find the solution for $P(z)$. It is expressed as

$$P(z) = \mathrm{PR}_0(\ln z) \cdot z^{a_2} \exp(-a_1 z),$$
$$a_2 = \frac{B}{\ln \xi}, \quad a_1 = \frac{\gamma}{\xi - 1}, \quad (a_{1,2} > 0). \tag{125}$$

In particular cases where $\mathrm{PR}_0(\ln z) = A_0$, we obtain from (111) the well-known χ^2-distribution.

(c). *The β-distribution.*

In papers [12, 13], it has been proved that the cumulative integral for the strongly correlated detrended sequences can be described by β-distribution. *How* to find the fractal (scaling properties) of any two random sequences compared if their β-distributions are kept invariant relative to linear transformations

$$x' = ax + b? \tag{126}$$

If we subject the initial β-distribution to the linear transformation, then we obtains

$$y = A(x - x_0)^{\alpha}(x_N - x)^{\beta} + B \to A\big(ax - (x_0 - b)\big)^{\alpha'}(x_N - b - ax)^{\beta'} + \tilde{B}$$
$$\equiv \tilde{A}(x - \tilde{x}_0)^{\alpha}(\tilde{x}_N - x)^{\beta} + \tilde{B}, \tag{127}$$

where

$$\tilde{A} = A \cdot a^{\alpha' + \beta'}, \quad \tilde{x}_0 = \frac{x_0 - b}{a}, \quad \tilde{x}_N = \frac{x_N - b}{a}, \quad \tilde{B} = B. \tag{128}$$

As one can see from (127), the β-distribution keeps its *invariant* properties as relatively linear transformations if the power-law exponents of two distributions compared satisfy the condition

$$\alpha + \beta = \alpha' + \beta' = \text{const} \equiv \text{inv}. \tag{129}$$

If the power-law exponents and the final points of location of two distribution are known from the fitting procedure (the linear fit of this function is shown in [13]), then the unknown values of linear transformation (a, b) from (126) can be found easily from the relationships (126)

$$\begin{cases} a\tilde{x}_0 + b = x_0 \\ a\tilde{x}_N + b = x_N \end{cases} \to \begin{pmatrix} a = \frac{x_N - x_0}{\tilde{x}_N - \tilde{x}_0} \\ b = \frac{x_0 \tilde{x}_N - \tilde{x}_0 x_N}{\tilde{x}_N - \tilde{x}_0} \end{pmatrix}. \tag{130}$$

So, we receive new possibilities for comparison of two random sequences having scaling properties in terms of the invariant properties of β-distribution. If we present this distribution in another form

$$P(z_1, z_2) = A z_1^{\alpha} z_2^{\beta},$$
$$z_1 = x - x_0, \quad z_2 = x_N - x, \tag{131}$$

then we can obtain easily the scaling equation relative to two variables for (131). It has the following form:

$$L(z_1\xi, z_2\xi) - L(z_1\xi, z_2) - L(z_1, z_2\xi) + L(z_1, z_2) = 0,$$
$$L(z_1, z_2) = \ln[P(z_1, z_2)], \quad z_1 + z_2 = x_N - x_0 = \text{const.} \tag{132}$$

So, the functional equations, formally containing two variables, can be used for investigation of the scaling properties of different fractal systems that cannot be described only in the frame of approach developed above.

In conclusion, we want to stress the basic results of the suggested approach outlined in Section 3.

1. This approach is applicable only in the cases where the structure-factor can be factorized and expressed in the form (117). For the other cases, the consideration of other approaches are necessary.
2. This approach is applicable when a current generation of the fractal considered can be expressed in the form of a set of star-vectors in **k**-space. Not all fractals can be expressed in this manner. Random fractals, for example, are needed to develop other methods for their consideration to establish the desired relationship between some class of fractals and the conventional fractional integrals in space.
3. From this consideration follows also that the fractal dimension cannot coincide with the power-law exponent figuring in the fractional integral. The log-periodic corrections (expressions (84) and (85)) that follow from the solutions of the functional equations are also possible. They appear in the case where the scaling parameter ξ is distributed over the denumerable set.
4. The accurate relationships between fractals and fractional integrals remain a "hot" spot for many researches working in the fractional calculus and fractal geometry field. From the results obtained in this paper, one can notice that the complex-conjugated part figuring in the fractional power-law exponent plays a *crucial* role. It is necessary to say that the power-law exponent with complex additive are presented in some papers, but the physical/geometrical origin of this additive was not clear. From the results obtained above, one can say that the complex-conjugated part is tightly associated with the discrete structure of the fractal process, and should be taken into account in fractional and kinetic equations that aim to describe self-similar processes in time-domain, at least. As for the spatial fractional integral, the finding of the accurate relationship for the given fractal in space remains an open problem. This problem can be divided in at least on two parts:
   1. Finding the proper fractional integral, based on the given fractal structure;
   2. Find a proper fractal for the fractional integral that is chosen for description of the self-similar process in space.

From our point of view, the general solution of this complex problem is absent, because each fractal in space can generate a specific fractional integral [14], but any efforts of researches actively working in this interesting field are most welcomed [3, 9].

Mathematical Appendix. The solutions of the functional Equation (80)

(See also some results obtained in paper [12]).

The solutions of this functional equation is closely related with well-known solutions of the difference equation of the k-th order with constant coefficients

$$Y_k = a_{k-1}Y_{k-1} + a_{k-2}Y_{k-2} + \cdots + a_0 Y_0. \tag{A1}$$

The solution of this equation (when all roots are different) can be written as

$$Y_k = K_1\lambda_1^k + \cdots + K_k\lambda_k^k. \tag{A2}$$

If one of the roots is *degenerate*, then the solution is written as (the integer value g defines the order of degeneracy)

$$Y_k = \left[\sum_{s=1}^{g}(C_s k^{s-1})\right]\cdot\lambda_g^k. \tag{A3}$$

For both cases, the desired roots are found from the polynomial

$$P(\lambda) = \lambda^k - a_{k-1}\lambda^{k-1} - a_{k-2}\lambda^{k-2} - \cdots - a_0 = 0. \tag{A4}$$

In complete analogy with these solutions, one can write the general solution of the functional Equation (16) for nondegenerate case (making the formal replacement $K_s \to \mathrm{PR}_s(\ln z)$, $k \to \ln(z)/\ln(\xi)$

$$\begin{aligned} L(z) &= \sum_{s=1}^{k}\mathrm{PR}_s(\ln z)\cdot(\lambda_s)^{\frac{\ln z}{\ln\xi}} \equiv \sum_{s=1}^{k}\mathrm{PR}_s(\ln z)\cdot\exp\left(\frac{\ln(\lambda_s)}{\ln\xi}\cdot\ln z\right)\\ &= \sum_{s=1}^{k}\mathrm{PR}_s(\ln z)\cdot z^{\nu_s},\quad \nu_s = \frac{\ln(\lambda_s)}{\ln\xi}, \end{aligned} \tag{A5}$$

and in the case where one of the roots is g-fold degenerate,

$$L_g(z) = \left[\sum_{r=1}^{g}\mathrm{PR}_r(\ln z)\cdot(\ln z)^{r-1}\right]\cdot z^{\nu_g},\quad \nu_g = \frac{\ln\lambda_q}{\ln\xi}. \tag{A6}$$

In expressions (A5) and (A6), the constants K_s are replaced by the log-periodic functions $\mathrm{PR}_r(\ln z \pm \ln \xi) = \mathrm{PR}_r(\ln z)$, which can be presented by the following decomposition to the Fourier series:

$$\mathrm{PR}_r(z) = A_0^{(r)} + \left[\sum_{k=1}^{\infty} Ac_k^{(r)} \cos\left(2\pi k \cdot \frac{\ln z}{\ln \xi}\right) + As_k^{(r)} \sin\left(2\pi k \cdot \frac{\ln z}{\ln \xi}\right)\right], \quad (A7)$$

$$r = 1, 2, \ldots, k, \ldots.$$

If one of the roots in (A6) is *negative*, then this root can replaced by its modulus value, and the periodic function can be changed for anti-periodic function having the following decomposition:

$$\mathrm{PR}_r^{(a)}(z) = \left[\sum_{k=1}^{\infty} Ac_k^{(r)} \cos\left(\pi k \cdot \frac{\ln z}{\ln \xi}\right) + As_k^{(r)} \sin\left(\pi k \cdot \frac{\ln z}{\ln \xi}\right)\right], \quad (A8)$$

$$r = 1, 2, \ldots, k, \quad \mathrm{PR}_r^{(a)}(\ln z \pm k \ln \xi) = (-1)^k \mathrm{PR}_r^{(a)}(\ln z).$$

The solution for the complex-conjugated roots is given by expression (22). Other similar functional equations that are reduced to this form are considered in paper [15].

Bibliography

[1] D. Baleanu, K. Diethelm, E. Scalas, and J. J. Trujillo, *Fractional Calculus Models and Numerical Methods*, Series on Complexity, Nonlinearity and Chaos, World Scientific, 2012.

[2] H. Bateman and A. Erdelyi, *Tables of Integral Transforms*, vol. 1, McGraw-Hill Company, New York, Toronto, London, 1954.

[3] S. Butera and M. Paola, A physically based connection between fractional calculus and fractal geometry, *Annals of Physics*, **350** (2014), 146–158.

[4] A. K. Golmankhaneh and D. Baleanu, New derivatives on the fractal subset of real-line, *Entropy*, (2016), 1–13.

[5] A. K. Golmankhaneha and D. Baleanu, Fractal calculus involving gauge function, *Communications in Nonlinear Science and Numerical Simulation*, **37** (2016), 125–130.

[6] A. Le Mehaute, R. R. Nigmatullin, and L. Nivanen, *Fleches du Temps et Geometrie Fractale*, Editions HERMES, Paris, 1998, p. 348. (In French).

[7] J. A. T. Machado, V. Kiryakova, and F. Mainardi, A poster about old history of fractional calculus, *Fractional Calculus and Applied Analysis*, **13**(4) (2010), 447–454.

[8] J. A. T. Machado, V. Kiryakova, and F. Mainardi, Recent history of fractional calculus, *Communications in Nonlinear Science and Numerical Simulation*, **16** (2011), 1140–1153.

[9] G. Maione, R. R. Nigmatullin, J. A. T. Machado, and J. Sabatier, Editorial paper: new challenges in fractional systems 2014. *Mathematical Problems in Engineering*, **2015** (2015), Article ID 870841, 3 pages. http://dx.doi.org/10.1155/2015/870841.

[10] R. R. Nigmatullin, Theory of dielectric relaxation in non-crystalline solids: from a set of micromotions to the averaged collective motion in the mesoscale region, *Physica B: Condensed Matter*, **358** (2005), 201–215.

[11] R. R. Nigmatullin, Fractional kinetic equations and universal decoupling of a memory function in mesoscale region, *Physica. A*, **363** (2006), 282–298.

[12] R. R. Nigmatullin, Strongly correlated variables and existence of the universal distribution function for relative fluctuations, *Physics of Wave Phenomena*, **16**(2) (2008), 119–145.

[13] R. R. Nigmatullin, Universal distribution function for the strongly-correlated fluctuations: general way for description of random sequences, *Communications in Nonlinear Science and Numerical Simulation*, **15** (2010), 637–647.

[14] R. R. Nigmatullin and D. Baleanu, New relationships connecting a class of fractal objects and fractional integrals in space, *Fractional Calculus and Applied Analysis*, **16**(4) (2013), 911–936.

[15] R. R. Nigmatullin and A. Le Mehaute, Is there a geometrical/physical meaning of the fractional integral with complex exponent? *Journal of Non-Crystalline Solids*, **351** (2005), 2888–2899.

[16] L. Pietronero and E. Tosatti, eds, *Proceedings of the 6th Trieste International Symposium on Fractals in Physics, ICTP*, Trieste, Italy, July 9–12, 1985, Chapter 8, Fractal in physics, N-Holland, 1986.

[17] I. Podlubny, *Fractional Differential Equations*, Academic Press, San Diego, 1999.

[18] S. G. Samko, A. A. Kilbas, and O. I. Marichev, *Integrals and Derivatives of the Fractional Order and Their Some Applications*, "Nauka and Tekhnika" Publishing House, 1987, p. 688. (In Russian).

[19] V. V. Uchaikin, *The Method of the Fractional Derivatives*, "Artishok" Publishing House, Ulianovsk, 2008, p. 512. (In Russian).

[20] X. J. Yang, D. Baleanu, and H. M. Srivastava, Local fractional similarity solution for the diffusion equation defined on Cantor sets, *Applied Mathematics Letters*, **47** (2015), 54–60.

Yuriy Povstenko

Thermodiffusion in a deformable solid: fractional calculus approach

Abstract: Methods of continuum mechanics and irreversible thermodynamics are used to describe heat conduction and diffusion in a deformable body. The balance equations and the equations of state are formulated. Nonlocal generalizations of the Fourier law in the theory of heat conduction and of the Fick equation in the theory of diffusion are analyzed. Theory of thermodiffusive stresses in a deformable solid is formulated, based on the heat conduction equation and the diffusion equation with derivatives of fractional order.

Keywords: Non-Fourier heat conduction, non-Fickian diffusion, fractional calculus, fractional thermoelasticity, thermal stresses, diffusive stresses, Duhamel–Neumann equation, generalized Beltrami–Michell equation

MSC 2000: 26A33, 35Q74, 35R11, 74F05, 80A20

1 Introduction

The beginning of the classical theory of heat conduction dates back to 1822 when Fourier [14] stated the linear dependence between the heat flux vector $\mathbf{q}$ and the temperature gradient ∇T, postulating the famous Fourier law

$$\mathbf{q} = -k\nabla T, \tag{1}$$

where ∇ is the gradient operator, and k is the thermal conductivity of a body. A few years later, Fourier's disciple Duhamel coupled the temperature field and the solid deformation [11], and pioneered studies on thermoelasticity. The Duhamel–Neumann equation [11, 27] states that stresses in a solid depend not only on strains, but also on the temperature field. Coupling of temperature and strains in the heat conduction equation was set up by Biot [3] using irreversible thermodynamics.

The Fourier law (1) is a phenomenological law, which states the proportionality of the heat flux to the gradient of the transported quantity. Likewise, the Fick law [13] in the theory of diffusion,

$$\mathbf{J} = -\kappa\nabla c, \tag{2}$$

Yuriy Povstenko, Institute of Mathematics and Computer Science, Faculty of Mathematics and Natural Sciences, Jan Długosz University in Czestochowa, Armii Krajowej 13/15, 42-200 Czestochowa, Poland, e-mail: j.povstenko@ajd.czest.pl

https://doi.org/10.1515/9783110571707-009

states the linear-dependence between the matter flux $\mathbf{J}$ and the concentration gradient ∇c, where κ is the diffusion conductivity.

The idea of matter redistribution during deformation was proposed by Gorsky [17]. This idea was used as the basis for the diffusion theory of elastic aftereffect, and was developed by Konobeevsky [21] and Fastov [12]. The modern theory of thermodiffusion in a deformable solid, based on the methods of continuum mechanics and irreversible thermodynamics, was formulated by Podstrigach (spelled also as Pidstrygach and Pidstryhach) [34–37], and was developed by Nowacki [30, 31].

The classical theory of heat conduction (based on the Fourier law (1)) and the standard theory of diffusion (based on the Fick law (2)) are quite acceptable for different physical situations. However, many theoretical and experimental studies of transport phenomena testify that in solids with complex internal structure, the classical Fourier law and Fick law are no longer accurate enough. This leads to formulation of nonclassical theories, in which the Fourier law and the Fick law, also the standard heat conduction and diffusion equations, are replaced by more general equations.

Each generalization of the heat conduction equation or the diffusion equation leads to the corresponding generalization of the theory of thermal or diffusive stresses. For example, thermoelasticity without energy dissipation proposed by Green and Naghdi [18] is based on the wave equation for temperature. The Cattaneo approach [5, 6] results in the telegraph equation for temperature, and leads to the generalized thermoelasticity of Lord and Shulman [23].

Fractional calculus (theory of integrals and derivatives of noninteger order) has many applications in physics, geophysics, geology, chemistry, engineering, bioengineering, medicine, and finance (see, for example, [24, 25, 33, 46, 51–53] and references therein).

Fractional thermoelasticity [38, 42, 47] starts from the heat conduction equation with differential operators of fractional order (see also [39–41], where this theory was formulated in terms of diffusive stresses). The fractional generalizations of the Cattaneo telegraph equation were used to formulate the corresponding time-fractional [43, 50, 55] and space-time-fractional [45, 48] theories of thermal stresses.

Below, the theory of thermodiffusive stresses, based on fractional calculus approach, is considered. The methods of continuum mechanics and irreversible thermodynamics are used to describe heat conduction and diffusion in a deformable body. The balance equations and the equations of state are formulated. Nonlocal generalizations of the Fourier law in the theory of heat conduction and the Fick law in the theory of diffusion are analyzed. The heat conduction equation and the diffusion equation with derivatives of fractional order are examined. The statements of problem in terms of displacements and in terms of stresses are discussed.

2 Balance equations

It is common practice to divide physical quantities which describe physical properties of a body into extensive and intensive quantities. An extensive physical quantity G depends on mass M of a material body, whereas an intensive one is independent of mass. Hence, an extensive quantity G can be written as an integral

$$G = \int_M g \, \mathrm{d}M = \int_V \rho g \, \mathrm{d}V, \tag{3}$$

where g is the density of G, and ρ is the mass density:

$$g = \frac{\mathrm{d}G}{\mathrm{d}M}, \quad \rho = \frac{\mathrm{d}M}{\mathrm{d}V}. \tag{4}$$

According to the Reynolds transport theorem [22, 26], the time rate of change of the quantity G is expressed as

$$\frac{\mathrm{d}G}{\mathrm{d}t} = \frac{\mathrm{d}}{\mathrm{d}t} \int_V \rho g \, \mathrm{d}V = \int_V \left[\frac{\mathrm{d}(\rho g)}{\mathrm{d}t} + \rho g \nabla \cdot \mathbf{v} \right] \mathrm{d}V, \tag{5}$$

where $\mathbf{v}$ is the velocity vector, and

$$\frac{\mathrm{d}g}{\mathrm{d}t} = \frac{\partial g}{\partial t} + \mathbf{v} \cdot \nabla g \tag{6}$$

is the material time derivative.

The change of the quantity G in a volume V can occur at the cost of the flux $\mathbf{J}^{(g)}$ through a surface A, which bounds a volume V, and the source of the quantity G with the density $\Theta^{(g)}$ within a volume V:

$$\frac{\mathrm{d}G}{\mathrm{d}t} = - \int_A \mathbf{n} \cdot \mathbf{J}^{(g)} \, \mathrm{d}A + \int_V \Theta^{(g)} \, \mathrm{d}V, \tag{7}$$

where $\mathbf{n}$ is the outer unit normal to the boundary surface A.

Taking into account the Gauss–Ostrogradsky formula, from (5) and (7) we get:

$$\int_V \left[\frac{\mathrm{d}(\rho g)}{\mathrm{d}t} + \rho g \nabla \cdot \mathbf{v} + \nabla \cdot \mathbf{J}^{(g)} - \Theta^{(g)} \right] \mathrm{d}V = 0. \tag{8}$$

Since a volume V is arbitrary, the integrand should vanish, and the general balance equation in the differential form is obtained:

$$\frac{\mathrm{d}(\rho g)}{\mathrm{d}t} + \rho g \nabla \cdot \mathbf{v} + \nabla \cdot \mathbf{J}^{(g)} - \Theta^{(g)} = 0. \tag{9}$$

The integral form of mass conservation is written as

$$\frac{\mathrm{d}M}{\mathrm{d}t} = \frac{\mathrm{d}}{\mathrm{d}t}\int_V \rho\,\mathrm{d}V = \int_V \left(\frac{\mathrm{d}\rho}{\mathrm{d}t} + \rho\nabla\cdot\mathbf{v}\right)\mathrm{d}V = 0 \tag{10}$$

or, in the local differential form known as the continuity equation,

$$\frac{\mathrm{d}\rho}{\mathrm{d}t} + \rho\nabla\cdot\mathbf{v} = 0. \tag{11}$$

Using (11), the general balance Equation (9) can be rewritten as

$$\rho\frac{\mathrm{d}g}{\mathrm{d}t} = -\nabla\cdot\mathbf{J}^{(g)} + \Theta^{(g)}. \tag{12}$$

If a solid solution is considered, then mass conservation of the solute substance reads

$$\rho\frac{\mathrm{d}c}{\mathrm{d}t} = -\nabla\cdot\mathbf{J}, \tag{13}$$

where c is the concentration of solute substance, and $\mathbf{J}$ is the matter flux.

According to the second Newton law for the material continuum, the time derivative of momentum in a material volume V bounded by a surface A equals the sum of volume and surface forces:

$$\frac{\mathrm{d}}{\mathrm{d}t}\int_V \rho\mathbf{v}\,\mathrm{d}V = \int_V \mathbf{F}_V\,\mathrm{d}V + \int_A \mathbf{F}_A\,\mathrm{d}A. \tag{14}$$

Here $\mathbf{F}_V$ is the volume force density, and $\mathbf{F}_A$ is the surface force density. In the following, the index V in the volume force density will be omitted.

The surface force density vector $\mathbf{F}_A$, at some point of a surface A, depends on the orientation of a surface element and is connected with the normal vector $\mathbf{n}$ by the linear dependence known as the Cauchy stress theorem:

$$\mathbf{F}_A(\mathbf{n}) = \mathbf{n}\cdot\boldsymbol{\sigma}, \tag{15}$$

where $\boldsymbol{\sigma}$ denotes the stress tensor. Hence,

$$\frac{\mathrm{d}}{\mathrm{d}t}\int_V \rho\mathbf{v}\,\mathrm{d}V = \int_V \mathbf{F}\,\mathrm{d}V + \int_A \mathbf{n}\cdot\boldsymbol{\sigma}\,\mathrm{d}A. \tag{16}$$

Taking into account the Reynolds transport theorem, the Gauss–Ostrogradsky formula, and the continuity Equation (11) results in the equation of motion

$$\rho\frac{\mathrm{d}\mathbf{v}}{\mathrm{d}t} = \mathbf{F} + \nabla\cdot\boldsymbol{\sigma}. \tag{17}$$

The balance of kinetic energy

$$\frac{1}{2}\rho\frac{\mathrm{d}\mathbf{v}^2}{\mathrm{d}t} = \mathbf{F}\cdot\mathbf{v} + (\nabla\cdot\boldsymbol{\sigma})\cdot\mathbf{v} \tag{18}$$

is obtained from (17) by multiplying both sides of the equation scalarly by $\mathbf{v}$.

The balance of moment of momentum with respect to the origin of the coordinate system is written as

$$\frac{\mathrm{d}}{\mathrm{d}t}\int_V \mathbf{r}\times\rho\mathbf{v}\,\mathrm{d}V = \int_V \mathbf{r}\times\mathbf{F}\,\mathrm{d}V + \int_A \mathbf{r}\times(\mathbf{n}\cdot\boldsymbol{\sigma})\mathrm{d}A, \tag{19}$$

where $\mathbf{r}$ denotes the radius-vector of a point, and $\times$ means the cross product. After using the Reynolds transport theorem and the Gauss–Ostrogradsky formula, Equation (19) is represented as

$$\int_V\left[\frac{\mathrm{d}}{\mathrm{d}t}(\mathbf{r}\times\rho\mathbf{v}) + (\mathbf{r}\times\rho\mathbf{v})\nabla\cdot\mathbf{v}\right]\mathrm{d}V = \int_V \mathbf{r}\times\mathbf{F}\,\mathrm{d}V - \int_V \nabla\cdot(\boldsymbol{\sigma}\times\mathbf{r})\mathrm{d}V \tag{20}$$

or, in the differential form,

$$\mathbf{r}\times\rho\frac{\mathrm{d}\mathbf{v}}{\mathrm{d}t} - \mathbf{r}\times\mathbf{F} + \nabla\cdot(\boldsymbol{\sigma}\times\mathbf{r}) = 0. \tag{21}$$

Multiplying both sides of the equation of motion (17) vectorialy by the radius-vector $\mathbf{r}$ and using the relation

$$\nabla\cdot(\boldsymbol{\sigma}\times\mathbf{r}) = -\mathbf{r}\times(\nabla\cdot\boldsymbol{\sigma}) + \boldsymbol{\epsilon}:\boldsymbol{\sigma}, \tag{22}$$

we get

$$\boldsymbol{\epsilon}:\boldsymbol{\sigma} = 0. \tag{23}$$

Here $\boldsymbol{\epsilon}$ is the antisymmetric Levi-Civita tensor, and the symbol : denotes the double inner product. Equation (23) means that the stress tensor $\boldsymbol{\sigma}$ is symmetric:

$$\boldsymbol{\sigma} = \boldsymbol{\sigma}^{\mathrm{T}}. \tag{24}$$

The total energy of a body is the sum of the internal energy and kinetic energy

$$\int_V \rho\varepsilon\,\mathrm{d}V = \int_V \rho u\,\mathrm{d}V + \int_V \frac{1}{2}\rho\mathbf{v}^2\,\mathrm{d}V, \tag{25}$$

where ε is the density of the total energy, and u is the density of the internal energy.

The change of the total energy is determined by the work of volume and surface forces and by the heat flux $\mathbf{q}$:

$$\frac{\mathrm{d}}{\mathrm{d}t}\int_V \rho\varepsilon\,\mathrm{d}V = \int_V \mathbf{F}\cdot\mathbf{v}\,\mathrm{d}V + \int_A \mathbf{n}\cdot\boldsymbol{\sigma}\cdot\mathbf{v}\,\mathrm{d}A - \int_A \mathbf{n}\cdot\mathbf{q}\,\mathrm{d}A. \tag{26}$$

It follows from Equations (25) and (26) that

$$\rho \frac{\mathrm{d}u}{\mathrm{d}t} + \rho \frac{\mathrm{d}}{\mathrm{d}t}\left(\frac{1}{2}\mathbf{v}^2\right) = \mathbf{F} \cdot \mathbf{v} + \nabla \cdot (\boldsymbol{\sigma} \cdot \mathbf{v}) - \nabla \cdot \mathbf{q}. \tag{27}$$

Taking into account the balance of kinetic energy (18) and the relation

$$\nabla \cdot (\boldsymbol{\sigma} \cdot \mathbf{v}) = (\nabla \cdot \boldsymbol{\sigma}) \cdot \mathbf{v} + \boldsymbol{\sigma} : \nabla \mathbf{v}, \tag{28}$$

valid for a symmetric tensor $\boldsymbol{\sigma}$, we obtain the balance equation of the internal energy

$$\rho \frac{\mathrm{d}u}{\mathrm{d}t} = \boldsymbol{\sigma} : \nabla \mathbf{v} - \nabla \cdot \mathbf{q}. \tag{29}$$

3 Equations of state

The principle of local (cellular) equilibrium [15, 20], which summarizes the first and second laws of thermodynamics, leads to the Gibbs equation for a deformable solid solution

$$\mathrm{d}u = T\mathrm{d}s + \frac{1}{\rho}\boldsymbol{\sigma} : \mathrm{d}\mathbf{e} + \varphi\, \mathrm{d}c, \tag{30}$$

where s is the entropy density, T is the temperature, φ stands for the chemical potential of the soluble substance, and $\mathbf{e}$ denotes the linear strain tensor

$$\mathbf{e} = \frac{1}{2}(\nabla \mathbf{u} + \mathbf{u}\nabla) \tag{31}$$

with $\mathbf{u}$ being the displacement vector.

According to Gibbs equation (30), the internal energy density depends on the entropy density, strain tensor, and concentration, that is, $u = u(s, \mathbf{e}, c)$, and the equations of state have the following form:

$$T = \left(\frac{\partial u}{\partial s}\right)_{\mathbf{e},c}, \quad \boldsymbol{\sigma} = \rho\left(\frac{\partial u}{\partial \mathbf{e}}\right)_{s,c}, \quad \varphi = \left(\frac{\partial u}{\partial c}\right)_{s,\mathbf{e}}. \tag{32}$$

It is convenient to introduce the density of the grand potential $\omega(T, \mathbf{e}, \varphi)$ by the Legendre transform

$$\omega = u - Ts - \varphi c \tag{33}$$

with the corresponding Gibbs equation

$$\mathrm{d}\omega = -s\mathrm{d}T + \frac{1}{\rho}\boldsymbol{\sigma} : \mathrm{d}\mathbf{e} - c\,\mathrm{d}\varphi \tag{34}$$

and the equations of state

$$s = -\left(\frac{\partial \omega}{\partial T}\right)_{\mathbf{e},\varphi}, \quad \boldsymbol{\sigma} = \rho\left(\frac{\partial \omega}{\partial \mathbf{e}}\right)_{T,\varphi}, \quad c = -\left(\frac{\partial \omega}{\partial \varphi}\right)_{T,\mathbf{e}}. \tag{35}$$

In the framework of linear theory, it is suggested that $\frac{\mathrm{d}g}{\mathrm{d}t} \sim \frac{\partial g}{\partial t}$, the mass density is assumed to be constant $\rho = \text{const}$, and the components of the strain tensor $\mathbf{e}$, the temperature change $\theta = T - T_0$, compared with the reference temperature T_0, and the chemical potential change $\phi = \varphi - \varphi_0$, compared with the reference chemical potential φ_0, are anticipated to be small.

For an elastic body with thermodiffusion, the temperature T, the components of the strain tensor $\mathbf{e}$, and the chemical potential of soluble substance φ define the state of a body. A single-valued functions of these state variables expressing other state variables are called the equations of state. For an isotropic body, with an accuracy of quadratic terms and under the assumption of absence of initial stresses, the grand potential density ω has the following series development:

$$\begin{aligned} \omega = {} & \frac{\mu}{\rho}\,\mathbf{e} : \mathbf{e} + \frac{\lambda}{2\rho}(\operatorname{tr}\mathbf{e})^2 - \frac{\beta_\theta K}{\rho}(\operatorname{tr}\mathbf{e})\theta - \frac{C}{2T_0}\theta^2 \\ & - \frac{\beta_\phi K}{\rho}(\operatorname{tr}\mathbf{e})\phi - \frac{D}{2}\phi^2 - \xi\theta\phi, \end{aligned} \tag{36}$$

which leads to the equations of state

$$\boldsymbol{\sigma} = \rho\left(\frac{\partial \omega}{\partial \mathbf{e}}\right)_{\theta,\phi} = 2\mu\mathbf{e} + (\lambda \operatorname{tr}\mathbf{e} - \beta_\theta K\theta - \beta_\phi K\phi)\mathbf{I}, \tag{37}$$

$$s = -\left(\frac{\partial \omega}{\partial \theta}\right)_{\mathbf{e},\phi} = \frac{C}{T_0}\theta + \frac{\beta_\theta K}{\rho}\operatorname{tr}\mathbf{e} + \xi\phi, \tag{38}$$

$$c = -\left(\frac{\partial \omega}{\partial \phi}\right)_{\mathbf{e},\theta} = D\phi + \frac{\beta_\phi K}{\rho}\operatorname{tr}\mathbf{e} + \xi\theta. \tag{39}$$

Here $\operatorname{tr}\mathbf{e} = \nabla \cdot \mathbf{u}$, λ and μ are Lamé constants, $K = \lambda + 2\mu/3$ is the bulk modulus, β_θ denotes the thermal coefficient of volumetric expansion, β_ϕ denotes the diffusive coefficient of volumetric expansion, C is the heat capacity, D is the mass capacity, and $\mathbf{I}$ stands for the unit tensor.

Equations (37)–(39) can be solved for strains

$$\mathbf{e} = \frac{1}{2\mu}\left[\boldsymbol{\sigma} - \frac{\nu}{1+\nu}(\operatorname{tr}\boldsymbol{\sigma})\mathbf{I}\right] + \left(\frac{\beta_\theta}{3}\theta + \frac{\beta_\phi}{3}\phi\right)\mathbf{I}, \tag{40}$$

temperature

$$\theta = \frac{T_0}{C'}s - \frac{\beta_s K}{\rho}\operatorname{tr}\mathbf{e} - \xi_c c, \tag{41}$$

and chemical potential

$$\phi = \frac{1}{D'}c - \frac{\beta_c K}{\rho}\,\mathrm{tr}\,\mathbf{e} - \xi_s s, \tag{42}$$

where ν is the Poisson ratio, and new coefficients C', D', β_s, β_c, ξ_c, and ξ_s can be easily calculated in terms of C, D, β_θ, β_ϕ, and ξ.

4 The constitutive equations for fluxes

For brevity, in what follows the interrelation between heat conduction and diffusion in the constitutive equations for the heat flux $\mathbf{q}$ and the matter flux $\mathbf{J}$ will be neglected. The consideration will be restricted to the dependence of the heat flux $\mathbf{q}$ on the temperature gradient $\nabla\theta$ and the dependence of the matter flux $\mathbf{J}$ on the chemical potential gradient $\nabla\phi$. Under this assumption, in linear approximation, the balance of entropy takes the form

$$\rho T_0 \frac{\partial s}{\partial t} = -\nabla \cdot \mathbf{q}. \tag{43}$$

For materials with time nonlocality, the heat flux at a point $\mathbf{x}$ at time t depends on the history of temperature gradient. To calculate the heat flux at time t, the temperature gradient should be integrated from a "starting point" to the present time with some kernel (the weight function), which characterizes memory. Mathematical models for describing memory effects in solids were intensively discussed in the literature (see, for example, [1, 2, 8–10, 49, 54]).

The general time-nonlocal constitutive equation for the heat flux was considered in [19]. Choosing 0 as a "starting point", we get [28, 29]:

$$\mathbf{q}(t) = -k_\theta \int_0^t K_\theta(t-\tau)\,\nabla\theta(\tau)\,\mathrm{d}\tau. \tag{44}$$

The entropy balance Equation (43), the equation of state for entropy (38), and the constitutive Equation (44) for the heat flux result in the general time-nonlocal heat conduction equation for thermodiffusion in an elastic solid:

$$\frac{\partial\theta}{\partial t} + \gamma_\theta \frac{\partial(\mathrm{tr}\,\mathbf{e})}{\partial t} + \eta_\theta \frac{\partial\phi}{\partial t} = a_\theta \int_0^t K_\theta(t-\tau)\,\Delta\theta(\tau)\,\mathrm{d}\tau, \tag{45}$$

where the coefficient $\gamma_\theta = \beta_\theta K T_0/(\rho C)$ describes the effect of deformation on heat conduction, $\eta_\theta = T_0\xi/C$, and $a_\theta = k_\theta/(\rho C)$ is the thermal diffusivity coefficient.

The classical Fourier law

$$\mathbf{q}(t) = -k_\theta \nabla\theta(t) \tag{46}$$

and the parabolic heat conduction equation

$$\frac{\partial\theta}{\partial t} + \gamma_\theta \frac{\partial(\operatorname{tr}\mathbf{e})}{\partial t} + \eta_\theta \frac{\partial\phi}{\partial t} = a_\theta \Delta\theta \tag{47}$$

are obtained from (44) and (45) for "instantaneous memory" with the kernel being the Dirac delta function, $K_\theta(t-\tau) = \delta(t-\tau)$.

"Full sclerosis" corresponds to the choice of the kernel as the time derivative of Dirac's delta: $K_\theta(t-\tau) = -\delta'(t-\tau)$ or

$$\mathbf{q}(t) = -k_\theta \frac{\partial}{\partial t} \nabla\theta(t), \tag{48}$$

thus leading to the corresponding Helmholtz equation for temperature

$$\theta + \gamma_\theta \operatorname{tr}\mathbf{e} + \eta_\theta \phi = a_\theta \Delta\theta. \tag{49}$$

"Full memory" [18, 28] means that there is no fading of memory and the kernel is constant: $K_\theta(t-\tau) = 1$. As a result, we get the wave equation for temperature

$$\frac{\partial^2\theta}{\partial t^2} + \gamma_\theta \frac{\partial^2(\operatorname{tr}\mathbf{e})}{\partial t^2} + \eta_\theta \frac{\partial^2\phi}{\partial t^2} = a_\theta \Delta\theta. \tag{50}$$

The time-nonlocal dependence between the heat flux vector $\mathbf{q}$ and the temperature gradient $\nabla\theta$ with the "long-tail" power kernel [38, 42, 44, 47] can be interpreted in terms of the Riemann–Liouville fractional integrals and derivatives

$$\mathbf{q}(t) = -k_\theta\, {}^{\mathrm{RL}}D_t^{1-\alpha} \nabla\theta(t), \quad 0 < \alpha \le 1, \tag{51}$$

$$\mathbf{q}(t) = -k_\theta\, I_t^{\alpha-1} \nabla\theta(t), \quad 1 < \alpha \le 2. \tag{52}$$

In (51) and (52), we have used the index t to emphasize that differentiation and integration is carried out with respect to time t. The constitutive Equations (51) and (52) lead to the time-fractional heat conduction equation with the Caputo fractional derivative

$${}^{C}D_t^{\alpha}\theta + \gamma_\theta\, {}^{C}D_t^{\alpha}(\operatorname{tr}\mathbf{e}) + \eta_\theta\, {}^{C}D_t^{\alpha}\phi = a_\theta \Delta\theta, \quad 0 < \alpha \le 2. \tag{53}$$

"Short-tail memory" described by the exponential kernel (see [7])

$$\mathbf{q}(t) = -\frac{k_\theta}{\zeta_\theta} \int_0^t \exp\left(-\frac{t-\tau}{\zeta_\theta}\right) \nabla\theta(\tau)\, \mathrm{d}\tau \tag{54}$$

with a nonnegative constant ζ_θ results in the corresponding telegraph equation for temperature

$$\frac{\partial\theta}{\partial t}+\zeta_\theta\frac{\partial^2\theta}{\partial t^2}+\gamma_\theta\frac{\partial(\mathrm{tr}\,\mathbf{e})}{\partial t}+\zeta_\theta\gamma_\theta\frac{\partial^2(\mathrm{tr}\,\mathbf{e})}{\partial t^2}+\eta_\theta\frac{\partial\phi}{\partial t}+\zeta_\theta\eta_\theta\frac{\partial^2\phi}{\partial t^2}=a_\theta\Delta\theta. \tag{55}$$

In the case of "middle-tail memory" [45]

$$\mathbf{q}(t)=-\frac{k_\theta}{\zeta_\theta}\int_0^t (t-\tau)^{\beta-2}E_{\beta-\alpha,\beta-1}\left[-\frac{(t-\tau)^{\beta-\alpha}}{\zeta_\theta}\right]\nabla\theta(\tau)\,\mathrm{d}\tau, \tag{56}$$

where $E_{\alpha,\beta}(z)$ is the Mittag-Leffler function in two parameters α and β, described by the series representation [16]

$$E_{\alpha,\beta}(z)=\sum_{k=0}^{\infty}\frac{z^k}{\Gamma(\alpha k+\beta)},\quad \alpha>0,\quad \beta>0,\quad z\in C,$$

we get the time-fractional telegraph equation for temperature

$$\begin{aligned}&{}^C D_t^\alpha\theta+\zeta_\theta\,{}^C D_t^\beta\theta+\gamma_\theta\,{}^C D_t^\alpha(\mathrm{tr}\,\mathbf{e})+\zeta_\theta\gamma_\theta\,{}^C D_t^\beta(\mathrm{tr}\,\mathbf{e})\\&+\eta_\theta\,{}^C D_t^\alpha\phi+\zeta_\theta\eta_\theta\,{}^C D_t^\beta\phi=a_\theta\Delta\theta,\quad 0<\alpha\le 1,\quad 1<\beta\le 2.\end{aligned} \tag{57}$$

Similar analysis can be carried out for the matter flux $\mathbf{J}$. The general time-nonlocal generalization of the Fick law is written as

$$\mathbf{J}(t)=-k_\phi\int_0^t K_\phi(t-\tau)\,\nabla\,\phi(\tau)\,\mathrm{d}\tau. \tag{58}$$

The equation of mass conservation of soluble substance (13), the equation of state for concentration (39), and the constitutive Equation (58) for the matter flux result in the general time-nonlocal diffusion equation for thermodiffusion in an elastic solid:

$$\frac{\partial\phi}{\partial t}+\gamma_\phi\frac{\partial(\mathrm{tr}\,\mathbf{e})}{\partial t}+\eta_\phi\frac{\partial\theta}{\partial t}=a_\phi\int_0^t K_\phi(t-\tau)\,\Delta\phi(\tau)\,\mathrm{d}\tau, \tag{59}$$

where $\gamma_\phi=\beta_\phi K/(\rho D)$, $\eta_\phi=\xi/D$, and $a_\phi=k_\phi/(\rho D)$.

The classical Fick law in terms of chemical potential

$$\mathbf{J}(t)=-k_\phi\nabla\phi(t) \tag{60}$$

and the parabolic diffusion equation

$$\frac{\partial\phi}{\partial t}+\gamma_\phi\frac{\partial(\mathrm{tr}\,\mathbf{e})}{\partial t}+\eta_\phi\frac{\partial\theta}{\partial t}=a_\phi\Delta\phi \tag{61}$$

are obtained from (58) and (59) for "instantaneous memory" with the kernel being the Dirac delta: $K_\phi(t-\tau) = \delta(t-\tau)$.

"Full sclerosis" conforms to the choice of the kernel as the time derivative of Dirac's delta: $K_\phi(t-\tau) = -\delta'(t-\tau)$ or

$$\mathbf{J}(t) = -k_\phi \frac{\partial}{\partial t} \nabla\phi(t), \tag{62}$$

which results in the corresponding Helmholtz equation for the chemical potential

$$\phi + \gamma_\phi \operatorname{tr}\mathbf{e} + \eta_\phi \theta = a_\phi \Delta\phi. \tag{63}$$

For "full memory," the kernel is constant: $K_\varphi(t-\tau) = 1$. Hence, the chemical potential satisfies the wave equation

$$\frac{\partial^2\phi}{\partial t^2} + \gamma_\phi \frac{\partial^2(\operatorname{tr}\mathbf{e})}{\partial t^2} + \eta_\phi \frac{\partial^2\theta}{\partial t^2} = a_\phi \Delta\phi. \tag{64}$$

The time-nonlocal dependence between the matter flux vector $\mathbf{J}$ and the chemical potential gradient $\nabla\phi$ with the "long-tail" power kernel has the form

$$\mathbf{J}(t) = -k_\phi \, {}^{\mathrm{RL}}D_t^{1-\alpha} \nabla\phi(t), \quad 0 < \alpha \le 1, \tag{65}$$

$$\mathbf{J}(t) = -k_\phi \, I_t^{\alpha-1} \nabla\phi(t), \quad 1 < \alpha \le 2, \tag{66}$$

and leads to the time-fractional diffusion equation with the Caputo fractional derivative

$${}^{C}D_t^{\alpha}\phi + \gamma_\phi \, {}^{C}D_t^{\alpha}(\operatorname{tr}\mathbf{e}) + \eta_\phi \, {}^{C}D_t^{\alpha}\theta = a_\phi \Delta\phi, \quad 0 < \alpha \le 2. \tag{67}$$

In the case of "short-tail memory" with the exponential kernel

$$\mathbf{J}(t) = -\frac{k_\phi}{\zeta_\phi} \int_0^t \exp\left(-\frac{t-\tau}{\zeta_\phi}\right) \nabla\phi(\tau) \, \mathrm{d}\tau \tag{68}$$

with ζ_ϕ being a nonnegative constant, the corresponding telegraph equation for the chemical potential is obtained as

$$\frac{\partial\phi}{\partial t} + \zeta_\phi \frac{\partial^2\phi}{\partial t^2} + \gamma_\phi \frac{\partial(\operatorname{tr}\mathbf{e})}{\partial t} + \zeta_\phi\gamma_\phi \frac{\partial^2(\operatorname{tr}\mathbf{e})}{\partial t^2} + \eta_\phi \frac{\partial\theta}{\partial t} + \zeta_\phi\eta_\phi \frac{\partial^2\theta}{\partial t^2} = a_\phi \Delta\phi. \tag{69}$$

For "middle-tail memory,"

$$\mathbf{J}(t) = -\frac{k_\phi}{\zeta_\phi} \int_0^t (t-\tau)^{\beta-2} E_{\beta-\alpha,\beta-1}\left[-\frac{(t-\tau)^{\beta-\alpha}}{\zeta_\phi}\right] \nabla\phi(\tau) \, \mathrm{d}\tau, \tag{70}$$

and we arrive at the time-fractional telegraph equation for chemical potential

$$\begin{aligned} &{}^{C}D_t^{\alpha}\phi + \zeta_\phi \, {}^{C}D_t^{\beta}\phi + \gamma_\phi \, {}^{C}D_t^{\alpha}(\operatorname{tr}\mathbf{e}) + \zeta_\phi\gamma_\phi \, {}^{C}D_t^{\beta}(\operatorname{tr}\mathbf{e}) \\ &+ \eta_\phi \, {}^{C}D_t^{\alpha}\theta + \zeta_\phi\eta_\phi \, {}^{C}D_t^{\beta}\theta = a_\phi \Delta\phi, \quad 0 < \alpha \le 1, \quad 1 < \beta \le 2. \end{aligned} \tag{71}$$

5 Statement of thermodiffusion problem for a deformable solid

The thermodiffusion problem for a deformable solid can be stated in terms of displacements, or in terms of stresses.

Substitution of the equation of state (37) into the motion Equation (17), taking into account the geometrical relations (31), leads to the equation of motion in terms of displacements

$$\mu \Delta \mathbf{u} + (\lambda + \mu) \nabla (\nabla \cdot \mathbf{u}) = \rho \frac{\partial^2 \mathbf{u}}{\partial t^2} - \mathbf{F} + \beta_\theta K \, \nabla \theta + \beta_\phi K \, \nabla \phi. \tag{72}$$

This equation should be considered with the constitutive equation for stresses

$$\boldsymbol{\sigma} = 2\mu \mathbf{e} + (\lambda \operatorname{tr} \mathbf{e} - \beta_\theta K \theta - \beta_\phi K \phi) \mathbf{I}, \tag{73}$$

the geometrical relations

$$\mathbf{e} = \frac{1}{2} (\nabla \mathbf{u} + \mathbf{u} \nabla), \tag{74}$$

and the chosen equations describing heat conduction and diffusion processes.

The statement of the problem in terms of stresses is described by the equation of motion

$$\nabla \cdot \boldsymbol{\sigma} = \rho \dot{\mathbf{v}} - \mathbf{F}, \tag{75}$$

where the superdot denotes the derivative with respect to time, the constitutive equation for strains

$$\mathbf{e} = \frac{1}{2\mu} \left[\boldsymbol{\sigma} - \frac{\nu}{1+\nu} (\operatorname{tr} \boldsymbol{\sigma}) \mathbf{I} \right] + \left(\frac{\beta_\theta}{3} \theta + \frac{\beta_\phi}{3} \phi \right) \mathbf{I}, \tag{76}$$

the geometrical relations

$$\mathbf{e} = \frac{1}{2} (\nabla \mathbf{u} + \mathbf{u} \nabla) \equiv \operatorname{Def} \mathbf{u}, \tag{77}$$

the corresponding heat conduction equation, and the diffusion equation.

For an arbitrary vector $\mathbf{u}$, the gradient operator ∇ fulfills the well-known equations

$$\nabla \times \nabla \mathbf{u} = 0 \quad \text{and} \quad \mathbf{u} \nabla \times \nabla = 0. \tag{78}$$

For a second-order tensor $\mathbf{a}$, the incompatibility operator Inc is introduced as

$$\operatorname{Inc} \mathbf{a} \equiv \nabla \times \mathbf{a} \times \nabla. \tag{79}$$

Due to (77)–(79),

$$\operatorname{Inc}\operatorname{Def}\mathbf{u} = 0. \tag{80}$$

Hence, the compatibility condition for the strain tensor $\mathbf{e}$ has the following form:

$$\operatorname{Inc}\mathbf{e} = 0. \tag{81}$$

As the problem is formulated in terms of stresses, Equation (81) should be also rewritten in terms of the stress tensor $\boldsymbol{\sigma}$. To do this, the incompatibility operator Inc should be applied to the strain tensor $\mathbf{e}$ calculated from the constitutive Equation (76), thus obtaining

$$\nabla \times \boldsymbol{\sigma} \times \nabla - \nabla \times \left[\left(\frac{\nu}{1+\nu}\operatorname{tr}\boldsymbol{\sigma} - \frac{2\mu\beta_\theta}{3}\theta - \frac{2\mu\beta_\phi}{3}\phi\right)\mathbf{I}\right] \times \nabla = 0. \tag{82}$$

For a symmetric tensor $\boldsymbol{\sigma}$, the following formula is valid [4]:

$$\begin{aligned}\nabla \times \boldsymbol{\sigma} \times \nabla = {} & (\nabla \cdot \boldsymbol{\sigma} \cdot \nabla)\mathbf{I} + \Delta\boldsymbol{\sigma} - \nabla(\nabla \cdot \boldsymbol{\sigma}) - (\boldsymbol{\sigma} \cdot \nabla)\nabla \\ & + \nabla\nabla(\operatorname{tr}\boldsymbol{\sigma}) - [\Delta(\operatorname{tr}\boldsymbol{\sigma})]\mathbf{I}.\end{aligned} \tag{83}$$

It can be easily calculated that

$$\nabla \times (\theta\mathbf{I}) \times \nabla = \nabla\nabla\theta - \Delta\theta\mathbf{I}. \tag{84}$$

Therefore,

$$\begin{aligned}&(\nabla \cdot \boldsymbol{\sigma} \cdot \nabla)\mathbf{I} + \Delta\boldsymbol{\sigma} - \nabla(\nabla \cdot \boldsymbol{\sigma}) - (\boldsymbol{\sigma} \cdot \nabla)\nabla + \frac{1}{1+\nu}\nabla\nabla(\operatorname{tr}\boldsymbol{\sigma}) \\ &- \frac{1}{1+\nu}\Delta(\operatorname{tr}\boldsymbol{\sigma})\mathbf{I} + \frac{2\mu\beta_\theta}{3}(\nabla\nabla\theta - \Delta\theta\mathbf{I}) + \frac{2\mu\beta_\phi}{3}(\nabla\nabla\phi - \Delta\phi\mathbf{I}) = 0.\end{aligned} \tag{85}$$

Taking the trace of Equation (85), we get

$$\nabla \cdot \boldsymbol{\sigma} \cdot \nabla = \frac{1-\nu}{1+\nu}\Delta(\operatorname{tr}\boldsymbol{\sigma}) + \frac{4\mu\beta_\theta}{3}\Delta\theta + \frac{4\mu\beta_\phi}{3}\Delta\phi. \tag{86}$$

Substituting (86) into (85), we have

$$\begin{aligned}&\Delta\boldsymbol{\sigma} - \nabla(\nabla \cdot \boldsymbol{\sigma}) - (\boldsymbol{\sigma} \cdot \nabla)\nabla + \frac{1}{1+\nu}\nabla\nabla(\operatorname{tr}\boldsymbol{\sigma}) - \frac{\nu}{1+\nu}[\Delta(\operatorname{tr}\boldsymbol{\sigma})]\mathbf{I} \\ &+ \frac{2\mu\beta_\theta}{3}(\nabla\nabla\theta + \Delta\theta\mathbf{I}) + \frac{2\mu\beta_\phi}{3}(\nabla\nabla\phi + \Delta\phi\mathbf{I}) = 0.\end{aligned} \tag{87}$$

This compatibility equation is valid independently of the equation of motion. Taking into account Equation (75), we obtain

$$\begin{aligned}&\Delta\boldsymbol{\sigma} + \frac{1}{1+\nu}\nabla\nabla(\operatorname{tr}\boldsymbol{\sigma}) - \frac{\nu}{1+\nu}[\Delta(\operatorname{tr}\boldsymbol{\sigma})]\mathbf{I} \\ &+ \frac{2\mu\beta_\theta}{3}(\nabla\nabla\theta + \Delta\theta\mathbf{I}) + \frac{2\mu\beta_\phi}{3}(\nabla\nabla\phi + \Delta\phi\mathbf{I}) \\ &= \nabla(\rho\dot{\mathbf{v}}) + (\rho\dot{\mathbf{v}})\nabla - \nabla\mathbf{F} - \mathbf{F}\nabla.\end{aligned} \tag{88}$$

The trace of Equation (88) gives

$$\Delta(\mathrm{tr}\,\boldsymbol{\sigma}) = \frac{1+\nu}{1-\nu}\left[\nabla\cdot(\rho\dot{\mathbf{v}}) - \nabla\cdot\mathbf{F} - \frac{4\mu\beta_\theta}{3}\Delta\theta - \frac{4\mu\beta_\phi}{3}\Delta\phi\right]. \tag{89}$$

Finally, we get the generalized Beltrami–Michell compatibility equation

$$\begin{aligned}\Delta\boldsymbol{\sigma} &+ \frac{1}{1+\nu}\nabla\nabla(\mathrm{tr}\,\boldsymbol{\sigma}) + \frac{2\mu\beta_\theta}{3}\left(\nabla\nabla\theta + \frac{1+\nu}{1-\nu}\Delta\theta\mathbf{I}\right)\\ &+ \frac{2\mu\beta_\phi}{3}\left(\nabla\nabla\phi + \frac{1+\nu}{1-\nu}\Delta\phi\mathbf{I}\right)\\ &= \nabla(\rho\dot{\mathbf{v}}) + (\rho\dot{\mathbf{v}})\nabla - \nabla\mathbf{F} - \mathbf{F}\nabla + \frac{\nu}{1-\nu}[\nabla\cdot(\rho\dot{\mathbf{v}}) - \nabla\cdot\mathbf{F}]\mathbf{I}.\end{aligned} \tag{90}$$

Different particular cases of Equation (90) can be found in the literature (see, for example, [4, 32]).

In what follows, most attention will be concentrated on thermoelasticity theory based on the heat conduction Equation (53) and the diffusion Equation (67):

$${}^C D_t^\alpha\theta + \gamma_\theta\,{}^C D_t^\alpha(\mathrm{tr}\,\mathbf{e}) + \eta_\theta\,{}^C D_t^\alpha\phi = a_\theta\Delta\theta, \quad 0 < \alpha \le 2, \tag{91}$$

$${}^C D_t^\alpha\phi + \gamma_\phi\,{}^C D_t^\alpha(\mathrm{tr}\,\mathbf{e}) + \eta_\phi\,{}^C D_t^\alpha\theta = a_\phi\Delta\phi, \quad 0 < \alpha \le 2. \tag{92}$$

For dynamic problems, initial values of displacement and velocity vectors are imposed:

$$t = 0: \quad \mathbf{u}(\mathbf{x}, t) = \mathbf{u}_0(\mathbf{x}), \tag{93}$$

$$t = 0: \quad \frac{\partial\mathbf{u}(\mathbf{x}, t)}{\partial t} = \mathbf{v}_0(\mathbf{x}). \tag{94}$$

In the quasi-static statement of the problem, which is characterized by neglecting the inertia term in (72), initial values of mechanical quantities are not considered. From a physical point of view, neglecting the inertia term in (72) means that no account has been taken of mechanical oscillations. The quasistatic statement of the thermodiffusion problem in a deformable solid is possible if the relaxation time of mechanical oscillations is considerably less than the relaxation times of heat conduction and diffusion processes.

For bounded domains, the boundary conditions should be imposed. The first fundamental boundary problem in terms of stresses is characterized by the conditions of traction at the boundary Σ

$$\mathbf{n}\cdot\boldsymbol{\sigma}(\mathbf{x}, t)|_\Sigma = \mathbf{S}(\mathbf{x}_\Sigma, t), \tag{95}$$

where $\mathbf{x}_\Sigma$ is a point at the surface Σ.

The second fundamental boundary problem incorporates the given condition for displacements

$$\mathbf{u}(\mathbf{x}, t)|_\Sigma = \mathbf{U}(\mathbf{x}_\Sigma, t). \tag{96}$$

For a mixed boundary problem, on a part of the boundary, the condition on tractions is imposed, whereas on the remainder part of the boundary the condition on displacements is prescribed.

In the case of classical parabolic heat conduction and diffusion equations, the Cauchy problem implies that the initial values of the temperature and chemical potential are given:

$$t = 0: \quad \theta(\mathbf{x}, t) = \theta_0(\mathbf{x}), \tag{97}$$

$$t = 0: \quad \phi(\mathbf{x}, t) = \phi_0(\mathbf{x}). \tag{98}$$

For more general equations with time-fractional derivatives of the order $1 < \alpha \leq 2$, the initial values of the first time derivative of temperature and of the first time derivative of chemical potential should also be imposed:

$$t = 0: \quad \frac{\partial \theta(\mathbf{x}, t)}{\partial t} = \chi_0(\mathbf{x}), \tag{99}$$

$$t = 0: \quad \frac{\partial \phi(\mathbf{x}, t)}{\partial t} = \psi_0(\mathbf{x}). \tag{100}$$

If the heat conduction and diffusion Equations (91) and (92) are considered in a bounded domain, the corresponding boundary conditions should be established. The Dirichlet boundary condition (the boundary condition of the first kind) specifies the temperature at the surface of a body:

$$\theta(\mathbf{x}, t)|_\Sigma = g(\mathbf{x}_\Sigma, t). \tag{101}$$

$$\phi(\mathbf{x}, t)|_\Sigma = g(\mathbf{x}_\Sigma, t). \tag{102}$$

For equations with the Caputo time-fractional derivatives the Neumann boundary condition (the boundary condition of the second kind) should be formulated in terms of the prescribed boundary values of the heat flux $\mathbf{q}$ and the matter flux $\mathbf{J}$. For example, for Equations (91) and (92), taking into account (51), (52), and (65), (66), the Neumann boundary conditions read:

$$ {}^{\mathrm{RL}}D_t^{1-\alpha} \frac{\partial \theta}{\partial n}\bigg|_\Sigma = g_\theta(\mathbf{x}_\Sigma, t), \quad 0 < \alpha \leq 1, \tag{103}$$

$$ I_t^{\alpha-1} \frac{\partial \theta}{\partial n}\bigg|_\Sigma = g_\theta(\mathbf{x}_\Sigma, t), \quad 1 < \alpha \leq 2, \tag{104}$$

$$ {}^{\mathrm{RL}}D_t^{1-\alpha} \frac{\partial \phi}{\partial n}\bigg|_\Sigma = g_\phi(\mathbf{x}_\Sigma, t), \quad 0 < \alpha \leq 1, \tag{105}$$

$$ I_t^{\alpha-1} \frac{\partial \phi}{\partial n}\bigg|_\Sigma = g_\phi(\mathbf{x}_\Sigma, t), \quad 1 < \alpha \leq 2, \tag{106}$$

where $\partial/\partial n$ denotes differentiation along the outward-drawn normal at the boundary surface Σ.

To formulate the correct physical Robin boundary conditions, we recall the Newton conditions of the convective heat exchange between a body and the environment with the temperature θ_{e}, and the convective mass exchange between a body and the environment with the chemical potential ϕ_{e}:

$$\mathbf{q} \cdot \mathbf{n}|_{\Sigma} = h_{\theta}(\theta|_{\Sigma} - \theta_{\mathrm{e}}), \tag{107}$$

$$\mathbf{J} \cdot \mathbf{n}|_{\Sigma} = h_{\phi}(\phi|_{\Sigma} - \phi_{\mathrm{e}}), \tag{108}$$

where h_{θ} is the convective heat transfer coefficient, and h_{ϕ} is the convective mass transfer coefficient.

The corresponding boundary conditions for the heat conduction Equation (91) and the diffusion Equation (92) are the following:

$$\left(h_{\theta}\theta + k_{\theta}\, {}^{\mathrm{RL}}D_t^{1-\alpha} \frac{\partial \theta}{\partial n} \right)\bigg|_{\Sigma} = g_{\theta}(\mathbf{x}_{\Sigma}, t), \quad 0 < \alpha \le 1, \tag{109}$$

$$\left(h_{\theta}\theta + k_{\theta} I_t^{\alpha-1} \frac{\partial \theta}{\partial n} \right)\bigg|_{\Sigma} = g_{\theta}(\mathbf{x}_{\Sigma}, t), \quad 1 < \alpha \le 2, \tag{110}$$

$$\left(h_{\phi}\phi + k_{\phi}\, {}^{\mathrm{RL}}D_t^{1-\alpha} \frac{\partial \phi}{\partial n} \right)\bigg|_{\Sigma} = g_{\phi}(\mathbf{x}_{\Sigma}, t), \quad 0 < \alpha \le 1, \tag{111}$$

$$\left(h_{\phi}\phi + k_{\phi} I_t^{\alpha-1} \frac{\partial \phi}{\partial n} \right)\bigg|_{\Sigma} = g_{\phi}(\mathbf{x}_{\Sigma}, t), \quad 1 < \alpha \le 2, \tag{112}$$

where $g_{\theta}(\mathbf{x}_{\Sigma}, t) = h_{\theta}\theta_{\mathrm{e}}(\mathbf{x}_{\Sigma}, t)$, $g_{\phi}(\mathbf{x}_{\Sigma}, t) = h_{\phi}\phi_{\mathrm{e}}(\mathbf{x}_{\Sigma}, t)$.

If the surfaces of two solids are in perfect thermal contact, the temperatures on the contact surface and the heat fluxes through the contact surface are the same for both solids, and we obtain the boundary conditions of the fourth kind. For example, if heat conduction in the first solid is described by Equation (91) with the time-derivative of order $0 < \alpha \le 2$, and, in the second solid, is described by the same equation with the time-derivative of the order $0 < \beta \le 2$, then

$$\theta_1|_{\Sigma} = \theta_2|_{\Sigma}, \tag{113}$$

$$k_{\theta 1}\, {}^{\mathrm{RL}}D_t^{1-\alpha} \frac{\partial \theta_1}{\partial n}\bigg|_{\Sigma} = k_{\theta 2}\, {}^{\mathrm{RL}}D_t^{1-\beta} \frac{\partial \theta_2}{\partial n}\bigg|_{\Sigma}, \quad 0 < \alpha \le 1, \quad 0 < \beta \le 1, \tag{114}$$

$$k_{\theta 1}\, I_t^{\alpha-1} \frac{\partial \theta_1}{\partial n}\bigg|_{\Sigma} = k_{\theta 2}\, I_t^{\beta-1} \frac{\partial \theta_2}{\partial n}\bigg|_{\Sigma}, \quad 1 < \alpha \le 2, \quad 1 < \beta \le 2, \tag{115}$$

where subscripts 1 and 2 refer to solids 1 and 2, respectively, and n is the common normal at the contact surface.

The similar reasoning for the time-fractional diffusion Equation (92) leads to the boundary conditions of perfect matter contact:

$$\phi_1|_{\Sigma} = \phi_2|_{\Sigma}, \tag{116}$$

$$k_{\phi 1}\,{}^{\mathrm{RL}}D_t^{1-\alpha}\frac{\partial \phi_1}{\partial n}\bigg|_{\Sigma} = k_{\phi 2}\,{}^{\mathrm{RL}}D_t^{1-\beta}\frac{\partial \phi_2}{\partial n}\bigg|_{\Sigma}, \quad 0<\alpha\le 1, \quad 0<\beta\le 1, \tag{117}$$

$$k_{\phi 1}\,I_t^{\alpha-1}\frac{\partial \phi_1}{\partial n}\bigg|_{\Sigma} = k_{\phi 2}\,I_t^{\beta-1}\frac{\partial \phi_2}{\partial n}\bigg|_{\Sigma}, \quad 1<\alpha\le 2, \quad 1<\beta\le 2. \tag{118}$$

It should be emphasized that for other generalized heat conduction equations presented above, the proper boundary conditions should be formulated in terms of the corresponding heat and matter fluxes (not in terms of the normal derivative of temperature and the normal derivative of chemical potential only).

For formulation of the moving interface boundary conditions and the conditions of nonperfect thermal and diffusive contact, the interested reader is referred to the book [47].

The proposed theories of thermodiffusion in a deformable solid, based on the heat conduction equation and diffusion equation with derivatives of fractional-order, interpolate the classical theories of thermal and diffusive stresses associated with the parabolic heat conduction and diffusion equations, the theories of thermal, and diffusive stressses without energy dissipation of Green and Naghdi [18], based on the wave equations for temperature and chemical potential, and the theory of Lord and Shulman [23], involving the telegraph equations for temperature and chemical potential of soluble substance.

Bibliography

[1] H.-D. Alber, *Materials With Memory: Initial-Boundary Value Problems for Constitutive Equations With Internal Variables*, Springer, Berlin, 1998.

[2] G. Amendola, M. Fabrizio, and J. M. Golden, *Thermodynamics of Materials With Memory: Theory and Applications*, Springer, New York, 2012.

[3] M. A. Biot, Thermoelasticity and irreversible thermodynamics, *Journal of Applied Physics*, **27** (1956), 240–253.

[4] V. I. Blokh, *Theory of Elasticity*, Kharkov University Press, Kharkov, 1964 (in Russian).

[5] C. Cattaneo, Sulla conduzione del calore, *Atti Del Seminario Matematico E Fisico Dell'Università Di Modena*, **3** (1948), 83–101.

[6] C. Cattaneo, Sur une forme de l'équation de la chaleur éliminant le paradoxe d'une propagation instantanée, *Comptes Rendus de L'Académie Des Sciences*, **247** (1958), 431–433.

[7] D. S. Chandrasekharaiah, Thermoelasticity with second sound: a review, *Applied Mechanics Reviews*, **39** (1986), 355–376.

[8] B. D. Coleman and V. J. Mizel, A general theory of dissipation in materials with memory, *Archive for Rational Mechanics and Analysis*, **27** (1968), 255–274.

[9] B. D. Coleman and V. J. Mizel, On the general theory of fading memory, *Archive for Rational Mechanics and Analysis*, **29** (1968), 18–31.

[10] W. A. Day, *The Thermodynamics of Simple Materials With Fading Memory*, Springer, Berlin, 1972.

[11] J.-M.-C. Duhamel, Second mémoire sur les phénomènes thermo-mécaniques, *Journal de L'École Polytechnique*, **15** (1837), 1–57.

[12] N. S. Fastov, Thermodynamics of irreversible processes in elastic deformation of bodies, in B. Ya. Lyubov and O. P. Maksimova (eds.) *Problems of Physical Metallurgy*, pp. 550–576, Metallurgizdat, Moscow, 1958 (in Russian).

[13] A. Fick, Über Diffusion, *Annalen der Physik*, **94** (1855), 59–86.

[14] J. B. J. Fourier, *Théorie analytique de la chaleur*, Firmin Didot, Paris, 1822.

[15] P. Germain, *Cours de mécanique des milieux continus, tome I: théorie générale*, Masson, Paris, 1973.

[16] R. Gorenflo, A. A. Kilbas, F. Mainardi, and S. V. Rogosin, *Mittag-Leffler Functions, Related Topics and Applications*, Springer, Berlin, 2014.

[17] W. S. Gorsky, Theorie der elastischen Nachwirkung in ungeordneten Misch-Kristallen (elastische Nachwirkung zweiter Art), *Physikalische Zeitschrift der Sowjetunion*, **8** (1935), 457–471.

[18] A. E. Green and P. M. Naghdi, Thermoelasticity without energy dissipation, *Journal of Elasticity*, **31** (1993), 189–208.

[19] M. E. Gurtin and A. C. Pipkin, A general theory of heat conduction with finite wave speeds, *Archive for Rational Mechanics and Analysis*, **31** (1968), 113–126.

[20] I. Gyarmati, *Non-Equilibrium Thermodynamics. Field Theory and Variational Principles*, Springer, Berlin, 1970.

[21] S. T. Konobeevsky, On the theory of phase transformations. II. Diffusion in solid solutions under stress distribution, *Journal of Experimental and Theoretical Physics*, **13** (1949), 200–214 (in Russian).

[22] L. G. Leal, *Advanced Transport Phenomena: Fluid Mechanics and Convective Transport Processes*, Cambridge University Press, Cambridge, 2007.

[23] H. W. Lord and Y. Shulman, A generalized dynamical theory of thermoelasticity, *Journal of the Mechanics and Physics of Solids*, **15** (1967), 299–309.

[24] R. L. Magin, *Fractional Calculus in Bioengineering*, Begell House Publishers, Connecticut, 2006.

[25] F. Mainardi, *Fractional Calculus and Waves in Linear Viscoelasticity: an Introduction to Mathematical Models*, Imperial College Press, London, 2010.

[26] J. E. Marsden and A. Tromba, *Vector Calculus*, 5th Ed., W. H. Freeman, New York, 2003.

[27] F. Neumann, *Vorlesungen über die Theorie des Elasticität des festen Körper und des Lichtäthers*, Teubner, Leipzig, 1895.

[28] R. R. Nigmatullin, To the theoretical explanation of the "universal response", *Physica Status Solidi B. Basic Solid State Physics*, **123** (1984), 739–745.

[29] R. R. Nigmatullin, On the theory of relaxation for systems with "remnant" memory, *Physica Status Solidi B. Basic Solid State Physics*, **124** (1984), 389–393.

[30] W. Nowacki, Dynamical problems of thermodiffusion in solids, *Bulletin de L'Académie Polonaise Des Sciences. Série Des Sciences Techniques*, **23** (1974), 55–64, 129–135, 257–266.

[31] W. Nowacki, Dynamic problems of thermodiffusion in elastic solids, *Proceedings of Vibration Problems*, **15** (1974), 105–128.

[32] W. Nowacki, *Thermoelasticity*, 2nd Ed., PWN–Polish Scientific Publishers, Warsaw and Pergamon Press, Oxford, 1986.

[33] I. Podlubny, *Fractional Differential Equations*, Academic Press, San Diego, 1999.

[34] Ya. S. Podstrigach, Differential equations of thermodiffusion problem in isotropic deformable solid, *Dopovidi Natsionalnoi Akademii Nauk Ukraini*, **2** (1961), 169–172 (in Ukrainian).

[35] Ya. S. Podstrigach, Diffusional theory of deformation of isotropic continuum, *Aspects of the Mechanics of Real Solids*, **2** (1964), 71–99 (in Russian).

[36] Ya. S. Podstrigach, Diffusional theory of anelasticity of metals, *Journal of Applied Mechanics and Technical Physics*, **6** (1965), 56–60.

[37] Ya. S. Podstrigach and V. S. Pavlina, Differential equations of thermodynamic processes in n-component solid solutions, *Soviet Materials Science*, **1** (1966), 259–264.

[38] Y. Povstenko, Fractional heat conduction equation and associated thermal stresses, *Journal of Thermal Stresses*, **28** (2005), 83–102.

[39] Y. Povstenko, Stresses exerted by a source of diffusion in a case of a non-parabolic diffusion equation, *International Journal of Engineering Science*, **43** (2005), 977–991.

[40] Y. Povstenko, Two-dimensional axisymmetric stresses exerted by instantaneous pulses and sources of diffusion in an infinite space in a case of time-fractional diffusion equation, *International Journal of Solids and Structures*, **44** (2007), 2324–2348.

[41] Y. Povstenko, Fundamental solutions to three-dimensional diffusion-wave equation and associated diffusive stresses, *Chaos, Solitons and Fractals*, **36** (2008), 961–972.

[42] Y. Povstenko, Thermoelasticity which uses fractional heat conduction equation, *Journal of Mathematical Sciences*, **162** (2009), 296–305.

[43] Y. Povstenko, Fractional Cattaneo-type equations and generalized thermoelasticity, *Journal of Thermal Stresses*, **34** (2011), 97–114.

[44] Y. Povstenko, Non-axisymmetric solutions to time-fractional diffusion-wave equation in an infinite cylinder, *Fractional Calculus & Applied Analysis. An International Journal for Theory and Applications*, **14** (2011), 418–435.

[45] Y. Povstenko, Theories of thermal stresses based on space-time-fractional telegraph equations, *Computers & Mathematics with Applications*, **64** (2012), 3321–3328.

[46] Y. Povstenko, *Linear Fractional Diffusion-Wave Equation for Scientists and Engineers*, Birkhäuser, New York, 2015.

[47] Y. Povstenko, *Fractional Thermoelasticity*, Springer, New York, 2015.

[48] Y. Povstenko, Theories of thermoelasticity based on space-time-fractional Cattaneo-type equation, in I. Podlubny, M. B. Vinagre Jara, Y. Q. Chen, V. Felin Batlle, I. Tejado Balsera (eds.) *Proceedings of the 4th IFAC Workshop on Fractional Differentiation and Its Applications*, Badajoz, Spain, 18–20 Oct 2010, Article No. FDA10-014.

[49] Yu. N. Rabotnov, *Elements of Hereditary Solid Mechanics*, Mir Publishers, Moscow, 1980.

[50] H. H. Sherief, A. M. A. El-Sayed, and A. M. Abd El-Latief, Fractional order theory of thermoelasticty, *International Journal of Solids and Structures*, **47** (2010), 269–275.

[51] V. E. Tarasov, *Fractional Dynamics. Applications of Fractional Calculus to Dynamics Of Particles, Fields and Media*, Higher Education Press, Beijing and Springer, Berlin, 2010.

[52] J. A. Tenreiro Machado, And I say to myself: "What a fractional world", *Fractional Calculus & Applied Analysis. An International Journal for Theory and Applications*, **14** (2011), 635–654.

[53] V. V. Uchaikin, *Fractional Derivatives for Physicists and Engineers*, Springer, Berlin, 2013.

[54] C.-C. Wang, The principle of fading memory, *Archive for Rational Mechanics and Analysis*, **18** (1965), 343–366.

[55] H. M. Youssef, Theory of fractional order generalized thermoelasticity, *Journal of Heat Transfer*, **132** (2010), 061301-1-7.

Elias C. Aifantis

Fractional generalizations of gradient mechanics

Abstract: This short chapter provides a fractional generalization of gradient mechanics, an approach (originally advanced by the author in the mid 80's) that has gained world-wide attention in the last decades due to its capability of modeling pattern forming instabilities and size effects in materials, and eliminating undesired elastic singularities. It is based on the incorporation of higher-order gradients (in the form of Laplacians) in the classical constitutive equations multiplied by appropriate internal lengths accounting for the geometry/topology of underlying micro/nano structures. This review will focus on the fractional generalization of the gradient elasticity equations (GradEla), an extension of classical elasticity, to incorporate the Laplacian of Hookean stress, by replacing the standard Laplacian by its fractional counterpart. On introducing the resulting fractional constitutive equation into the classical static equilibrium equation for the stress, a fractional differential equation is obtained, whose fundamental solutions are derived by using the Green's function procedure. As an example, Kelvin's problem is analyzed within the aforementioned setting. Then, an extension to consider constitutive equations for a restrictive class of nonlinear elastic deformations and deformation theory of plasticity is pursued. Finally, the methodology is applied for extending the author's higher-order diffusion theory from the integer to the fractional case.

Keywords: Fractional gradient elasticity, Thompson's problem, fractional Helmholtz equation, fractional gradient plasticity, higher-order fractional diffusion, steady-state fractional diffusion

MSC 2000: 35R11, 26A33

1 Introduction

This contribution concerns the fractional generalization of the author's gradient elasticity and higher-order diffusion. Both of these theories were introduced three decades ago to model deformation and transport problems in media with micro/nanostruc-

Acknowledgement: Support of the Ministry of Education and Science of Russian Federation under grant no. 14.Z50.31.0039 is acknowledged.

Elias C. Aifantis, Aristotle University of Thessaloniki, Thessaloniki, 54124, Greece; and Michigan Technological University, Houghton, MI 49931, USA; and Togliatti State University, Togliatti 445020, Russia, e-mail: mom@mom.gen.auth.gr; Tel.: +30-2310995921

https://doi.org/10.1515/9783110571707-010

tures. A new Laplacian term was added in the standard constitutive equations of Hookean elasticity and Fickean diffusion to interpret experimental data that could not be modeled by classical theories. Among the new results were the elimination of undesirable elastic singularities in dislocation lines and crack tips, and new robust continuum models for grain boundary diffusion. All these problems were successfully and efficiently addressed by incorporating internal lengths in the standard constitutive equations of elasticity and diffusion as scalar multipliers of newly introduced Laplacian terms of the persistent constitutive variables, to account for "weakly" nonlocal effects. The resulting internal length gradient (ILG) framework and its applications to various areas of material mechanics are reviewed in a recent article by the author [5], where extensive bibliography can also be found. In the same article, a brief account of fractional and fractal generalization of the ILG framework is given.

We expand on the aforementioned review by providing an updated summary of the fractional generalization of the ILG framework, focusing on static elasticity with a few related remarks on plasticity and steady-state diffusion. In that connection, it is noted that the basic balance laws for the mass and momentum are assumed to retain their classical (integer) form. For stationary deformation and steady-state diffusion problems, these laws lead to the standard balance equations

$$\operatorname{div}\boldsymbol{\sigma} = 0 \quad \text{or} \quad \sigma_{ij,j} = 0 \tag{1}$$

for the stress tensor and

$$\operatorname{div}\boldsymbol{j} = 0 \quad \text{or} \quad j_{i,i} = 0 \tag{2}$$

for the diffusive flux vector.

The standard constitutive equations of the ILG framework for $\boldsymbol{\sigma}$ and $\boldsymbol{j}$ are of the form

$$\begin{aligned} \boldsymbol{\sigma} &= \lambda(\operatorname{tr}\boldsymbol{\varepsilon})\mathbf{1} + 2\mu\boldsymbol{\varepsilon} - l_\varepsilon^2\nabla^2[\lambda(\operatorname{tr}\boldsymbol{\varepsilon})\mathbf{1} + 2\mu\boldsymbol{\varepsilon}], \\ \sigma_{ij} &= \lambda\varepsilon_{kk}\delta_{ij} + 2\mu\varepsilon_{ij} - l_\varepsilon^2\nabla^2[\lambda\varepsilon_{kk}\delta_{ij} + 2\mu\varepsilon_{ij}], \end{aligned} \tag{3}$$

and

$$\boldsymbol{j} = -D\nabla(\rho - l_d^2\nabla^2\rho), \quad j_i = -D(\rho - l_d^2\nabla^2\rho)_{,i}, \tag{4}$$

respectively. The classical elastic moduli (λ, μ) and diffusivity (D) have their usual meaning, the quantities ε_{ij} and ρ denote strain and concentration, respectively; whereas the newly introduced parameters l_ε and l_d are deformation-induced and diffusion-induced internal lengths (ILs) accounting for "weakly" nonlocal effects.

The fractional generalization of the above equations consists of replacing the standard (integer) Laplacian Δ in Equations (3) and (4) with a fractional one of the

Riesz form $(-^R\Delta)^{\alpha/2}$ or the Caputo form ${}^C\Delta_W^\alpha$. Then, the corresponding fractional generalization of Equation (3) reads

$$\sigma_{ij} = (\lambda\varepsilon_{kk}\delta_{ij} + 2\mu\varepsilon_{ij}) - l_\varepsilon^2(\alpha)(-^R\Delta)^{\alpha/2}[\lambda\varepsilon_{kk}\delta_{ij} + 2\mu\varepsilon_{ij}], \tag{5}$$

where $(-^R\Delta)^{\alpha/2}$ is the fractional generalization of the Laplacian in the Riesz form, and

$$\sigma_{ij} = (\lambda\varepsilon_{kk}\delta_{ij} + 2\mu\varepsilon_{ij}) - l_\varepsilon^2(\alpha)^C\Delta_W^\alpha[\lambda\varepsilon_{kk}\delta_{ij} + 2\mu\varepsilon_{ij}], \tag{6}$$

where ${}^C\Delta_W^\alpha$ is the fractional Laplacian in the Caputo form [26].

Equations (5) and (6) are fractional generalizations of the original GradEla model. Similar equations can be written for the fractional generalization of Equation (4). They read

$$\boldsymbol{j} = -D\nabla[\rho - l_d^2(\alpha)\{(-^R\Delta)^{\alpha/2}\rho\}], \quad j_i = -D[\rho - l_d^2(\alpha)\{(-^R\Delta)^{\alpha/2}\rho\}]_{,i}, \tag{7}$$

and

$$\boldsymbol{j} = -D\nabla[\rho - l_d^2(\alpha)\{^C\Delta_W^\alpha\rho\}], \quad j_i = -D[\rho - l_d^2(\alpha)\{^C\Delta_W^\alpha\rho\}]_{,i}, \tag{8}$$

respectively.

On introducing the aforementioned fractional gradient constitutive equations into the nonfractional balance laws given by Equations (1) and (2), the corresponding partial differential equations of fractional order are obtained, which need to be solved with the aid of appropriate boundary conditions for finite domains. To dispense with the complication of higher-order fractional boundary conditions, we consider infinite domains and derive fundamental solutions for the respective problems by employing a fractional extension of the Green's function method. The above is illustrated in detail in the next section (Section 2) by considering the classical Thomson (Lord Kelvin) elasticity problem in the framework of fractional GradEla. In Section 3 we give a brief account on preliminaries of fractional gradient nonlinear elasticity or deformation theory of plasticity. We note that both of these sections are an update of the fractional considerations of [5], based on the detailed elaborations contained in the initial articles by Tarasov and the author [25–27], and subsequent further discussions by Tarasov [20–24]. In Section 4 we return to the problem of fractional Laplacian and obtain fundamental solutions for a general fractional equation of Helmholtz type, which turns out to govern both gradient elasticity and higher-order diffusion theory. Since the basics of fractional deformation have been outlined in Sections 2 and 3, we show in Section 4 how these basic results are directly applicable to fractional diffusion problems. In this connection, it is noted that solutions of fractional GradEla problems are reduced to solutions of an inhomogenous Helmholtz equation, which is also the governing equation of fractional generalization of electrostatics with Debye screening [28].

2 Gradient elasticity (GradEla): revisiting Kelvin's problem

To shed light on the implications of fractional GradEla, a specific 3D configuration with spherical symmetry is considered below. The model of Equation (5) is employed due to the fact that definite results are available for the fractional Laplacian of Riesz type. The corresponding most general fractional GradEla governing equation is of the form [25]

$$c_\alpha((-\Delta)^{\alpha/2}u)(\boldsymbol{r}) + c_\beta((-\Delta)^{\beta/2}u)(\boldsymbol{r}) = f(\boldsymbol{r}) \quad (\alpha > \beta), \tag{9}$$

where $\boldsymbol{r} \in \mathbb{R}^3$ and $r = |\boldsymbol{r}|$ are dimensionless, $(-\Delta)^{\alpha/2}$ is the Riesz fractional Laplacian of order α, with the same for the symbols characterized by β, and the (fractional) gradient coefficients (c_α, c_β) are material constants related to the elastic moduli and the internal length, respectively. The rest of the symbols have their usual meaning: u denotes displacement and $f(\boldsymbol{r})$ body force. For $\alpha > 0$ and suitable functions $u(\boldsymbol{r})$, the Riesz fractional derivative can be defined in terms of the Fourier transform $\mathcal{F}$ by

$$((-\Delta)^{\alpha/2}u)(\boldsymbol{r}) = \mathcal{F}^{-1}(|\boldsymbol{k}|^\alpha(\mathcal{F}u)(\boldsymbol{k})), \tag{10}$$

where $\boldsymbol{k}$ denotes the wave vector. If $\alpha = 4$ and $\beta = 2$, then we have the well-known GradEla equation

$$c_2\Delta u(\boldsymbol{r}) - c_4\Delta^2 u(\boldsymbol{r}) + f(\boldsymbol{r}) = 0, \tag{11}$$

where $c_2 = (\lambda + 2\mu)$ and $c_4 = \pm(\lambda + 2\mu)\, l_s$ for spherically symmetric problems. Equation (11) is a fractional partial differential equation with a solution of the form

$$u(\boldsymbol{r}) = \int_{\mathbb{R}^3} G_{\alpha,\beta}(\boldsymbol{r} - \boldsymbol{r}')f(\boldsymbol{r}')\, d^3\boldsymbol{r}', \tag{12}$$

with the Green-type function $G_{\alpha,\beta}$ given by

$$G_{\alpha,\beta}(\boldsymbol{r}) = \int_{\mathbb{R}^3} \frac{1}{c_\alpha|\boldsymbol{k}|^\alpha + c_\beta|\boldsymbol{k}|^\beta} e^{i\boldsymbol{k}\cdot\boldsymbol{r}}\, d^3\boldsymbol{k} = \frac{1}{(2\pi)^{3/2}\sqrt{|\boldsymbol{r}|}} \int_0^\infty \frac{\lambda^{3/2}J_{1/2}(\lambda|\boldsymbol{r}|)}{c_\alpha\lambda^\alpha + c_\beta\lambda^\beta}\, d\lambda, \tag{13}$$

where $J_{1/2}(z) = \sqrt{2/\pi z}\,\sin(z)$ is the Bessel function of the first kind, and the dot denotes inner product.

To proceed further, we consider Thomson's problem of an applied point load f_0, that is,

$$f(\boldsymbol{r}) = f_0\delta(\boldsymbol{r}) = f_0\delta(x)\delta(y)\delta(z). \tag{14}$$

Then, the displacement field $u(\boldsymbol{r})$ has a simple form given by the particular solution

$$u(\boldsymbol{r}) = f_0 G_{\alpha,\beta}(\boldsymbol{r}), \tag{15}$$

with the Green's function given by Equation (13), that is,

$$u(\boldsymbol{r}) = \frac{f_0}{2\pi^2|\boldsymbol{r}|} \int_0^\infty \frac{\lambda \sin(\lambda|\boldsymbol{r}|)}{c_\alpha \lambda^\alpha + c_\beta \lambda^\beta}\, d\lambda, \quad (\alpha > \beta). \tag{16}$$

It turns out that the asymptotic form of the solution given by Equation (16) for $0 < \beta < 2$ and $\alpha \neq 2$ reads

$$u(\boldsymbol{r}) \approx \frac{f_0 \Gamma(2-\beta)\sin(\pi\beta/2)}{2\pi^2 c_\beta} \frac{1}{|\boldsymbol{r}|^{3-\beta}} \quad (|\boldsymbol{r}| \to \infty). \tag{17}$$

This asymptotic behavior as $|\boldsymbol{r}| \to \infty$ does not depend on the parameter α, and (as will be seen below) the corresponding asymptotic behavior as $|\boldsymbol{r}| \to 0$ does not depend on the parameter β , where $\alpha > \beta$. It follows that the displacement field at large distances from the point of load application is determined only by the term $(-\Delta)^{\beta/2}$, where $\beta < \alpha$. This can be interpreted as a fractional nonlocal "deformation" counterpart of the classical elasticity result based on Hooke's law. We can also note the existence of a maximum for the quantity $u(\boldsymbol{r}) \cdot |\boldsymbol{r}|$ in the case $0 < \beta < \alpha < 2$. Indeed, these observations become clear by considering, in detail, the following two special cases that emerge.

A) Sub-GradEla model: $\alpha = 2$; $0 < \beta < 2$. In this case, Equation (9) becomes

$$c_2 \Delta u(\boldsymbol{r}) - c_\beta((-\Delta)^{\beta/2} u)(\boldsymbol{r}) + f(\boldsymbol{r}) = 0 \quad (0 < \beta < 2). \tag{18}$$

The order of the fractional Laplacian $(-\Delta)^{\beta/2}$ is less than the order of the first term related to the usual Hooke's law. For example, one can consider the square of the Laplacian, that is, $\beta = 1$. In general, the parameter β defines the order of the power-law nonlocality. The particular solution of Equation (18) in the present case, reads

$$u(\boldsymbol{r}) = \frac{f_0}{2\pi^2|\boldsymbol{r}|} \int_0^\infty \frac{\lambda \sin(\lambda|\boldsymbol{r}|)}{c_2 \lambda^2 + c_\beta \lambda^\beta}\, d\lambda \quad (0 < \beta < 2). \tag{19}$$

The following asymptotic behavior for Equation (19) can be derived in the form

$$u(\boldsymbol{r}) = \frac{f_0}{2\pi^2|\boldsymbol{r}|} \int_0^\infty \frac{\lambda \sin(\lambda|\boldsymbol{r}|)}{c_2 \lambda^2 + c_\beta \lambda^\beta} \approx \frac{C_0(\beta)}{|\boldsymbol{r}|^{3-\beta}} + \sum_{k=1}^\infty \frac{C_k(\beta)}{|\boldsymbol{r}|^{(2-\beta)(k+1)+1}} \quad (|\boldsymbol{r}| \to \infty), \tag{20}$$

where

$$C_0(\beta) = \frac{f_0\Gamma(2-\beta)\sin(\pi\beta/2)}{2\pi^2 c_\beta},$$

$$C_k(\beta) = -\frac{f_0 c_2^k}{2\pi^2 c_\beta^{k+1}} \int_0^\infty z^{(2-\beta)(k+1)-1} \sin(z)\, dz. \tag{21}$$

As a result, the displacement field generated by the force that is applied at a point in the fractional gradient elastic continuum described by the fractional Laplacian $(-\Delta)^{\beta/2}$ with $0 < \beta < 2$ is given by

$$u(\mathbf{r}) \approx \frac{C_0(\beta)}{|\mathbf{r}|^{3-\beta}} \quad (0 < \beta < 2), \tag{22}$$

for large distances ($|\mathbf{r}| \to \infty$).

B) Super-GradEla model: $\alpha > 2$ and $\beta = 2$. In this case, Equation (9) becomes

$$c_2 \Delta u(\mathbf{r}) - c_\alpha((-\Delta)^{\alpha/2} u)(\mathbf{r}) + f(\mathbf{r}) = 0 \quad (\alpha > 2). \tag{23}$$

The order of the fractional Laplacian $(-\Delta)^{\alpha/2}$ is greater than the order of the first term related to the usual Hooke's law. The parameter $\alpha > 2$ defines the order of the power-law nonlocality of the elastic continuum. If $\alpha = 4$, then Equation (23) reduces to Equation (11). The case can be viewed as corresponding as closely as possible ($\alpha \approx 4$) to the usual gradient elasticity model of Equation (11). The asymptotic behavior of the displacement field $u(\mathbf{r})$ for $\mathbf{r} \to 0$ in the case of super-gradient elasticity is given by

$$u(\mathbf{r}) \approx \frac{f_0 \Gamma((3-\alpha)/2)}{2^\alpha \pi^2 \sqrt{\pi} c_\alpha \Gamma(\alpha/2)} \frac{1}{|\mathbf{r}|^{3-\alpha}} \quad (2 < \alpha < 3),$$

$$u(\mathbf{r}) \approx \frac{f_0}{2\pi\alpha c_2^{1-3/\alpha} c_\alpha^{3/\alpha} \sin(3\pi/\alpha)} \quad (\alpha > 3). \tag{24}$$

Note that the above asymptotic behavior does not depend on the parameter β and that the corresponding relation of Equation (24) does not depend on c_β. The displacement field $u(\mathbf{r})$ for short distances away from the point of load application is determined only by the term with $(-\Delta)^{\alpha/2}$ ($\alpha > \beta$), that is, the fractional counterpart of the usual extra non-Hookean term of gradient elasticity. More details for the above results can be found in [20, 22, 25–27].

3 Fractional gradient plasticity

In this section we provide an introductory account of fractional deformation theory of plasticity (as opposed to the flow theory of plasticity) by simply elaborating on a specific constitutive equation, which could also be viewed as a very special form of nonlinear elasticity.

The corresponding (nonlinear) fractional gradient constitutive equation involves scalar measures of the stress and strain tensors; that is, their second invariants, as these quantities enter in both theories of nonlinear elasticity and plasticity. In plasticity theory, in particular, we employ the second invariants of the deviatoric stress and plastic strain tensors. The effective (equivalent) stress σ is defined by the equation

$$\sigma = \sqrt{(1/2)\sigma'_{ij}\sigma'_{ij}}, \quad \sigma'_{ij} = \sigma'_{ij} - \frac{1}{3}\sigma'_{kk}\delta_{ij}, \tag{25}$$

where σ_{ij} is the stress tensor. The effective (equivalent) plastic strain is defined as

$$\varepsilon = \int dt \sqrt{2\dot{\varepsilon}_{ij}\dot{\varepsilon}_{ij}}, \tag{26}$$

where $\dot{\varepsilon}_{ij}$ is the plastic strain rate tensor, which is assumed to be traceless in order to satisfy plastic incompressibility.

Motivated by the above, we propose the following form of nonlinear fractional differential equation for the scalar quantities σ and ε, which can be used as a basis for future tensorial formulation of nonlinear elasticity and plasticity theories

$$\sigma(\mathbf{r}) = E\,\varepsilon(\mathbf{r}) + c(\alpha)\left((-\Delta)^{\alpha/2}\varepsilon\right)(\mathbf{r}) + \eta\, K(\varepsilon(\mathbf{r})) \quad (\alpha > 0), \tag{27}$$

where $K(\varepsilon(\mathbf{r}))$ is a nonlinear function, which describes the usual (homogeneous) material's response (linear hardening plasticity), $c(\alpha)$ is an internal parameter that measures the nonlocal character of deformation mechanisms, E is a shear-like elastic modulus, η is a small parameter of nonlinearity, and $(-\Delta)^{\alpha/2}$ is the fractional Laplacian in the Riesz form. As a simple example of the nonlinear function, we can consider

$$K(\varepsilon) = \varepsilon^{\beta}(\mathbf{r}) \quad (\beta > 0). \tag{28}$$

Equation (27), where $K(\varepsilon(\boldsymbol{r}))$ is defined by Equation (28), is the fractional Ginzburg–Landau equation. For various choices of the parameters (E, η, β) characterizing the homogenous material response, different models of nonlinear elastic and plastic behavior may result.

Let us derive a particular solution of Equation (27) with $K(\varepsilon(\mathbf{r})) = 0$. To solve the linear fractional differential equation

$$\sigma(\mathbf{r}) = E\,\varepsilon(\mathbf{r}) + c(\alpha)\left[(-\Delta)^{\alpha/2}\varepsilon\right](\mathbf{r}), \tag{29}$$

we apply the Fourier method and the fractional Green function method. Using Theorem 5.22 of Kilbas et al. ([8], pages 342 and 343; see also [19]) for the case $E \neq 0$ and

$\alpha > (n-1)/2$, we see that Equation (29) is solvable, and its particular solution is given by the equation

$$\varepsilon(\mathbf{r}) = G_{n,\alpha} * \sigma = \int_{\mathbb{R}^n} G_{n,\alpha}(\mathbf{r}-\mathbf{r}')\,\sigma(\mathbf{r}')\,d^n\mathbf{r}', \tag{30}$$

where the symbol $*$ denotes convolution, and $G_{n,\alpha}(\mathbf{r})$ is defined by

$$G_{n,\alpha}(\mathbf{r}) = \frac{|\mathbf{r}|^{(2-n)/2}}{(2\pi)^{n/2}} \int_0^\infty \frac{\lambda^{n/2} J_{(n-2)/2}(\lambda|\mathbf{r}|)}{c(\alpha)\,\lambda^\alpha + E}\,d\lambda, \tag{31}$$

where $n = 1, 2, 3$, $\alpha > (n-1)/2$, and $J_{(n-2)/2}$ is the Bessel function of the first kind.

Let us consider a deformation of unbounded fractional nonlocal continuum, where the stress is applied to an infinitesimally small region in this continuum. In this case, we can assume that the strain $\varepsilon(\boldsymbol{r})$ is induced by a point stress $\sigma(\boldsymbol{r})$ at the origin of coordinates

$$\sigma(\mathbf{r}) = \sigma_0\delta(\mathbf{r}), \tag{32}$$

that is, the particular solution is proportional to the Green's function. As a result, the stress field is

$$\varepsilon(\mathbf{r}) = \frac{1}{2\pi^2}\frac{\sigma_0}{|\mathbf{r}|} \int_0^\infty \frac{\lambda\,\sin(\lambda|\mathbf{r}|)}{E + c(\alpha)\,\lambda^\alpha}\,d\lambda. \tag{33}$$

3.1 Perturbation of linearized fractional deformations by nonlinear hardening

Suppose that $\varepsilon(\boldsymbol{r}) = \varepsilon_0(\boldsymbol{r})$ is the solution of Equation (27) with $\eta = 0$, that is, is the solution of the linear equation

$$\sigma(\mathbf{x}) = E\,\varepsilon_0(\mathbf{r}) + c(\alpha)\,((-\Delta)^{\alpha/2}\varepsilon_0)(\mathbf{r}). \tag{34}$$

The solution of this fractional differential equation may be written in the form

$$\varepsilon(\mathbf{r}) = \varepsilon_0(\mathbf{r}) + \eta\,\varepsilon_1(\mathbf{r}) + \cdots. \tag{35}$$

This means that we consider perturbations to the strain field $\varepsilon_0(\boldsymbol{r})$ of the fractional gradient deformation state, which are caused by weak nonlinear hardening effects. The first-order approximation with respect to η gives the equation

$$E\,\varepsilon_1(\mathbf{r}) + c(\alpha)\,((-\Delta)^{\alpha/2}\varepsilon_1)(\mathbf{r}) + K(\varepsilon_0(\mathbf{r})) = 0, \tag{36}$$

which is equivalent to the linear equation

$$\sigma_{\text{eff}}(\mathbf{r}) = E\,\varepsilon_1(\mathbf{r}) + c(\alpha)\left((-\Delta)^{\alpha/2}\varepsilon_1\right)(\mathbf{r}), \tag{37}$$

where the effective stress $\sigma_{\text{eff}}(\mathbf{x})$ is defined by the equation

$$\sigma_{\text{eff}}(\mathbf{r}) = -K(\varepsilon_0(\mathbf{r})). \tag{38}$$

Equation (36) gives a particular solution in the form

$$\varepsilon(\mathbf{r}) = \varepsilon_0(\mathbf{r}) + \varepsilon_1(\mathbf{r}) = G_{n,\alpha} * \sigma + \eta\, G_{n,\alpha} * \sigma_{\text{eff}}, \tag{39}$$

where the convolution operation and $G_{n,\alpha}$ are defined by Equations (30) and (31). Upon substitution of Equation (38) into Equation (39), we obtain

$$\varepsilon(\mathbf{r}) = G_{n,\alpha} * \sigma - \eta\, G_{n,\alpha} * K(G_{n,\alpha} * \sigma). \tag{40}$$

For a “point stress” of the form given by Equation (32), Equation (39) can be written in the form

$$\varepsilon(\mathbf{r}) = \sigma_0\, G_{n,\alpha}(\mathbf{r}) - \eta\left(G_{n,\alpha} * K(\sigma_0\, G_{n,\alpha})\right)(\mathbf{r}), \tag{41}$$

which, for $K(\cdot)$ given by Equation (28), results in

$$\varepsilon(\mathbf{r}) = \sigma_0\, G_{n,\alpha}(\mathbf{r}) - \eta\, \sigma_0^{\beta}\left(G_{n,\alpha} * (G_{n,\alpha})^{\beta}\right)(\mathbf{r}). \tag{42}$$

3.2 Perturbation of plasticity by fractional gradient nonlocality

Let us now consider an equilibrium state by setting ε_0 = const. (that is, $(-\Delta)^{\alpha/2}\varepsilon_0 = 0$) and $\sigma(\mathbf{r}) = \sigma$ = const. in Equation (27), that means

$$E\,\varepsilon_0 + \eta\, K(\varepsilon_0) = \sigma. \tag{43}$$

This, for the case where the function K is defined by Equation (28) with $\beta = 3$ becomes

$$E\,\varepsilon_0 + \eta\,\varepsilon_0^3 = \sigma. \tag{44}$$

For $\sigma \neq 0$, there is no solution $\varepsilon_0 = 0$. For $E > 0$ and the weak scalar stress field $\sigma \ll \sigma_c$ with respect to the critical value $\sigma_c = \sqrt{E/\eta}$, there exists only one solution

$$\varepsilon_0 \approx \sigma/E. \tag{45}$$

For negative stiffness materials ($E < 0$) and $\sigma = 0$, we have three solutions

$$\varepsilon_0 \approx \pm\sqrt{|E|/\eta}, \quad \varepsilon_0 = 0. \tag{46}$$

For $\sigma < (2\sqrt{3}/9)\sigma_c$, there also exist three solutions. For $\sigma \gg \sigma_c$, we can neglect the first term ($E \approx 0$),

$$\eta\,\varepsilon_0^3 \approx \sigma, \tag{47}$$

and obtain

$$\varepsilon_0 \approx (\sigma/\eta)^{1/3}. \tag{48}$$

In general, the equilibrium values ε_0 are solutions of the nonlinear algebraic relation given by Equation (43).

Let us consider a deviation $\varepsilon_1(\boldsymbol{r})$ of the field from the equilibrium value $\varepsilon_0(\boldsymbol{r})$. For this purpose, we will seek a solution in the form

$$\varepsilon(\mathbf{r}) = \varepsilon_0 + \varepsilon_1(\mathbf{r}). \tag{49}$$

In general, the stress field is not constant, that is, $\sigma(\mathbf{x}) \neq \sigma$. In a first approximation, we obtain the equation

$$\sigma(\mathbf{r}) = c(\alpha)\,((-\Delta)^{\alpha/2}\varepsilon_1)(\mathbf{r}) + (E + \eta\,K_\varepsilon'(\varepsilon_0))\varepsilon_1(\mathbf{r}), \tag{50}$$

where $K_\varepsilon' = \partial K(\varepsilon)/\partial\varepsilon$. Equation (50) is equivalent to the linear fractional differential equation

$$\sigma(\mathbf{x}) = E_{\text{eff}}\,\varepsilon_1(\mathbf{r}) + c(\alpha)\,((-\Delta)^{\alpha/2}\varepsilon_1)(\mathbf{r}), \tag{51}$$

with the effective modulus E_{eff} defined by

$$E_{\text{eff}} = E + \eta\,K_\varepsilon'(\varepsilon_0). \tag{52}$$

For the case $K(\varepsilon) = \varepsilon^\beta$, we have

$$E_{\text{eff}} = E + \beta\,\eta\,\varepsilon_0^{\beta-1}. \tag{53}$$

A particular solution of Equation (50) can be written in the form of Equation (30), where we use E_{eff} instead of E for the "point stress" (see Equations (32)–(33)). Equation (40) gives

$$\varepsilon_1(\mathbf{r}) = \frac{1}{2\pi^2}\frac{\sigma_0}{|\mathbf{r}|}\int_0^\infty \frac{E + E_{\text{eff}} + 2c(\alpha)\,\lambda^\alpha}{(c(\alpha)\,\lambda^\alpha + E)\,(c(\alpha)\,\lambda^\alpha + E_{\text{eff}})}\,\sin(\lambda|\mathbf{r}|)\,d\lambda. \tag{54}$$

For the case $\alpha = 2$, the field $\varepsilon_1(\mathbf{r})$ is given by the equation

$$\varepsilon_1(\mathbf{r}) = \frac{\sigma_0}{4\pi c(\alpha)\,|\mathbf{r}|}\,e^{-|\mathbf{r}|/r_c}, \tag{55}$$

where r_c is defined by

$$r_c^2 = \frac{c(\alpha)}{E + \eta K_\varepsilon'(\varepsilon_0)}. \tag{56}$$

It should be noted that an analogous situation exists in classical theory of electric fields. In the electrodynamics, the field $\varepsilon_1(\mathbf{r})$ describes the Coulomb potential with the Debye screening. For a fractional differential field equation ($\alpha \neq 2$), we have a power-law type of screening that is described in [28]. The electrostatic potential for media with power-law spatial dispersion differs from the Coulomb potential by the factor

$$C_{\alpha,0}(|\mathbf{r}|) = \frac{2}{\pi} \int_0^\infty \frac{\lambda \sin(\lambda|\mathbf{r}|)}{E_{\text{eff}} + c(\alpha)\lambda^\alpha} d\lambda. \tag{57}$$

Note that the Debye potential differs from the Coulomb potential by the exponential factor $C_D(|\mathbf{x}|) = \exp(-|\mathbf{r}|/r_D)$.

4 Fractional Helmholtz equation

On introducing the fractional GradEla constitutive relation given by Equation (5) into the equilibrium relation given by Equation (1), we obtain

$$[1 + l_\varepsilon^\alpha(-\Delta)^{\alpha/2}][\lambda\nabla \operatorname{tr}\boldsymbol{\varepsilon} + 2\mu \operatorname{div}\boldsymbol{\varepsilon}] = 0, \tag{58}$$

where the notation $l_\varepsilon^2(\alpha) \equiv l_\varepsilon^\alpha$ and $(-{}^R\Delta)^{\alpha/2} \equiv (-\Delta)^{\alpha/2}$ was used for simplicity. Noting the fact that the operators ∇ and $(-\Delta)^{\alpha/2}$ commute and that the second bracket in Equation (58) is also zero by replacing $\boldsymbol{\varepsilon}$ with $\boldsymbol{\varepsilon}_0$, where $\boldsymbol{\varepsilon}_0$ denotes the solution of the corresponding equation for classical elasticity (that is, $\lambda\nabla \operatorname{tr}\boldsymbol{\varepsilon} + 2\mu \operatorname{div}\boldsymbol{\varepsilon} = 0$), we can easily deduce that the solution of Equation (58) satisfies the reduced fractional partial differential equation

$$[1 + l_\varepsilon^\alpha(-\Delta)^{\alpha/2}]\boldsymbol{\varepsilon} = \boldsymbol{\varepsilon}_0, \tag{59}$$

which for the case $\alpha = 2$ reduces to the inhomogeneous Helmholtz equation derived for the nonfractional GradEla (Ru–Aifantis theorem [18]) and was used successfully to derive nonsingular solutions for dislocations and cracks [2, 4, 7, 9, 10]. It turns out that compatible displacements $\boldsymbol{u}$ ($\varepsilon_{ij} = (1/2)[u_{i,j} + u_{j,i}]$) also obey Equation(59), and the same holds for corresponding fields in electrostatics with Debye screening [28], and for steady-state higher-order diffusion problems [1, 3].

It is thus critical to derive fundamental solutions for Equation(59); that is, for the equation

$$[1 + l_\varepsilon^\alpha(-\Delta)^{\alpha/2}]G_\alpha(\boldsymbol{r}) = \delta(\boldsymbol{r}), \tag{60}$$

where $G_\alpha(\boldsymbol{r})$ denotes the fundamental solution, $\delta(\boldsymbol{r})$ denotes the delta function, and $\boldsymbol{r}$ is the radial coordinate in a 3D space.

To compute the fundamental solution of Equation (60) with the natural boundary condition $G_\alpha(\boldsymbol{r}) \to 0$ as $\boldsymbol{r} \to \infty$, we employ the method of Fourier transforms. Using the properties of the Fourier transform of the Riesz fractional Laplacian for every "well-behaved" scalar function $f(\boldsymbol{r})$,

$$\mathcal{F}(\,(-\Delta)^{\alpha/2} f(\boldsymbol{r})\,)(\boldsymbol{k}) = |\boldsymbol{k}|^\alpha \mathcal{F}(f(\boldsymbol{r}))(\boldsymbol{k})\,, \tag{61}$$

and the well-known transform of the delta function $\mathcal{F}(\delta(\boldsymbol{r}))(\boldsymbol{k}) = 1$, we obtain the following algebraic equation for the fundamental solution:

$$[\,1 + l_\varepsilon^\alpha |\boldsymbol{k}|^\alpha\,]\, G_\alpha(\boldsymbol{k}) = 1, \tag{62}$$

which gives

$$G_\alpha(\boldsymbol{k}) = \frac{1}{1 + l_\varepsilon^\alpha |\boldsymbol{k}|^\alpha}. \tag{63}$$

Consequently, the fundamental solution of Equation (60) in the physical space is obtained through inversion of Equation (63):

$$G_\alpha(\boldsymbol{r}) = \frac{1}{(2\pi)^3} \int_{-\infty}^{\infty} \frac{1}{1 + l_\varepsilon^\alpha |\boldsymbol{k}|^\alpha}\, e^{i\boldsymbol{k}\cdot\boldsymbol{r}}\, d^3\boldsymbol{k}. \tag{64}$$

To simplify Equation (64), we perform the change of variables $\boldsymbol{k} \to l_\varepsilon^{-1}\boldsymbol{k}$, which results in a factor of l_ε^{-3} and a change in scale $\boldsymbol{r} \to \boldsymbol{r}/l_\varepsilon$. Therefore, for simplicity, we omit those factors and restore them at the end result.

The integral given by Equation (64) is defined in a three-dimensional Euclidean space and can be analytically computed in spherical coordinates by applying the well-known relationship (see, for example, Lemma 25.1 of Samko et al. [19])

$$\frac{1}{(2\pi)^3} \int_{-\infty}^{\infty} f(|\boldsymbol{k}|) e^{i\boldsymbol{k}\cdot\boldsymbol{r}}\, d^3\boldsymbol{k} = \frac{1}{(2\pi)^{3/2}\sqrt{|\boldsymbol{r}|}} \int_0^\infty k^{3/2} f(k)\, J_{1/2}(k|\boldsymbol{r}|)\, dk. \tag{65}$$

In Equation (65), k denotes the magnitude of the wave vector, and $J_{1/2} = \sqrt{2/(\pi z)}\,\sin(z)$ denotes the Bessel function of order 1/2. Introduction of Equation (65) in Equation (64), by omitting the scaling factors, results in

$$\begin{aligned} G_\alpha(\boldsymbol{r}) &= \frac{1}{(2\pi)^{3/2}\sqrt{|\boldsymbol{r}|}} \int_0^\infty \frac{k^{3/2}}{1 + k^\alpha} J_{1/2}(k|\boldsymbol{r}|)\, dk \\ &= \frac{1}{2\pi^2 |\boldsymbol{r}|} \int_0^\infty \frac{k}{1 + k^\alpha} \sin(k|\boldsymbol{r}|)\, dk. \end{aligned} \tag{66}$$

The integral in Equation (66) can be computed using the convolution property of the Mellin transform, defined in [6] by the relationship

$$\mathcal{M}(f(x))(s) = \int_0^\infty f(x)\, x^{s-1}\, dx. \tag{67}$$

Its inverse is given by

$$f(x) = \frac{1}{2\pi i} \int_{\gamma - i\infty}^{\gamma + i\infty} f(s)\, x^{-s}\, ds, \tag{68}$$

where the path of integration is a vertical strip separating the poles of $\mathcal{M}(f(x))(s)$ defined in $\gamma_1 < \mathrm{Re}(s) < \gamma_2$. For more details about the Mellin transform, we refer the reader to [13]. Here we only use the basic results

$$\begin{aligned} \mathcal{M}\left(\frac{1}{1+x^\alpha}\right)(s) &= \frac{1}{\alpha}\Gamma\left(\frac{s}{\alpha}\right)\Gamma\left(1-\frac{s}{\alpha}\right), \\ \mathcal{M}(x^{3/2} J_{1/2}(x))(s) &= 2^{1/2+s}\, \frac{\Gamma(1+\frac{s}{2})}{\Gamma(\frac{1}{2}-\frac{s}{2})}, \end{aligned} \tag{69}$$

where we made use of the Mellin transform of the Bessel function (see also Section 6.8 of [6]),

$$\mathcal{M}(J_\sigma(2\sqrt{u}))(s) = \frac{\Gamma(\frac{\sigma}{2}+s)}{\Gamma(\frac{\sigma}{2}+1-s)}. \tag{70}$$

Consequently, Equation (66) can be evaluated using the above results and performing the inverse Mellin transform by computing the Mellin–Barnes integral

$$G_\alpha(\mathbf{r}) = \frac{1}{2\alpha\pi^{3/2}\,|\mathbf{r}|^2}\, \frac{1}{2\pi i} \int_{\gamma-i\infty}^{\gamma+i\infty} \frac{\Gamma(\frac{1}{\alpha}-\frac{s}{\alpha})\,\Gamma(1-\frac{1}{\alpha}+\frac{s}{\alpha})\Gamma(1+\frac{s}{2})}{\Gamma(\frac{1}{2}-\frac{s}{2})} \left(\frac{|\mathbf{r}|}{2}\right)^{-s} ds. \tag{71}$$

The Mellin–Barnes integral representation of Equation (71) can be expressed in terms of the corresponding Fox-H function of fractional analysis (see, for example, [11, 12, 14–16]):

$$G_\alpha(\mathbf{r}) = \frac{1}{2\alpha\pi^{3/2}\,|\mathbf{r}|^2}\, H^{2,1}_{1,3}\left[\frac{|\mathbf{r}|}{2}\,\middle|\, \begin{matrix} (1-\frac{1}{\alpha},\frac{1}{\alpha}) \\ (1-\frac{1}{\alpha},\frac{1}{\alpha}),(1,\frac{1}{2}),(\frac{1}{2},\frac{1}{2}) \end{matrix}\right]. \tag{72}$$

The integral (71) has poles at the points $s = 1-\alpha(\nu-1)$ and $s = 1-(3+2\nu)$, $\nu \in \mathbb{N}$. To evaluate it, we first apply the residue theorem to the poles of $\Gamma(1-\frac{1-s}{\alpha})$, since they correspond to the singularity near the origin $r \approx 0$.

After a change of variables $s \to s + 1 - \alpha$, (71) becomes [11, 12, 14]

$$G_\alpha(\boldsymbol{r}) = \frac{1}{\alpha\, 2^\alpha\, \pi^{3/2}\, |\boldsymbol{r}|^{3-\alpha}} \frac{1}{2\pi i} \int_{\gamma-i\infty}^{\gamma+i\infty} \frac{\Gamma(1-\frac{s}{\alpha})\,\Gamma(\frac{s}{\alpha})\Gamma(\frac{3}{2}-\frac{\alpha}{2}+\frac{s}{2})}{\Gamma(\frac{\alpha}{2}-\frac{s}{2})} \left(\frac{|\boldsymbol{r}|}{2}\right)^{-s} ds. \tag{73}$$

Next, we perform the other change of variables $s \to \alpha s$ in Equation (73), which results in

$$G_\alpha(\boldsymbol{r}) = \frac{1}{2^\alpha\, \pi^{3/2}\, |\boldsymbol{r}|^{3-\alpha}} \frac{1}{2\pi i} \int_{\gamma-i\infty}^{\gamma+i\infty} \frac{\Gamma(1-s)\,\Gamma(s)\Gamma(\frac{3}{2}-\frac{\alpha}{2}(1-s))}{\Gamma(\frac{\alpha}{2}(1-s))} \left(\frac{|\boldsymbol{r}|}{2}\right)^{-\alpha s} ds, \tag{74}$$

which provides an alternative representation in terms of the Fox-H function, that is,

$$G_\alpha(\boldsymbol{r}) = \frac{1}{2^\alpha\, \pi^{3/2}\, |\boldsymbol{r}|^{3-\alpha}} H^{2,1}_{1,3}\left[\left(\frac{|\boldsymbol{r}|}{2}\right)^\alpha \,\middle|\, \begin{matrix} (0,1) \\ (0,1), (\frac{3}{2}-\frac{\alpha}{2}, \frac{\alpha}{2}), (1-\frac{\alpha}{2}, \frac{\alpha}{2}) \end{matrix}\right]. \tag{75}$$

The contour integral in Equation (75) can be evaluated using the method of residues from complex analysis, by closing the contour encircling all poles at $s = -\nu$ and then applying the Cauchy residue theorem:

$$G_\alpha(\boldsymbol{r}) = \frac{1}{2^\alpha\, \pi^{3/2}\, |\boldsymbol{r}|^{3-\alpha}} \sum_{\nu=0}^{\infty} \lim_{s\to-\nu} \left\{(s+\nu)\Gamma(s) \frac{\Gamma(1-s)\,\Gamma(\frac{3}{2}-\frac{\alpha}{2}(1-s))}{\Gamma(\frac{\alpha}{2}(1-s))} \left(\frac{|\boldsymbol{r}|}{2}\right)^{-\alpha s}\right\}. \tag{76}$$

Equation (76) can be evaluated using the relation

$$\lim_{s\to-\nu} (s+\nu)\,\Gamma(s) = \lim_{s\to-\nu} \frac{\Gamma(s+\nu+1)}{s(s+1)..(s+\nu-1)} = \frac{(-1)^\nu}{\nu\,!}. \tag{77}$$

This gives

$$G_\alpha(\boldsymbol{r}) = \frac{1}{2^\alpha\, \pi^{3/2}\, |\boldsymbol{r}|^{3-\alpha}} \sum_{\nu=0}^{\infty} \frac{(-1)^\nu}{\nu\,!} \frac{\Gamma(1+\nu)\,\Gamma(\frac{3}{2}-\frac{\alpha}{2}(1+\nu))}{\Gamma(\frac{\alpha}{2}(1+\nu))} \left(\frac{|\boldsymbol{r}|}{2}\right)^{\alpha\nu}, \tag{78}$$

which can be simplified by noting that $\Gamma(1+\nu) = \nu!$ for $\nu \in \mathbb{N}$. The final result is

$$G_\alpha(\boldsymbol{r}) = \frac{1}{2^\alpha\, \pi^{3/2}\, |\boldsymbol{r}|^{3-\alpha}} \sum_{\nu=0}^{\infty} \frac{\Gamma(\frac{3}{2}-\frac{\alpha}{2}(1+\nu))}{\Gamma(\frac{\alpha}{2}(1+\nu))} (-1)^\nu \left(\frac{|\boldsymbol{r}|}{2}\right)^{\alpha\nu}. \tag{79}$$

An asymptotic expression near the origin is obtained from the dominating term of Equation (79) as $r \to 0$, that is,

$$G_\alpha(\boldsymbol{r}) \approx \frac{\Gamma(\frac{3}{2}-\frac{\alpha}{2})}{2^\alpha\, \pi^{3/2}\Gamma(\frac{\alpha}{2})} \frac{1}{|\boldsymbol{r}|^{3-\alpha}} \quad (\boldsymbol{r} \to 0). \tag{80}$$

This asymptotic form cancels the singularity of the fundamental solution of corresponding classical theories. To see this, one can compute the contributions from the poles of $\Gamma(1+s/2)$ in Equation (71), which correspond to nonsingular asymptotic behavior near the origin, using the same techniques. The result is

$$G_\alpha(\boldsymbol{r}) = \frac{1}{2^\alpha \pi^{3/2} |\boldsymbol{r}|^{3-\alpha}} \sum_{\nu=0}^{\infty} \frac{\Gamma(\frac{3}{2} - \frac{\alpha}{2}(1+\nu))}{\Gamma(\frac{\alpha}{2}(1+\nu))} (-1)^\nu \left(\frac{|\boldsymbol{r}|}{2}\right)^{\alpha\nu} + \frac{2}{\alpha(4\pi)^{3/2}} \sum_{\nu=0}^{\infty} \frac{\Gamma(\frac{3+2\nu}{\alpha})\Gamma(1-\frac{1}{\alpha}(3+2\nu))}{\Gamma(\frac{3}{2}+\nu)} \frac{(-1)^\nu}{\nu!} \left(\frac{|\boldsymbol{r}|}{2}\right)^{2\nu}. \tag{81}$$

In the special case $\alpha \to 2$, Equation (81) reduces to the Green's function of the classical Helmholtz equation, that is,

$$G_\alpha(\boldsymbol{r}) = \frac{1}{4\pi|\boldsymbol{r}|} e^{-|\boldsymbol{r}|}. \tag{82}$$

It is easily proved that Equations (80), (81) give the same results as Equation (64) of [28], since the latter solves the same mathematical equation (that is, the fractional inhomogeneous Helmholtz equation (their Equation (59)) for a different physical problem, the problem of a point charge.

5 Fractional higher-order diffusion

On introducing the fractional diffusion constitutive relation given by Equation (7) into the classical (nonfractional) mass balance law

$$\frac{\partial \rho}{\partial t} + \operatorname{div} \boldsymbol{j} = 0 \quad \text{or} \quad \rho_{,t} + j_{i,i} = 0, \tag{83}$$

we obtain the fractional high-order diffusion equation

$$\frac{\partial \rho}{\partial t} = D\Delta\rho + D l_d^\alpha \nabla \cdot \{(-\Delta)^{\alpha/2} \nabla \rho\}, \tag{84}$$

along with the auxiliary conditions $\rho(\boldsymbol{r},0) = \delta(\boldsymbol{r})$, $\rho(\boldsymbol{r},t) \to 0$ as $|\boldsymbol{r}| \to \infty$, and $\delta(\boldsymbol{r})$ denoting, as usual, the delta function. (The notation $l_d^2(\alpha) \equiv l_d^\alpha$ and $(-{}^R\Delta)^{\alpha/2} \equiv (-\Delta)^{\alpha/2}$ was used for simplicity.)

To solve Equation (84), we employ the method of Fourier transform and exploit the properties of the Riesz fractional Laplacian, along with the symmetry of the problem. This gives

$$\frac{\partial \rho(\boldsymbol{k},t)}{\partial t} = -D|\boldsymbol{k}|^2 \rho(\boldsymbol{k},t) - D\, l_d^\alpha |\boldsymbol{k}|^\alpha |\boldsymbol{k}|^2 \rho(\boldsymbol{k},t), \tag{85}$$

where $\boldsymbol{k}$ denotes the wave vector. Equation (85) is a first-order ordinary differential equation with respect to time with the initial condition $\rho(\boldsymbol{k},0) = \mathcal{F}(\delta(\boldsymbol{r})) = 1$. Its solution is

$$\rho(\boldsymbol{k},t) = \exp(-Dt|\boldsymbol{k}|^2)\exp(-D_\alpha t|\boldsymbol{k}|^{\alpha+2}), \tag{86}$$

where we defined $D_\alpha \equiv Dl_d^\alpha$. The solution of Equation (86) in configuration space is obtained by inversion of the Fourier transform

$$\rho(\boldsymbol{r},t) = \frac{1}{(2\pi)^3}\int_{-\infty}^{\infty}\exp(-Dt|\boldsymbol{k}|^2)\exp(-D_\alpha t|\boldsymbol{k}|^{\alpha+2})\exp(i\boldsymbol{k}\cdot\boldsymbol{r})\,d^3\boldsymbol{k}. \tag{87}$$

Equation (87) is the inverse Fourier transform of the product of two independent terms, and can be expressed as the convolution of the corresponding solutions in the physical space using the following well-known property of the Fourier transform:

$$\mathcal{F}((f*g)(\boldsymbol{r},t))(\boldsymbol{k}) = \mathcal{F}(f(\boldsymbol{r},t))(\boldsymbol{k})\,\mathcal{F}(g(\boldsymbol{r},t))(\boldsymbol{k}), \tag{88}$$

$$(f*g)(\boldsymbol{r},t) = \int_{-\infty}^{\infty} f(\boldsymbol{r}-\boldsymbol{r}',t)g(\boldsymbol{r}',t)\,d^3\boldsymbol{r}'. \tag{89}$$

Using Equation (88), we recognize Equation (87) as the convolution

$$\rho(\boldsymbol{r},t) = (G_2 * G_{\alpha+2})(\boldsymbol{r},t), \tag{90}$$

where we defined the set of functions G_α as

$$\begin{aligned} G_\alpha(\boldsymbol{r},t) &= \mathcal{F}^{-1}\{\exp(-D_\alpha t|\boldsymbol{k}|^\alpha)\} \\ &= \frac{1}{(2\pi)^3}\int_{-\infty}^{\infty}\exp(-D_\alpha t|\boldsymbol{k}|^\alpha)\exp(i\boldsymbol{k}\cdot\boldsymbol{r})\,d^3\boldsymbol{k}. \end{aligned} \tag{91}$$

Equation (91) is the fundamental solution (that is, the Green's function) for the fractional diffusion equation

$$\frac{\partial G_\alpha(\boldsymbol{r},t)}{\partial t} = -D_\alpha(-\Delta)^{\alpha/2}G_\alpha(\boldsymbol{r},t). \tag{92}$$

The corresponding fundamental solution of Equation (84) is then deduced from Equation (91) through the convolution property of Equation (90).

Applying Equation (65) to the fundamental solution $G_\alpha(\boldsymbol{r})$ of Equation (91), we obtain

$$\begin{aligned} G_\alpha(\boldsymbol{r},t) &= \frac{1}{(2\pi)^{3/2}\sqrt{|\boldsymbol{r}|}}\int_0^\infty k^{3/2}\exp(-D_\alpha t\,k^\alpha)J_{1/2}(k|\boldsymbol{r}|)\,dk \\ &= \frac{1}{2\pi^2|\boldsymbol{r}|}\int_0^\infty k\exp(-D_\alpha t\,k^\alpha)\sin(k|\boldsymbol{r}|)\,dk. \end{aligned} \tag{93}$$

The integral in Equation (93) can be computed using the convolution property of the Mellin transform, as in the previous section. The final result is the following series expansion expression [17]:

$$G_\alpha(\boldsymbol{r},t) = \frac{2}{\alpha(4\pi)^{3/2}(D_\alpha t)^{3/\alpha}} \sum_{\nu=0}^{\infty} \frac{(-1)^\nu}{\nu!} \frac{\Gamma(\frac{3}{\alpha}+\frac{2\nu}{\alpha})}{\Gamma(\frac{3}{2}+\nu)} \left(\frac{|\boldsymbol{r}|^2}{4(D_\alpha t)^{2/\alpha}}\right)^\nu . \tag{94}$$

Equation (94) can be represented in terms of the Wright's function ${}_1\Psi_1$ as

$$G_\alpha(\boldsymbol{r},t) = \frac{2}{\alpha(4\pi)^{3/2}(D_\alpha t)^{3/\alpha}}\ {}_1\Psi_1\left[\begin{matrix}(\frac{3}{\alpha},\frac{2}{\alpha})\\ (\frac{3}{2},1)\end{matrix};-\frac{|\boldsymbol{r}|^2}{4(D_\alpha t)^{\frac{2}{\alpha}}}\right]. \tag{95}$$

The generalized Wright's function is defined by the following series [8, 19]:

$${}_p\Psi_q(z) = {}_p\Psi_q\left[\begin{matrix}(a_1,A_1) & \dots & (a_p,A_p)\\ (b_1,B_1) & \dots & (b_q,B_q)\end{matrix};z\right] = \sum_{\nu=0}^{\infty} \frac{\prod_{j=1}^{p}\Gamma(a_j+A_j\nu)}{\prod_{j=1}^{q}\Gamma(b_j+B_j\nu)}\frac{z^\nu}{\nu!}. \tag{96}$$

It is easy to check that when $\alpha = 2$, the series expansion reduces to the Green's function of the ordinary diffusion equation in 3-dimensional space. This is readily seen by letting $\alpha \to 2$ in Equations (94) and (95), resulting in the expression

$$\begin{aligned} G_2(\boldsymbol{r},t) &= \frac{1}{(4\pi D t)^{3/2}}\ {}_1\Psi_1\left[\begin{matrix}(\frac{3}{2},1)\\ (\frac{3}{2},1)\end{matrix};-\frac{|\boldsymbol{r}|^2}{4Dt}\right] \\ &= \frac{1}{(4\pi D t)^{3/2}} \sum_{\nu=0}^{\infty}\frac{(-1)^\nu}{\nu!}\left(\frac{|\boldsymbol{r}|^2}{4Dt}\right)^\nu \\ &= \frac{1}{(4\pi D t)^{3/2}} \exp\left(-\frac{|\boldsymbol{r}|^2}{4Dt}\right). \end{aligned} \tag{97}$$

Consequently, the fundamental solution of the second-order fractional diffusion Equation (84), denoted as $G(\boldsymbol{r},t)$, is obtained through convolution of Equation (88), with $G_\alpha(\boldsymbol{r},t)$ given by Equation (95) and (97) for $\alpha = 2$, that is,

$$G(\boldsymbol{r},t) = \int_{-\infty}^{\infty} G_{\alpha+2}(\boldsymbol{r}-\boldsymbol{r}',t)G_2(\boldsymbol{r}',t)\,d^3\boldsymbol{r}'. \tag{98}$$

We can extend Equation (83) to include distributed sources (for example, chemical reaction or trapping) with density/concentration rate $q(\boldsymbol{r},t)$. In this particular case, the classical mass balance law becomes

$$\frac{\partial\rho}{\partial t} + \operatorname{div}\boldsymbol{j} = q, \tag{99}$$

and the corresponding inhomogeneous fractional diffusion equation reads

$$\frac{\partial\rho}{\partial t} = D\Delta\rho + Dl_d^\alpha\nabla\cdot\{(-\Delta)^{\alpha/2}\nabla\rho\} + q. \tag{100}$$

Using the Fourier transform method, we can obtain the fundamental solution of Equation (100) as follows:

$$\rho(\boldsymbol{r},t) = \int_0^t \int_{-\infty}^{\infty} G(\boldsymbol{r}-\boldsymbol{r}',t-\tau)q(\boldsymbol{r}',\tau)\,d^3\boldsymbol{r}'\,d\tau, \tag{101}$$

where $G(\boldsymbol{r},t)$ is given by Equation (98). For the special case of a unit point source $q(\boldsymbol{r},t) = \delta(\boldsymbol{r})\delta(t)$, it is readily seen that Equation (101) reduces to the fundamental solution $G(\boldsymbol{r},t)$.

The fractional diffusion equation admits steady-state solutions, under the presence of external sources/sinks with density/rate $q(\boldsymbol{r})$. The governing equation for this time independent configuration is

$$D\Delta\rho + Dl_d^{\alpha}\nabla\cdot\{(-\Delta)^{\alpha/2}\nabla\rho\} + q = 0. \tag{102}$$

Equation (102) can be generalized to a higher-order steady-state fractional diffusion equation of the form

$$D_{\alpha}((-\Delta)^{\alpha/2}\rho)(\boldsymbol{r}) + D_{\beta}((-\Delta)^{\beta/2}\rho)(\boldsymbol{r}) = q(\boldsymbol{r}) \quad (\alpha > \beta), \tag{103}$$

where (α,β) denote arbitrary positive fractional orders, and (D_{α}, D_{β}) are the corresponding fractional diffusion coefficients. Equation (103) can be derived by considering a fractional extension of the conservation law given by Equation (99), along with the constitutive relation given by Equation (7), and/or a further fractional extension for its classical gradient (∇) part.

Equation (103) is a fractional partial differential equation, whose solution reads

$$\rho(\boldsymbol{r}) = \int_{\mathbb{R}^3} G_{\alpha,\beta}(\boldsymbol{r}-\boldsymbol{r}')q(\boldsymbol{r}')\,d^3\boldsymbol{r}', \tag{104}$$

with the Green-type function $G_{\alpha,\beta}(\boldsymbol{r})$ given by

$$G_{\alpha,\beta}(\boldsymbol{r}) = \int_{\mathbb{R}^3} \frac{1}{D_{\alpha}|\boldsymbol{k}|^{\alpha} + D_{\beta}|\boldsymbol{k}|^{\beta}} e^{i\boldsymbol{k}\cdot\boldsymbol{r}}\,d^3\boldsymbol{k} = \frac{1}{(2\pi)^{3/2}\sqrt{|\boldsymbol{r}|}} \int_0^{\infty} \frac{\lambda^{3/2}J_{1/2}(\lambda|\boldsymbol{r}|)}{D_{\alpha}\lambda^{\alpha} + D_{\beta}\lambda^{\beta}}\,d\lambda. \tag{105}$$

Let us now consider the particular problem of a unit point source located at the origin of the form

$$q(\boldsymbol{r}) = q_0\delta(\boldsymbol{r}) = q_0\delta(x)\delta(y)\delta(z). \tag{106}$$

Upon substitution of Equation (106) into Equation (104), we obtain the particular solution

$$\rho(\boldsymbol{r}) = q_0 G_{\alpha,\beta}(\boldsymbol{r}), \tag{107}$$

with the Green function $G_{\alpha,\beta}(\mathbf{r})$ given by Equation (105). Then by using the particular expression for the Bessel function of the first kind, we obtain

$$\rho(\mathbf{r}) = \frac{q_0}{2\pi^2|\mathbf{r}|} \int_0^\infty \frac{\lambda \sin(\lambda|\mathbf{r}|)}{c_\alpha \lambda^\alpha + c_\beta \lambda^\beta} d\lambda \quad (\alpha > \beta). \tag{108}$$

Two distinct modes of diffusion arise, depending on the particular form of the fractional parameters (α, β), which are discussed in detail below:
A) Sub-GradDiffusion model: $\alpha = 2$; $0 < \beta < 2$. In this case, Equation (102) becomes

$$D\Delta\rho(\mathbf{r}) - D_\beta((-\Delta)^{\beta/2}\rho)(\mathbf{r}) + q(\mathbf{r}) = 0 \quad (0 < \beta < 2). \tag{109}$$

The order of the fractional Laplacian $(-\Delta)^{\beta/2}$ is less than the order of the first term related to the usual Fick's law. The parameter β defines the order of the power-law nonlocality. The particular solution of Equation (109) reads

$$\rho(\mathbf{r}) = \frac{q_0}{2\pi^2|\mathbf{r}|} \int_0^\infty \frac{\lambda \sin(\lambda|\mathbf{r}|)}{D\lambda^2 + D_\beta \lambda^\beta} d\lambda \quad (0 < \beta < 2). \tag{110}$$

The following asymptotic behavior for Equation (110) can be derived:

$$\rho(\mathbf{r}) = \frac{q_0}{2\pi^2|\mathbf{r}|} \int_0^\infty \frac{\lambda \sin(\lambda|\mathbf{r}|)}{D\lambda^2 + D_\beta \lambda^\beta} \approx \frac{C_0(\beta)}{|\mathbf{r}|^{3-\beta}} + \sum_{k=1}^{\infty} \frac{C_k(\beta)}{|\mathbf{r}|^{(2-\beta)(k+1)+1}} \quad (|\mathbf{r}| \to \infty), \tag{111}$$

where

$$C_0(\beta) = \frac{q_0 \Gamma(2-\beta)\sin(\pi\beta/2)}{2\pi^2 D_\beta},$$

$$C_k(\beta) = -\frac{q_0 D^k}{2\pi^2 D_\beta^{k+1}} \int_0^\infty z^{(2-\beta)(k+1)-1} \sin(z)\, dz. \tag{112}$$

As a result, the density of the diffusive species generated by the source that is concentrated at a single point in space, for large distances from the source, is given asymptotically by the expression

$$\rho(\mathbf{r}) \approx \frac{C_0(\beta)}{|\mathbf{r}|^{3-\beta}} \quad (0 < \beta < 2) \tag{113}$$

for large distances ($|\mathbf{r}| \to \infty$).
B) Super-GradDiffusion: $\alpha > 2$ and $\beta = 2$. In this case, Equation (102) becomes

$$D\Delta\rho(\mathbf{r}) - D_\alpha((-\Delta)^{\alpha/2}\rho)(\mathbf{r}) + q(\mathbf{r}) = 0, \quad (\alpha > 2). \tag{114}$$

The order of the fractional Laplacian $(-\Delta)^{\alpha/2}$ is greater than the order of the first term related to the usual Fick's law. The asymptotic density $\rho(\boldsymbol{r})$ for $\boldsymbol{r} \to 0$, in this case, is given by

$$\rho(\boldsymbol{r}) \sim \frac{q_0 \Gamma((3-\alpha)/2)}{2^\alpha \pi^2 \sqrt{\pi} D_\alpha \Gamma(\alpha/2)} \frac{1}{|\boldsymbol{r}|^{3-\alpha}} \quad (2 < \alpha < 3),$$
$$\rho(\boldsymbol{r}) \approx \frac{q_0}{2\pi\alpha D^{1-3/\alpha} D_\alpha^{3/\alpha} \sin(3\pi/\alpha)} \quad (\alpha > 3). \tag{115}$$

Note that the above asymptotic behavior does not depend on the parameter β and that the corresponding relation of Equation (115) does not depend on D_β. The density $\rho(\boldsymbol{r})$ for short distances away from the point of source application is determined only by the term with $(-\Delta)^{\alpha/2}$ $(\alpha > \beta)$.

Finally, and especially for the case of more complicated boundary value problems, we mention that the steady-state diffusion of Equation (102) can be factored as

$$D\nabla \cdot \nabla\{1 + l_d^\alpha(-\Delta)^{\alpha/2}\}\rho + q = 0. \tag{116}$$

By defining the "classical" operator $L^0 \equiv D\nabla \cdot \nabla$, and likewise its "fractional gradient" counterpart $L^\alpha \equiv 1 + l_d^\alpha(-\Delta)$, we can prove that L^α satisfies the classical steady-state Fickean diffusion equation. This is a direct consequence of the fact that the operators L^0 and L^α commute. Therefore, we arrive at the following "operator-split" scheme

$$(1 + l_d^\alpha(-\Delta)^{\alpha/2})\rho = \rho_0; \quad D\nabla \cdot \nabla\rho_0 + q = 0. \tag{117}$$

Equation (117) is the fractional counterpart of the Ru–Aifantis theorem [18] for the steady-state fractional higher-order diffusion equation.

Bibliography

[1] E. C. Aifantis, On the problem of diffusion in solids, *Acta Mechanica*, **37** (1980), 265–296.

[2] E. C. Aifantis, On the gradient approach—relation to Eringen's nonlocal theory, *International Journal of Engineering Science*, **49** (2011), 1367–1377.

[3] E. C. Aifantis, Gradient Nanomechanics: applications to deformation, fracture, and diffusion in nanopolycrystals, *Metallurgical and Materials Transactions. A, Physical Metallurgy and Materials Science*, **42** (2011), 2985–2998.

[4] E. C. Aifantis, On non-singular GRADELA crack fields, *Theoretical and Applied Mechanics Letters*, **4** (2014), 051005.

[5] E. C. Aifantis, Internal length gradient (ILG) material mechanics across scales and disciplines, *Advances in Applied Mechanics*, **49** (2016), 1–110.

[6] H. Bateman and A. Erdelyi, *Tables of Integral Transforms, Volume 1*, McGraw-Hill, New York, 1954.

[7] M. Yu. Gutkin and E. C. Aifantis, Dislocations and disclinations in gradient elasticity, *Physica Status Solidi B. Basic Solid State Physics*, **214** (1999), 245–284.

[8] A. Kilbas, M. Srivastava, and J. Trujillo, *Theory and Applications of Fractional Differential Equations*, Elsevier, 2006.

[9] M. Lazar and G. Maugin, Dislocations in gradient elasticity revisited, *Proceedings of the Royal Society A. Mathematical, Physical and Engineering Sciences*, **462** (2006), 3465–3480.

[10] M. Lazar, G. Maugin, and E. C. Aifantis, On dislocations in a special class of generalized elasticity, *Physica Status Solidi*, **242** (2005), 2365–2390.

[11] Y. Luchko and V. Kiryakova, The Mellin integral transform in fractional calculus, *Fractional Calculus & Applied Analysis. An International Journal for Theory and Applications*, **16** (2013), 405–430.

[12] F. Mainardi, Y. Luchko, and G. Pagnini, The fundamental solution of the space-time fractional diffusion equation, *Fractional Calculus & Applied Analysis. An International Journal for Theory and Applications*, **4** (2001), 153–192.

[13] O. I. Marichev, *Handbook of Integral Transforms of Higher Transcendental Functions: Theory and Algorithmic Tables*, Ellis Horwood, 1982.

[14] A. Mathai, *The H-Function: Theory and Applications*, Springer-Verlag, New York, 2010.

[15] R. Metzler and J. Klafter, The random walk's guide to anomalous diffusion: a fractional dynamics approach, *Physics Reports*, **339** (2000), 1–77.

[16] R. Metzler and T. F. Nonnenmacher, Space- and time-fractional diffusion and wave equations, fractional Fokker–Planck equations, and physical motivation, *Chemical Physics*, **284** (2002), 67–90.

[17] K. Parisis and E. C. Aifantis, Fractional generalization of higher-order diffusion, arXiv:1808.03241 [physics.class–ph], 2018.

[18] C. Q. Ru and E. C. Aifantis, A simple approach to solve boundary-value problems in gradient elasticity, *Acta Mechanica*, **101** (1993), 59–68.

[19] S. Samko, A. Kilbas, and O. Marichev, *Integrals and Derivatives of Fractional Order and Applications*, Gordon and Breach, New York, 1993.

[20] V. E. Tarasov, Lattice model of fractional gradient and integral elasticity: long-range interaction of Grünwald–Letnikov–Riesz type, *Mechanics of Materials*, **70** (2014), 106–114.

[21] V. E. Tarasov, Lattice with long-range interaction of power-law type for fractional non-local elasticity, *International Journal of Solids and Structures*, **51** (2014), 2900–2907.

[22] V. E. Tarasov, Three-dimensional lattice approach to fractional generalization of continuum gradient elasticity, *Progress in Fractional Differentiation and Applications*, **1** (2015), 243–258.

[23] V. E. Tarasov, Three-dimensional lattice models with long-range interactions of Grünwald–Letnikov type for fractional generalization of gradient, *Meccanica*, **51** (2016), 125–138.

[24] V. E. Tarasov, Fractional mechanics of elastic solids: continuum aspects, *Journal of Engineering Mechanics*, **143** (2017), D4016001–8.

[25] V. E. Tarasov and E. C. Aifantis, Toward fractional gradient elasticity, *Journal of Mechanical Behaviour of Materials*, **23** (2014), 41–46.

[26] V. E. Tarasov and E. C. Aifantis, Non-standard extensions of gradient elasticity: fractional non-locality, memory and fractality, *Communications in Nonlinear Science and Numerical Simulation*, **22** (2015), 197–227.

[27] V. E. Tarasov and E. C. Aifantis, On fractional and fractal formulations of gradient linear and nonlinear elasticity, arxiv:1808.04452, 2018, 37 pages.

[28] V. E. Tarasov and J. Trujillo, Fractional power-law spatial dispersion in electro-dynamics, *Annals of Physics*, **334** (2013), 1–23.

Jun Li and Martin Ostoja-Starzewski

Application of fractional calculus to fractal media

Abstract: This chapter is a survey of continuum-type mechanics of porous media having a generally anisotropic fractal geometry. The approach relies on expressing the global balance laws in terms of fractional integrals based on product measures and, then, converting them to integer-order integrals in conventional (Euclidean) space. Via localization, this allows development of local balance laws of fractal media: conservation of mass, microinertia, linear momentum, angular momentum, and energy; also the second law of thermodynamics. The product measure formulation, together with the angular momentum balance, directly leads to a generally asymmetric Cauchy stress and, hence, to a micropolar (rather than classical) mechanics of fractal materials. The continuum-thermodynamic development follows the lines of thermomechanics with internal variables.

Keywords: Anisotropic fractals, product measure, continuum mechanics, balance laws, micropolar continuum

MSC 2000: 26A33, 28A80, 74B99

1 Introduction

The present article outlines an approach to continuum-type mechanics of heterogeneous porous media of fractal type, that is, those lacking a clear separation of length scales. The basic ideas, extending the dimensional regularization concepts and proceeding in the vein of a field theory, hark back to the pioneering work by Tarasov [17–19], who developed continuum-type equations of fractal porous media, and extended them to a range of problems in fields, such as electrodynamics, fluid mechanics, heat transfer, and wave motion [20]. The power of this approach stems from the fact that much of the framework of continuum mechanics/physics may be generalized and partial differential equations (with derivatives of integer order) may still be employed. Whereas the original formulation was based on the Riesz measure and re-

Acknowledgement: JL acknowledges the support by the start-up funds at University of Massachusetts Dartmouth. MO-S acknowledges the funding by the NSF (grant CMMI-1462749).

Jun Li, Department of Mechanical Engineering, University of Massachusetts, Dartmouth, MA 02747-2300, USA, e-mail: jun.li@umassd.edu
Martin Ostoja-Starzewski, Department of Mechanical Science and Engineering, Institute for Condensed Matter Theory and Beckman Institute, University of Illinois at Urbana-Champaign, Urbana, IL 61801, USA, e-mail: martinos@illinois.edu

https://doi.org/10.1515/9783110571707-011

stricted to isotropic and fluid-like media, the present authors introduced a framework of product measure [2, 9, 15, 16], grasping the anisotropy of fractal geometry (that is, different fractal dimensions in different directions) generally in solid materials. This modeling strategy has been classified as the "fractional-integral approach" in the comprehensive review article [20]; note that four other approaches to fractal media—each having its pros and cons—are also possible.

In this article, we review the key ideas of the product measure formulation, showing that it provides a unique and consistent expression of the line, surface, and volume transformation coefficients for mapping the fractal porous medium into an approximating homogenized continuum. The full mapping is based on the conservation of mass, microinertia, linear momentum, angular momentum, energy, and the second law of thermodynamics. The product measure formulation, together with the angular momentum balance, directly leads to a generally asymmetric Cauchy stress and, hence, to a micropolar (rather than classical) mechanics of fractal materials.

2 Anisotropic fractal structures

2.1 Product measures

The mass of a body occupying a region $\mathcal{W}$ in the Euclidean 3-space obeys a power law

$$m(R) = kR^D. \tag{1}$$

Here R is the length scale of measurement (or resolution), k is a proportionality constant, and D (typically < 3) is the fractal dimension. Note that (1) can be applied to a pre-fractal, that is, a fractal-type, physical object with lower and upper cutoffs. Next, Tarasov employed a fractional integral to represent the mass as

$$m(\mathcal{W}) = \int_{\mathcal{W}} \rho(\mathbf{r})dV_D = \int_{\mathcal{W}} \rho(\mathbf{r})c_3(D,R)dV_3, \tag{2}$$

where the first and second equalities, respectively, involve fractional (Riesz-type) integrals and conventional integrals, whereas the coefficient $c_3(D,R)$ provides a transformation between the two [17–19]. This formulation is now primarily of historical interest since it has several drawbacks:

1. It involves a fractional derivative that does not give zero when applied to a constant function (for example, frame translation or rigid body motion), which is a rather unphysical property.
2. The mechanics-type derivation of wave equations yields a different result from the variational-type derivation.

3. The 3d (three-dimensional) wave equation does not correctly reduce to the 1d wave equation.
4. It is limited to isotropic media.

Motivated by these considerations, we have replaced (2) by a more general power-law relation with respect to each coordinate (with the basic idea going back to [19])

$$m(R) \sim x_1^{\alpha_1} x_2^{\alpha_2} x_3^{\alpha_3}, \tag{3}$$

whereby the mass distribution is specified via a product measure

$$m(x_1, x_2, x_3) = \int\int_{W}\int \rho(x_1, x_2, x_3) d\mu(x_1) d\mu(x_2) d\mu(x_3). \tag{4}$$

Here the length measurement in each coordinate x_k is provided by

$$d\mu(x_k) = c_1^{(1)}(\alpha_k, x_k) dx_k, \tag{5}$$

where $c_1^{(1)}$ is called a *line transformation coefficient* in x_k. Relation (5) implies that the infinitesimal fractal volume element dV_D is the product

$$\begin{aligned} dV_D &= dl_{\alpha_1}(x_1) dl_{\alpha_2}(x_2) dl_{\alpha_3}(x_3) = c_1^{(1)} c_1^{(2)} c_1^{(3)} dx_1 dx_2 dx_3 = c_3 dV_3, \\ &\text{with } c_3 = c_1^{(1)} c_1^{(2)} c_1^{(3)}. \end{aligned} \tag{6}$$

By analogy to (5), c_3 is called a *volume transformation coefficient*. It follows from (6) that the total fractal dimension D of mass m is $\alpha_1 + \alpha_2 + \alpha_3$.

Equation (3) implies that the mass fractal dimension D equals $\alpha_1 + \alpha_2 + \alpha_3$, where each α_k plays the role of a fractal dimension in the direction x_k. Whereas it is noted that, the anisotropic fractal body's fractal dimension is not necessarily the sum of projected fractal dimensions, we quote from a book on mathematics of fractals [4]: "*Many fractals encountered in practice are not actually products, but are product-like.*"

Focusing on a cubic-shaped dV_D in Figure 1, for each surface element $S_d^{(k)}$ defined by a direction x_k normal to it, the *surface transformation coefficient* $c_2^{(k)}$ is set up in terms of its two in-plane line coefficients ($c_1^{(i)}$ and $c_1^{(j)}$) or, equivalently, in terms of the volume coefficient c_3 and the line transformation coefficient $c_1^{(k)}$ in direction normal to the surface

$$c_2^{(k)} = c_1^{(i)} c_1^{(j)} = c_3 / c_1^{(k)}, \quad i \neq j \quad \text{and} \quad i, j \neq k. \tag{7}$$

To express the $c_1^{(k)}$ coefficients, we note that a variety of forms could exist to represent the fractional power-law relation (3). In particular, there are two forms currently in use, namely, the form used by Tarasov [17–19] and Balankin [1] from the Riesz

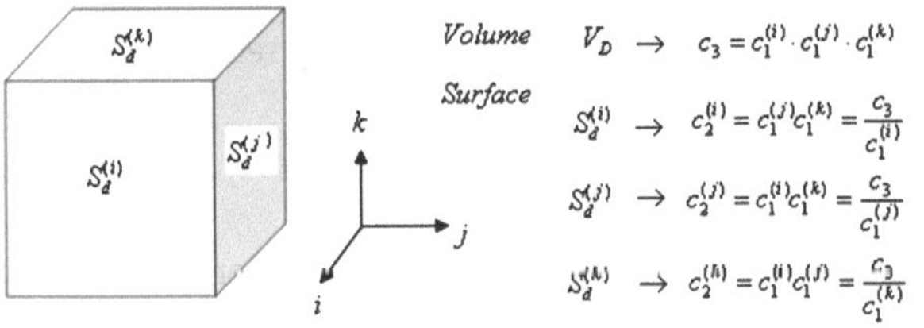

Figure 1: Roles of the transformation coefficients $c_1^{(i)}$, $c_2^{(k)}$, and c_3 in homogenizing a fractal body of volume dV_D, surface dS_d, and lengths dl_α into an Euclidean parallelepiped of volume dV_3, surface dS_2, and length dx.

fractional integral, or the one we utilized [2, 9, 16], which is according to a modified Riemann–Liouville fractional integral of Jumarie [8], so that

$$c_1^{(k)} = \alpha_k {x_k}^{\alpha_k - 1}, \quad k = 1, 2, 3, \quad \text{or} \tag{8}$$

$$c_1^{(k)} = \alpha_k (l_k - x_k)^{\alpha_k - 1}, \quad k = 1, 2, 3. \tag{9}$$

Here l_k is the total length (integral interval) along x_k. The density distributions in both forms are biased towards either the coordinate origin (8) or the boundary (9), which may not be the case in specific fractal media [11]. Besides, this formulation cannot differentiate between two fractal media having the same fractal dimension but different density distributions and mechanical behaviors.

To overcome the aforementioned issues, we propose a series expression of $c_1^{(k)}$:

$$c_1^{(k)} = \sum a_i |x_k - b_i|^{\alpha_k - 1}, \quad k = 1, 2, 3 \quad \text{and} \quad i = 1, 2, \ldots, N. \tag{10}$$

Here b_i refers to the coordinate of those points surrounded by dense particles in fractal media, and a_i represents their weight in the normalization of mass. It is expected that the above equation could capture the fractional size scaling and the density distributions in specific fractal media.

Note that the above expression of $c_1^{(k)}$ has a fractional length dimension, which is understandable, since in mathematics a fractal set has a finite measure only with respect to its Hausdorff dimension. In practice, it is convenient to work with dimensionless variables to avoid issues of physical dimensions. Thus, we suggest to replace x_k by x_k/l_0 and $|x_k - b_i|$ by $|x_k - b_i|/l_0$ in the fractional integration [9] (l_0 is a characteristic scale, for example, the mean pore size).

2.2 Tensor calculus on anisotropic fractals

To have a working continuum theory, a generalization of the Green–Gauss theorem is needed. Following [9, 16], it is formulated within the framework of product measures discussed above as

$$\begin{aligned} \int_{\partial \mathcal{W}} f_k n_k dS_d &= \int_{\mathcal{W}} (f_k c_2^{(k)}),_k \, c_3^{-1} dV_D \\ &= \int_{\mathcal{W}} f_{k,k} \, c_2^{(k)} \, c_3^{-1} dV_D = \int_{\mathcal{W}} \frac{f_{k,k}}{c_1^{(k)}} dV_D. \end{aligned} \tag{11}$$

This leads to the definition of the *fractal derivative* (in general, *fractal gradient*) operator

$$\nabla^D \phi = \mathbf{e}_k \nabla_k^D \phi \quad \text{or} \quad \nabla_k^D \phi = \frac{1}{c_1^{(k)}} \frac{\partial \phi}{\partial x_k} \quad \text{(no sum on } k\text{)}, \tag{12}$$

where $\mathbf{e}_k$ are the base vectors. This operator has three desirable properties:
(1) It is the "inverse" operator of a fractional integral.
(2) It satisfies the product rule: $\nabla_k^D(AB) = \nabla_k^D(A)B + A\nabla_k^D(B)$.
(3) Its operation on any constant is zero, which indeed is a property not possessed by the usual Riemann–Liouville fractional derivative. Consequently, the fractional generalization of Reynold's transport theorem is

$$\frac{d}{dt}\int_{\mathcal{W}} P\, dV_D = \int_{\mathcal{W}} \left[\frac{\partial}{\partial t}P + (Pv_k),_k\right] dV_D, \tag{13}$$

implying that the fractal material time derivative is the same as the conventional material time derivative (dP/dt):

$$\left(\frac{d}{dt}\right)_D P = \frac{d}{dt}P = \frac{\partial}{\partial t}P + P,_k v_k. \tag{14}$$

From a homogenization standpoint, the above developments allow interpretation of the fractal (intrinsically discontinuous and multiscale) medium as a continuum with a "fractal metric" embedded in the equivalent homogenized continuum model, that is,

$$dl_D = c_1 dx, \quad dS_d = c_2 dS_2, \quad dV_D = c_3 dV_3. \tag{15}$$

Here dl_D, dS_d, and dV_D represent the line, surface, and volume elements in the fractal body, whereas dx, dS_2, and dV_3, respectively, denote those in the homogenized model; see Figure 1. The coefficients c_1, c_2, c_3 provide relations between both pictures. Standard image analysis techniques (such as the "box method" or the "sausage method") allow a quantitative calibration of these coefficients for every direction and every cross-sectional plane.

Another advantage of the product measure formulation is the direct generalization of tensor calculus. To see this clearly, on account of (12), the *fractal divergence* of a vector field is

$$\operatorname{div}\mathbf{f} = \nabla^D \cdot \mathbf{f} \quad \text{or} \quad \nabla_k^D f_k = \frac{1}{c_1^{(k)}} \frac{\partial f_k}{\partial x_k}. \tag{16}$$

This leads to the *fractal curl* operator of a vector field

$$\operatorname{curl}\mathbf{f} = \nabla^D \times \mathbf{f} \quad \text{or} \quad e_{jki}\nabla_k^D f_i = e_{jki}\frac{1}{c_1^{(k)}} \frac{\partial f_i}{\partial x_k}, \tag{17}$$

where e_{jki} is the permutation symbol.

The following four fundamental identities of the conventional vector calculus can now be shown to carry over in terms of these new operators.

(i) The divergence of the curl of a vector field **f**:

$$\operatorname{div}\cdot\operatorname{curl}\mathbf{f} = \nabla_j^D \cdot e_{jki}\nabla_k^D f_i = \frac{1}{c_1^{(j)}}\frac{\partial}{\partial x_j}\left[e_{jki}\frac{1}{c_1^{(k)}}\frac{\partial f_i}{\partial x_k}\right] = 0 \tag{18}$$

(ii) The curl of the gradient of a scalar field ϕ:

$$\operatorname{curl}\times(\operatorname{grad}\phi) = e_{ijk}\nabla_j^D(\nabla_k^D\phi) = e_{ijk}\frac{1}{c_1^{(j)}}\frac{\partial}{\partial x_j}\left[\frac{1}{c_1^{(k)}}\frac{\partial\phi}{\partial x_k}\right] = 0. \tag{19}$$

In both cases above, we can pull $1/c_1^{(j)}$ out in front of the gradient because the coefficient $c_1^{(j)}$ is independent of x_j.

(iii) The divergence of the gradient of a scalar field ϕ is written in terms of the fractal gradient as

$$\operatorname{div}\cdot(\operatorname{grad}\phi) = \nabla_j^D\cdot\nabla_j^D\phi = \frac{1}{c_1^{(j)}}\frac{\partial}{\partial x_j}\left[\frac{1}{c_1^{(j)}}\frac{\partial\phi}{\partial x_j}\right] = \frac{1}{c_1^{(j)}}\left[\frac{\partial\phi,_j}{c_1^{(j)}}\right],_j, \tag{20}$$

which gives an explicit form of the *fractal Laplacian*.

(iv) The curl of the curl operating on a vector field **f**:

$$\operatorname{curl}\times(\operatorname{curl}\mathbf{f}) = e_{prj}\nabla_r^D(e_{jki}\nabla_k^D f_i) = \nabla_p^D(\nabla_r^D f_r) - \nabla_r^D\nabla_r^D f_p. \tag{21}$$

As a result, also other rules of vector calculus (for example, the Helmholtz decomposition) hold [15].

3 Continuum mechanics of fractal media

With the product measure, the fractional integral, the fractal derivative, the generalized Green–Gauss, and Reynolds theorems we can now develop continuum poromechanics of fractal media. In what follows, the field equations for fractal media will be formulated analogously to the field equations of classical continuum mechanics, but will be based on fractional integrals and expressed in terms of the fractal derivative (12). First, we specify the relationship between surface force $\mathbf{F}^S$ $(= F_k^S)$ and the Cauchy stress tensor σ_{kl} using fractional integrals as

$$F_k^S = \int_S \sigma_{lk} n_l dS_d, \tag{22}$$

where n_l are the components of the outward normal $\mathbf{n}$ to S. On account of $(15)_2$, this force becomes

$$F_k^S = \int_S \sigma_{lk} n_l dS_d = \int_S \sigma_{lk} n_l c_2^{(l)} dS_2. \tag{23}$$

To specify the strain, we observe, using $(15)_1$ and the definition of a fractal derivative (12), that

$$\frac{\partial}{\partial l_{\alpha_k}} = \frac{\partial x_k}{\partial l_{\alpha_k}} \frac{\partial}{\partial x_k} = \frac{1}{c_1^{(k)}} \frac{\partial}{\partial x_k} = \nabla_k^D . \tag{24}$$

Thus, for small deformations, we define the strain ε_{ij} in terms of the displacement u_k as

$$\varepsilon_{ij} = \frac{1}{2}(\nabla_j^D u_i + \nabla_i^D u_j) = \frac{1}{2}\left[\frac{1}{c_1^{(j)}} u_{i},_{j} + \frac{1}{c_1^{(i)}} u_{j},_{i}\right] \quad \text{(no sum in indicies).} \tag{25}$$

As shown in [9], this definition of strain results in the same equations governing the wave motion in linear elastic materials when derived by a variational approach as when derived by a mechanical approach; this is the case with 1d-, 2d-, and 3d-wave motions and the elastodynamics of beams [16].

In the following, we apply the balance laws for mass, linear, and angular momenta, energy, and entropy production to the fractal medium to derive the corresponding continuum equations.

3.1 Fractal conservation of mass and microinertia

Begin with the local equation for conservation of mass for a $d\mathcal{W}$ element

$$\rho d\mathcal{W} = \rho_0 d\mathcal{W}_0, \tag{26}$$

where ρ and ρ_0 is the density of the medium, respectively, in the current and reference state. For the entire fractal body $\mathcal{W}$, we have

$$\frac{d}{dt} \int_{\mathcal{W}} \rho dV_D = 0. \tag{27}$$

Since the above also holds for any arbitrary subset of $\mathcal{W}$, application of the fractional Reynolds transport theorem (13) yields

$$\frac{\partial \rho}{\partial t} + (v_k \rho),_{k} = \frac{d\rho}{dt} + \rho v_{k},_{k} = 0. \tag{28}$$

On account of (24), we find the *fractal conservation of mass* (that is, *fractal continuity equation*)

$$\frac{d\rho}{dt} + \rho c_1^{(k)} \nabla_k^D v_k = 0. \tag{29}$$

To determine the inertia tensor i_{kl} at any micropolar point at time t, we consider a rigid particle p having, at any time t, a volume element $\mathcal{P}_t$, whose angular momentum is

$$\sigma_A(t) = \int_{\mathcal{P}_t} (\mathbf{x} - \mathbf{x}_A) \times \mathbf{v}(\mathbf{x}, t) d\mu(\mathbf{x}). \tag{30}$$

Here $\mathbf{x}_A$ is the location of some fixed point A within $\mathcal{P}$. Taking $\mathbf{v}(\mathbf{x}, t)$ as a helicoidal vector field (for some vector $\boldsymbol{\omega} \in \mathbb{R}^3$) [21]

$$\mathbf{v}(\mathbf{x}, t) = \mathbf{v}(\mathbf{x}_A, t) + \boldsymbol{\omega} \times (\mathbf{x} - \mathbf{x}_A), \tag{31}$$

we find

$$\sigma_A(t) = mAG \times \mathbf{v}(\mathbf{x}, t) + \int_{\mathcal{P}_t} (\mathbf{x} - \mathbf{x}_A) \times \mathbf{v}(\mathbf{x}, t) d\mu(\mathbf{x}). \tag{32}$$

Here G is the center of mass of $\mathcal{P}$. Taking $A = O$ (the origin of the coordinate system), the second term on the right gives all the components of i_{kl} (diagonal and off-diagonal) as

$$i_{kl} = \int_{\mathcal{P}_0} [x_m x_m \delta_{kl} - x_k x_l] \rho(\mathbf{x}) dV_D. \tag{33}$$

The micropolar particle $\mathcal{P}$ is characterized by the vector Ξ_K in the material description and by the vector ξ_k in the spatial description, with the linear mapping from the first into the second: $\xi_k = \chi_{kK} \Xi_K$, here χ_{kK} is called the *microdeformation* [3]. Next, consider the local equation mapping the microinertia in the reference state (I_{KL}) into that in the current state (i_{kl})

$$i_{kl} = I_{KL} \chi_{kK} \chi_{lL}, \tag{34}$$

where i_{kl} and I_{KL} are the microinertia of the medium, respectively, in the current and reference states. For the entire fractal particle $\mathcal{P}$, we have

$$\frac{d}{dt} \int_{\mathcal{P}} \rho I_{KL} dV_D = 0, \tag{35}$$

which, in view of (14), results in a *fractal conservation of microinertia*

$$\frac{d}{dt} i_{kl} = i_{kr} v_{lr} + i_{lr} v_{kr}. \tag{36}$$

3.2 Fractal linear and angular momentum equations

Consider the balance law of linear momentum for $\mathcal{W}$

$$\frac{d}{dt}\int_{\mathcal{W}} \rho v_k dV_D = F_k^B + F_k^S, \tag{37}$$

where F_k^B is the body force, and F_k^S is the surface force given by (22). Introducing the body force density X_k, (37) can be written as

$$\frac{d}{dt}\int_{\mathcal{W}} \rho v_k dV_D = \int_{\mathcal{W}} X_k dV_D + \int_{\partial\mathcal{W}} \sigma_{lk} n_l dS_d. \tag{38}$$

Using the Reynolds transport theorem (13) and the continuity equation (29), the left-hand side is changed to

$$\begin{aligned}\frac{d}{dt}\int_{\mathcal{W}} \rho v_k dV_D &= \int_{\mathcal{W}} \left[\frac{\partial \rho v_k}{\partial t} + (v_k v_l \rho),_l\right] dV_D \\ &= \int_{\mathcal{W}} \rho\left[\frac{\partial v_k}{\partial t} + v_l v_{k},_l\right] dV_D = \int_{\mathcal{W}} \rho \frac{dv_k}{dt} dV_D.\end{aligned} \tag{39}$$

Next, by the Green–Gauss theorem (11) and localization, we obtain the *fractal linear momentum equation*

$$\rho \dot{v}_k = X_k + \nabla_l^D \sigma_{lk}. \tag{40}$$

The conservation of angular momentum in a fractal medium is stated as

$$\frac{d}{dt}\int_{\mathcal{W}} \rho e_{ijk} x_j v_k dV_D = \int_{\mathcal{W}} e_{ijk} x_j X_k dV_D + \int_{\partial\mathcal{W}} e_{ijk} x_j \sigma_{lk} n_l dS_d. \tag{41}$$

Using (40) and (11) yields

$$e_{ijk}\frac{\sigma_{jk}}{c_1^{(j)}} = 0. \tag{42}$$

Observe now that the presence of an anisotropic fractal structure is reflected by discrepancies in fractal dimensions α_i in different directions, which generally implies that $c_1^{(j)} \neq c_1^{(k)}$, $j \neq k$ [10]. Therefore, (42) indicates that the Cauchy stress is generally asymmetric in fractal media, suggesting that the micropolar effects should be accounted for and (41) should be augmented by the presence of couple-stresses. It is important to note here that a material may have an anisotropic fractal structure, yet be isotropic in terms of its constitutive laws.

In micropolar continuum mechanics, one introduces a couple-stress tensor μ_{ik} and a rotation vector φ_j augmenting, respectively, the Cauchy stress tensor τ_{ik} (thus denoted so as to distinguish it from the symmetric σ_{ik}) and the deformation vector u_i. The surface force and surface couple in the fractal setting can be specified by fractional integrals of $\boldsymbol{\tau}$ and $\boldsymbol{\mu}$, respectively, as

$$T_k^S = \int_{\partial\mathcal{W}} \tau_{ik} n_i dS_d, \quad M_k^S = \int_{\partial\mathcal{W}} \mu_{ik} n_i dS_d. \tag{43}$$

The above is consistent with the relation of force tractions and couple tractions to the force stresses and couple stresses on any surface element dS_d

$$t_k = \tau_{ik} n_i, \quad m_k = \mu_{ik} n_i. \tag{44}$$

Now, proceeding in a fashion similar as before, we obtain (40) and the *fractal angular momentum equation*

$$\frac{e_{ijk}}{c_1^{(j)}} \tau_{jk} + \nabla_j^D \mu_{ji} + Y_i = \frac{d}{dt}(i_{ij}\varphi_j). \tag{45}$$

In the above, Y_i is the body force couple density.

3.3 Fractal energy equation and second law of thermodynamics

Globally, the balance of energy in continuum (thermo)mechanics has the following form:

$$\frac{d}{dt}(\mathcal{K} + \mathcal{E}) = \mathcal{P} + \mathcal{H}, \tag{46}$$

where $\mathcal{K}$ is the kinetic energy, $\mathcal{E}$ the internal energy, $\mathcal{P}$ is the power, and $\mathcal{H}$ is the thermal energy supplied externally. More explicitly, we have

$$\begin{aligned} \mathcal{K} &= \int_{\mathcal{W}} \frac{1}{2}\rho v_i v_i dV_D, \quad \mathcal{E} = \int_{\mathcal{W}} \rho \frac{de}{dt} dV_D, \\ \mathcal{P} &= \int_{\mathcal{W}} (X_i v_i + Y_i w_i) dV_D + \int_{\partial\mathcal{W}} (t_i v_i + m_i w_i) dS_d, \\ \mathcal{H} &= -\int_{\partial\mathcal{W}} q_i n_i dV_D + \int_{\partial\mathcal{W}} \rho h dS_d, \end{aligned} \tag{47}$$

where e is the internal energy density, q_i is heat flux through the boundary of $\mathcal{W}$, h is the heat generation within $\mathcal{W}$, whereas v_k $(= \dot{u}_k)$ and w_k $(= \dot{\varphi}_k)$ are the deformation and rotation velocities of the continuum point $\mathcal{P}$, respective. As an aside, we note that, just like in conventional (nonfractal media) continuum mechanics, the balance equations of linear momentum (40) and angular momentum (45) can be consistently derived from the invariance of (46) with respect to rigid body translations ($v_i \to v_i + b_i$, $w_i \to w_i$) and rotations ($v_i \to v_i + e_{ijk}x_j\omega_k$, $w_i \to w_i + \omega_i$), respectively.

Substituting (47) into (46), with the help of (11), we obtain the local rate of change of internal energy

$$\dot{e} = \tau_{ji}(\nabla_j^D v_i - e_{kji}\varphi_k) + \mu_{ji}\nabla_i^D w_j - \nabla_i^D q_i + \rho h. \tag{48}$$

Since we are interested in poroelasticity, we restrict attention to small motions, that is, to the infinitesimal strain tensor and the curvature tensor in fractal media

$$\gamma_{ji} = \nabla_j^D u_i - e_{kji}\varphi_k, \quad \kappa_{ji} = \nabla_j^D \varphi_i. \tag{49}$$

Then, (48) gives the *fractal energy equation*

$$\rho\dot{e} = \tau_{ij}\dot{\gamma}_{ij} + \mu_{ij}\dot{\kappa}_{ij} - \nabla_i^D q_i + h\rho. \tag{50}$$

Assuming e to be a state function of γ_{ij} and κ_{ij} only (which is natural for an elastic solid) and assuming τ_{ij} and μ_{ij} not to be explicitly dependent on the temporal derivatives of γ_{ij} and κ_{ij}, we find

$$\tau_{ij} = \frac{\partial e}{\partial \gamma_{ij}}, \quad \mu_{ij} = \frac{\partial e}{\partial \kappa_{ij}}. \tag{51}$$

That is, just as in nonfractal continuum mechanics, also in fractal media, (τ_{ij}, γ_{ij}) and (μ_{ij}, κ_{ij}) are the conjugate pairs.

To derive the field equation of the second law of thermodynamics in a fractal medium $B(\omega)$, we begin with the global form of that law in the volume V_D, having an Euclidean boundary $\partial\mathcal{W}$, that is,

$$\dot{S} = \dot{S}^{(r)} + \dot{S}^{(i)} \quad \text{with } \dot{S}^{(r)} = \frac{\dot{Q}}{T}, \quad \dot{S}^{(i)} \geq 0, \tag{52}$$

where $\dot{S}$, $\dot{S}^{(r)}$, and $\dot{S}^{(i)}$ stand, respectively, for the total, reversible, and irreversible entropy production rates in V_D. Equivalently,

$$\dot{S} \geq \dot{S}^{(r)}. \tag{53}$$

Thus, we can write (53) as

$$\frac{d}{dt}\int_W \rho s\, dV_D = \dot{S} \geq \dot{S}^{(r)} = -\int_{\partial W} \frac{q_k n_k}{T} dS_d, \tag{54}$$

which, on account of (11), may be rewritten as

$$\int_W \rho \frac{d}{dt} s\, dV_D \geq -\int_W \nabla_k^D \left(\frac{q_k}{T}\right) dV_D, \tag{55}$$

so as to result in a local form of the second law

$$\rho \frac{ds}{dt} \geq -\nabla_k^D \left(\frac{q_k}{T} \right), \tag{56}$$

or, more explicitly,

$$\rho \frac{ds}{dt} \geq -\frac{\nabla_k^D q_k}{T} + \frac{q_k \nabla_k^D T}{T^2}. \tag{57}$$

Just as in thermomechanics of nonfractal bodies [22], we now introduce the rate of irreversible entropy production $\rho \dot{s}^{(i)}$, which—in view of (57)—gives

$$0 \leq \rho \dot{s}^{(i)} = \rho \dot{s} + \frac{\nabla_k^D q_k}{T} - \frac{q_k \nabla_k^D T}{T^2} + \rho h. \tag{58}$$

Here by s we denote specific entropies (that is, per unit mass). Next, we recall the classical relation between the free energy density ψ, the internal energy density e, the entropy s, and the absolute temperature T: $\psi = e - Ts$. This allows us to write for time rates of these quantities

$$\dot{\psi} = \dot{e} - s\dot{T} - T\dot{s}. \tag{59}$$

On the other hand, with ψ being a function of the strain γ_{ji} and curvature-torsion κ_{ji} tensors, the internal variables α_{ij} (strain-type) and ζ_{ij} (curvature-torsion-type), and temperature T, we have

$$\rho \dot{\psi} = \rho \frac{\partial \psi}{\partial \gamma_{ij}} \dot{\gamma}_{ij} + \rho \frac{\partial \psi}{\partial \alpha_{ij}} \dot{\alpha}_{ij} + \rho \frac{\partial \psi}{\partial \kappa_{ij}} \dot{\kappa}_{ij} + \rho \frac{\partial \psi}{\partial \zeta_{ij}} \dot{\zeta}_{ij} + \rho \frac{\partial \psi}{\partial T} \dot{T}. \tag{60}$$

In the above, we shall adopt the conventional relations giving the (external and internal) quasi-conservative Cauchy and Cosserat (couple) stresses and the entropy density as gradients of ψ:

$$\tau_{ij}^{(q)} = \rho \frac{\partial \psi}{\partial \varepsilon_{ij}}, \quad \beta_{ij}^{(q)} = \rho \frac{\partial \psi}{\partial \alpha_{ij}}, \quad \mu_{ij}^{(q)} = \rho \frac{\partial \psi}{\partial \kappa_{ij}}, \quad \eta_{ij}^{(q)} = \rho \frac{\partial \psi}{\partial \zeta_{ij}}, \quad s = -\frac{\partial \psi}{\partial T}. \tag{61}$$

This is accompanied by a split of total Cauchy and micropolar stresses into their quasi-conservative and dissipative parts

$$\tau_{ij} = \tau_{ij}^{(q)} + \tau_{ij}^{(d)}, \quad \mu_{ij} = \mu_{ij}^{(q)} + \mu_{ij}^{(d)}, \tag{62}$$

along with relations between the internal quasi-conservative and dissipative stresses

$$\beta_{ij}^{(q)} = -\beta_{ij}^{(d)}, \quad \eta_{ij}^{(q)} = -\eta_{ij}^{(d)}. \tag{63}$$

In view of (59)–(61), we obtain

$$\rho(\dot{\psi} + s\dot{T}) = \rho(\dot{e} - T\dot{s}) = \tau_{ij}^{(q)} \dot{\gamma}_{ij} + \beta_{ij}^{(q)} \alpha_{ij} + \mu_{ij}^{(q)} \dot{\kappa}_{ij} + \eta_{ij}^{(q)} \dot{\zeta}_{ij} + \rho h. \tag{64}$$

On account of the energy balance, this is equivalent to

$$T\rho\dot{s} = \tau_{ij}^{(d)}\dot{\gamma}_{ij} + \beta_{ij}^{(d)}\dot{\alpha}_{ij} + \mu_{ij}^{(d)}\dot{\kappa}_{ij} + \eta_{ij}^{(d)}\dot{\zeta}_{ij} - \nabla_k^D q_k + \rho h. \tag{65}$$

Recalling (58), we find the local form of the *fractal second law* in terms of time rates of strains and internal parameters

$$0 \le T\rho\dot{s}^{(i)} = \tau_{ij}^{(d)}\dot{\gamma}_{ij} + \beta_{ij}^{(d)}\dot{\alpha}_{ij} + \mu_{ij}^{(d)}\dot{\kappa}_{ij} + \eta_{ij}^{(d)}\dot{\zeta}_{ij} - \frac{q_k\nabla_k^D T}{T} + \rho h. \tag{66}$$

The above is a generalization of the Clausius–Duhem inequality to fractal dissipative media with internal parameters. Upon dropping the internal parameters (which is the case in nondissipative phenomena), the terms $\beta_{ij}^{(d)}\dot{\alpha}_{ij}$ and $\eta_{ij}^{(d)}\dot{\zeta}_{ij}$ vanish, implying that $\tau_{ij}^{(d)}$ reduces to the symmetric Cauchy stress and $\dot{\gamma}_{ij}$ reduces to the deformation rate $d_{ij} := v_{(i,j)}$. On the other hand, upon neglecting the micropolar effects (and thus reverting to classical continuum mechanics), the terms $\mu_{ij}^{(d)}\dot{\kappa}_{ij}$ and $\eta_{ij}^{(d)}\dot{\zeta}_{ij}$ vanish.

For nonfractal bodies, the stress tensor τ_{ij} reverts back to σ_{ij}, and (66) reduces to the simple well-known form [22]

$$0 \le T\rho\dot{s}^{(i)} = \sigma_{ij}^{(d)}\dot{\gamma}_{ij} - \frac{T,_k q_k}{T} + \rho h. \tag{67}$$

4 Conclusion

Continuum-type formulation of mechanics of fractal media is possible through a certain type of fractional calculus. This chapter shows how fundamental balance laws for generally anisotropic fractals can be developed using product measures. The method hinges on expressing the balance laws for fractal media in terms of fractional integrals and, then, converting them to integer-order integrals in conventional (Euclidean) space. An important aspect is the general lack of symmetry of the Cauchy stress tensor, necessitating the adoption of a micropolar continuum model. The entire approach has been applied to wave equations in several settings (1d and 3d wave motions, fractal Timoshenko beam, and elastodynamics) [5–7], extremum and variational principles [14], equations of turbulence in fractal media [13], and fractal planetary rings [12]. In all the cases, the derived equations for fractal media depend explicitly on fractal dimensions and reduce to conventional forms for continuous media with Euclidean geometries upon setting the dimensions to integers.

Bibliography

[1] A. S. Balankin, J. Bory-Reyes, and M. Shapiro, Towards a physics on fractals: differential vector calculus in three-dimensional continuum with fractal metric, *Physica A*, **444** (2016), 345–359.

[2] P. N. Demmie and M. Ostoja-Starzewski, Waves in fractal media, *Journal of Elasticity*, **104** (2011), 187–204.
[3] A. C. Eringen, *Microcontinuum Field Theories I. Foundations and Solids*, Springer-Verlag, New York, 1999.
[4] K. Falconer, *Fractal Geometry: Mathematical Foundations and Applications*, J. Wiley, 2003.
[5] H. Joumaa and M. Ostoja-Starzewski, On the wave propagation in isotropic fractal media, *Zeitschrift für Angewandte Mathematik und Physik*, **62** (2011), 1117–1129.
[6] H. Joumaa and M. Ostoja-Starzewski, Acoustic-elastodynamic interaction in isotropic fractal media, *The European Physical Journal Special Topics*, **222**(8) (2013), 1951–1960.
[7] H. Joumaa, M. Ostoja-Starzewski, and P. N. Demmie, Elastodynamics in micropolar fractal solids, *Mathematics and Mechanics of Solids*, **19**(2) (2014), 117–134.
[8] G. Jumarie, On the representation of fractional Brownian motion as an integral with respect to $(dt)^a$, *Applied Mathematics Letters*, **18** (2005), 739–748.
[9] J. Li and M. Ostoja-Starzewski, Fractal solids, product measures and fractional wave equations, *Proceedings of the Royal Society A. Mathematical, Physical and Engineering Sciences*, **465** (2009), 2521–2536. Errata, *ibid* (2010).
[10] J. Li and M. Ostoja-Starzewski, Micropolar continuum mechanics of fractal media, *International Journal of Engineering Science*, **49** (2011), 1302–1310.
[11] J. Li and M. Ostoja-Starzewski, Comment on "Hydrodynamics of fractal continuum flow" and "Map of fluid flow in fractal porous medium into fractal continuum flow", *Physical Review E*, **88**(5) (2013), 057001.
[12] A. Malyarenko and M. Ostoja-Starzewski, Fractal planetary rings: energy inequalities and random field model, *International Journal of Modern Physics B*, **31**(30) (2017), 1750236 (14 pages).
[13] M. Ostoja-Starzewski, On turbulence in fractal porous media, *Zeitschrift für Angewandte Mathematik und Physik*, **59**(6) (2008), 1111–1117.
[14] M. Ostoja-Starzewski, Extremum and variational principles for elastic and inelastic media with fractal geometries, *Acta Mechanica*, **205** (2009), 161–170.
[15] M. Ostoja-Starzewski, Electromagnetism on anisotropic fractal media, *Zeitschrift für Angewandte Mathematik und Physik*, **64**(2) (2013), 381–390.
[16] M. Ostoja-Starzewski and J. Li, Fractal materials, beams and fracture mechanics, *Zeitschrift für Angewandte Mathematik und Physik*, **60**(6) (2009), 1194–1205.
[17] V. E. Tarasov, Continuous medium model for fractal media, *Physics Letters A*, **336** (2005), 167–174.
[18] V. E. Tarasov, Fractional hydrodynamic equations for fractal media, *Annalen der Physik*, **318**(2) (2005), 286–307.
[19] V. E. Tarasov, *Fractional Dynamics: Applications of Fractional Calculus to Dynamics of Particles, Fields and Media*, Springer-Verlag, 2010.
[20] V. E. Tarasov, Continuum mechanics of fractal media, in *Encyclopedia of Continuum Mechanics*, Springer-Verlag, 2018, in press.
[21] R. Temam and A. Miranville, *Mathematical Modeling in Continuum Mechanics*, Cambridge University Press, 2005.
[22] H. Ziegler and Ch. Wehrli, The derivation of constitutive relations from the free energy and the dissipation functions, *Advances in Applied Mechanics*, **25** (1987), 183–238.

Francesco Paolo Pinnola and Massimiliano Zingales

Fractional-order constitutive equations in mechanics and thermodynamics

Abstract: This chapter is devoted to the application of fractional calculus in mechanics of materials and thermodynamics. The use of fractional calculus in mechanics is related to the definition of fractional-order constitutive equations leading to the class of fractional hereditariness. In this regard, a brief description of the classical rheological models of material hereditariness and a comparison with the fractional elements are reported. It is shown that a rheological hierarchy corresponding to the fractional-order stress-strain relation may be defined. Such a model provides a multi-scale mechanical picture of the power-law hereditariness and it leads toward an unique definition of material free energy. The chapter is also devoted to the investigation of the fractional-order Fourier equation. The analysis of the anomalous heat transfer has been conducted with the a multi-scale approach similar to that used in material hereditariness. Thermodynamic consistency of the model has been reported in terms of the irreversible entropy production.

Keywords: Power-law hereditariness, anomalous heat transfer, thermodynamics restrictions

PACS: 46, 46.35.+z, 83, 83.10.Gr, 83.60.-a, 83.60.Bc, 05.70.-a, 05.70.Ce, 44, 44.05.+e, 44.10.+i

1 Introduction

Recent applications of fractional-order calculus in mechanics and thermodynamics have been introduced in several fields of engineering and mathematical physics, including nonlocal mechanics [7, 9, 25, 26, 48], constitutive models [6, 10, 22, 23, 52, 62, 64], poromechanics [18, 32, 66], viscoelastic fluid behavior [41], random vibration [19, 65], heat transport [44, 72, 73], electrical behavior of capacitors [2, 35, 58], and Brownian motion [5, 11, 16, 20, 51, 67].

Despite the large application of fractional calculus in physics, the use of power-laws and fractional-order models have not been focused on the physical context be-

Francesco Paolo Pinnola, Department of Innovation Engineering, Università del Salento, Lecce, Italy, e-mail: francesco.pinnola@unisalento.it, http://orcid.org/0000-0002-2926-0020
Massimiliano Zingales, Department of Civil, Environmental, Aerospace and Materials Engineering, Università degli Studi di Palermo, Palermo, Italy; and Bio/NanoMechanics for Medical Sciences Laboratory (BNM2-Lab), Advanced Technology Network (ATeN)-Center, CHAB Pole, Palermo, Italy, e-mail: massimiliano.zingales@unipa.it, http://orcid.org/0000-0001-9093-9529

https://doi.org/10.1515/9783110571707-012

yond their introduction. The mechanics and the thermodynamics involved in the use of power-laws and fractional-order related operators will be discussed in this chapter to capture material hereditariness and anomalous thermal diffusion.

2 Fractional hereditary materials

Elasticity is the capability of solid to modify its shape and size under an external load/deformation following a completely reversible process. On the other hand, viscous behavior denotes a relation between the deformation rate and the internal stress of the fluids. A material that exhibits—at the same time—viscous and elastic properties in its mechanical behavior is known as viscoelastic or hereditary. Examples of this kind of materials are: polymers [8, 21, 36, 45, 56, 63], biological tissues and bones [17, 53, 54, 70], mortars [69], woods [38], asphalts and bitumen mixtures [1, 43], and some kinds of soils and rocks [4, 71]; in other words, any real material has hereditary behavior.

This section is devoted to the *fractional hereditariness*, often regarded as *fractional viscoelasticity*, that is, the description of the viscoelastic phenomenon through noninteger derivatives and integrals. Probably, this is the most resonant application of the fractional calculus in a physical problem that has involved several scientists from the first decades of the last century.

2.1 Preliminaries on linear viscoelasticity

The description of the mechanical behavior of viscoelastic materials is obtained through stress–strain relations that bond kinematic and static functions. In this section uniaxial stress–strain relations are considered.

2.1.1 Classical models

The elastic stress–strain relation is expressed by a linear time-independent law

$$\sigma = k\gamma, \tag{1}$$

where σ represents the stress (Pa), γ denotes the strain (dimensionless), and k is a characteristic coefficient of the material known as Young's modulus (Pa). Equation (1), known as *Hooke's law*, is commonly represented by a perfect spring with constant stiffness k (Pa). In such relation, if the imposed stress goes to zero, then the deformation becomes null without loss of energy. This means that the expended work during any loading process is totally recovered during the unloading phase.

The viscous model, describing behavior of several fluids, is the following linear time-dependent law:

$$\sigma(t) = \mu\dot{\gamma}(t), \tag{2}$$

where $\sigma(t)$ is the stress history, $\dot{\gamma}(t)$ is the strain rate ($\sec^{-1}$), and μ is a characteristic parameter of the fluid, known as viscosity (Pa·sec = 10 Poise). Equation (2), known as *Newton–Petroff' law*, shows that the stress $\sigma(t)$ depends on the time strain rate $\dot{\gamma}(t)$. The external energy is not accumulated by the fluid (that is unable to return to the initial configuration), and then it is entirely converted into heat. In other words, the viscous fluid flows in an irreversible way owing to external stress. Equation (2) is commonly represented by an ideal dashpot with viscosity as damping coefficient μ.

As it has been stated, any real material exhibits viscoelastic stress–strain relation, and both elastic and viscous characteristics appear in the mechanical behavior. In particular, it has been observed that if a real material is subjected to a constant stress for a certain time, the corresponding deformation history increases. This strain flow, due to constant stress, is known as *creep*. In contrast, if the applied strain is constant, the stress history decays, and this phenomenon is known as *relaxation*. Creep and relaxation cannot be modeled by Equation (1) or Equation (2), but a proper combination of these two laws can reproduce these two phenomena. Basically, combining these two laws means to assemble perfect springs and ideal dashpots. Following this approach, the simplest viscoelastic models are obtained by a spring and a dashpot connected in series or in parallel.

A perfect spring connected in series with a dashpot represent the *Maxwell* model. In this case, the stress–strain relation becomes

$$\dot{\sigma}(t) + \nu\sigma(t) = k\dot{\gamma}(t), \tag{3}$$

where

$$\nu = \frac{k}{\mu}, \tag{4}$$

k being the stiffness of the spring, and μ the viscous coefficient of the dashpot.

The other simple viscoelastic model is obtained by a spring and a dashpot connected to each other in parallel. In this way the *Kelvin–Voigt* model is obtained, and the stress–strain relation becomes

$$\dot{\gamma}(t) + \nu\gamma(t) = \frac{\sigma(t)}{\mu}. \tag{5}$$

By increasing the number of the elements (spring and dashpot) in the Kelvin–Voigt and/or in the Maxwell model, it is possible to obtain other complex models that are used to describe accurately the viscoelastic phenomenon, for example, standard linear solid, Zener, and Maxwell–Wiechert [14, 27, 39, 63]. For these complex models the stress–strain relations contain more than two coefficients (stiffness and viscosity). The number of parameters increases with the complexity of the model. In the general case, the stress–strain relation of these *multi-element models* can be expressed as

$$\sum_{k=0}^{n} a_k \frac{d^k}{dt^k} \sigma(t) = \sum_{k=0}^{m} b_k \frac{d^k}{dt^k} \gamma(t), \tag{6}$$

where the numbers of the involved parameters are $m + n$. Observe that Equations (3) and (5) are particular cases of Equation (6).

2.1.2 Integral formulation: creep and the relaxation

In linear viscoelasticity, there are two fundamental functions, that is, the *creep compliance* and the *relaxation modulus*. The first one is the response in terms of deformation history under of unitary stress history, whereas the second one is the stress history due to an imposed unitary strain history.

Creep compliance and relaxation modulus contain all informations about the viscoelastic behavior. For any real material, these functions can be determined by two experimental tests. In particular, the creep compliance, denoted by $\psi(t)$, is obtained with the aid of a load control test; imposing a unit step function (Heaviside) as stress history, and measuring the response in terms of strain history. That is,

$$\sigma(t) = \mathcal{H}(t) \Rightarrow \gamma(t) = \psi(t), \tag{7}$$

where $\mathcal{H}(t)$ is the unit step function. On the contrary, relaxation modulus, denoted by $\phi(t)$, is evaluated by imposing a unit step function as strain history in a displacement control test, and measuring the response in terms of stress history. That is,

$$\gamma(t) = \mathcal{H}(t) \Rightarrow \sigma(t) = \phi(t). \tag{8}$$

Observe that creep compliance is nothing else than the unit response function in terms of strain, and relaxation modulus is the unit response function in terms of stress. Both functions are positive for $t \geqslant 0$, whereas are null for $t < 0$, and $\psi(t)$ is a monotonically increasing function as $\phi(t)$ decreases monotonically.

Consider a creep test in which the imposed stress history is that one shown in Figure 1(a). That is, for $t = t_1$, the stress is $\sigma(t) = \sigma_1 \mathcal{H}(t - t_1)$, whereas for $t > t_2$, the total stress is $\sigma_1 + \sigma_2$, and then the imposed stress is

$$\sigma(t) = \sigma_1 \mathcal{H}(t - t_1) + \sigma_2 \mathcal{H}(t - t_2). \tag{9}$$

The creep compliance $\psi(t)$ is defined for $t \geqslant 0$, and then the strain response under the imposed stress in Equation (9), for all $t:\ t_1 < t < t_2$, is

$$\gamma(t) = \sigma_1 \psi(t - t_1), \tag{10}$$

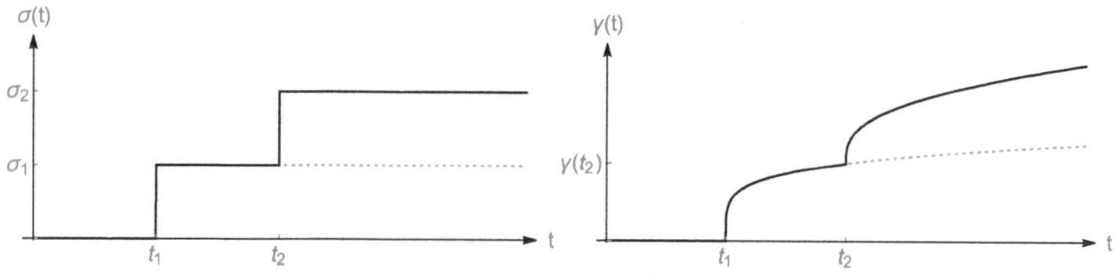

(a) $\sigma(t)$ (b) $\gamma(t)$

Figure 1: Imposed stress history and corresponded strain history.

whereas, for $t > t_2$, the strain history is

$$\gamma(t) = \sigma_1\psi(t - t_1) + (\sigma_2 - \sigma_1)\psi(t - t_2). \tag{11}$$

Equation (11) is valid if the system is linear, and since $\psi(t) \neq 0$ for $t \geqslant 0$, we have that $\psi(t - t_i)$, with $t_i = t_1, t_2$ is different from zero only if $t \geqslant t_1$. Equation (11) is the response in terms of strain due to a particular stress history in which there are two jumps. If the imposed stress history has n jumps in the time steps $t_1, t_2, \ldots, t_n$ of $\sigma_1, \sigma_2, \ldots, \sigma_n$, then the corresponding strain history becomes

$$\gamma(t) = \sum_{j=1}^{n}(\sigma_j - \sigma_{j-1})\psi(t - t_j) = \sum_{j=1}^{n}\Delta\sigma_j\psi(t - t_j); \tag{12}$$

this equation represents the *superposition principle* available in linear viscoelasticity. If the imposed stress history is a continuous law, then it is possible to discretize $\sigma(t)$ by considering time increments Δt and stress increments $\Delta\sigma$. When time increments $\Delta t \to 0$, the stress increment becomes infinitesimal, $\Delta\sigma \to d\sigma(t)$, and the summation in Equation (12) becomes an integral. That is,

$$\gamma(t) = \int_0^t d\sigma(\tau)\psi(t - \tau) = \int_0^t \dot{\sigma}(\tau)d\tau\psi(t - \tau). \tag{13}$$

If at $t = 0$ the imposed stress is different from zero, that is, $\sigma(0) = \sigma_0$, then the strain history is given as

$$\gamma(t) = \int_0^t \dot{\sigma}(\tau)\psi(t - \tau)d\tau + \sigma_0\psi(t). \tag{14}$$

Equation (13) is a convolution integral (or *faltung*), where the kernel is the creep compliance. It represents the relationship between the strain-history response at a certain time and the stress imposed by the whole past. In other words, the material has memory of the past, and the response in terms of the strain history is related to the entire imposed stress history. Equation (14) represents the *integral formulation of linear viscoelasticity*, introduced by *Ludwig Boltzmann* and *Vito Volterra*. By a similar approach the stress history due to an imposed strain history can be found by using the relaxation modulus. That is,

$$\sigma(t) = \int_0^t \dot{\gamma}(\tau)\phi(t-\tau)d\tau + \gamma_0\phi(t), \tag{15}$$

where $\gamma(0) = \gamma_0$ is the initial value of the imposed deformation at $t = 0$. In Equation (15) the stress is obtained from the imposed strain by a convolution integral in which the kernel is the relaxation modulus.

Obviously, the two functions $\phi(t)$ and $\psi(t)$ must be related to each other. In particular, Equations (14) and (15) for $\sigma_0 = 0$ and $\gamma_0 = 0$ in the Laplace domain lead to

$$\hat{\Psi}(s)\hat{\Phi}(s) = s^{-2}, \tag{16}$$

where $\hat{\Psi}(s)$ and $\hat{\Phi}(s)$ are the Laplace transforms of the creep compliance and relaxation modulus, respectively. Equation (16) represents the link between the two functions in the Laplace domain.

Relaxation modulus of Maxwell model can be obtained by placing $\gamma(t) = \mathcal{H}(t)$ and $\sigma(0) = 0$ in Equation (3). That is,

$$\phi(t) = ke^{-\nu t}. \tag{17}$$

The Laplace relation in Equation (16) allows providing the creep compliance of the Maxwell model from Equation (17) as

$$\psi(t) = \frac{\nu}{k}t + \frac{1}{k}. \tag{18}$$

Observe that for Maxwell model the relaxation modulus is a decaying exponential function that can reproduce real experimental evidence, but the creep compliance is a ramp that is unrealistic, since every real material does not exhibit a linear trend during the creep test.

Just as in the previous case, the creep compliance of the Kelvin–Voigt model can be evaluated from Equation (5) by placing $\sigma(t) = \mathcal{H}(t)$ and $\gamma(0) = 0$. That is,

$$\psi(t) = \frac{1}{k}(1 - e^{-\nu t}) \tag{19}$$

from Equation (19), and Equation (16) the relaxation modulus of the Kelvin–Voigt model is obtained as

$$\phi(t) = k\left[1 - \frac{\delta(t)}{\nu}\right], \tag{20}$$

where $\delta(t)$ is the Dirac delta function. In this case, the limitation of the Kelvin–Voigt model is manifested: it is able to describe the creep phenomenon but fails in describing relaxation. In contrast, the Maxwell model well approximates the relaxation tests, but it is unable to describe that of the creep. To overcome these problems, it is possible to use the multi-element models in Equation (6) or consider another approach. In particular, a more accurate description that is able to capture creep and relaxation phenomena through a simple stress–strain relation, is obtained by fractional-order models.

2.2 Fractional-order stress–strain relation

An efficient way to describe the viscoelastic phenomenon is achieved by introducing fractional-order operators in the stress–strain relation. The results of such a fractional viscoelastic model show high degree of agreement with several experimental evidences.

2.2.1 Spring-pot model

The process that has led to the mechanical description of viscoelastic phenomenon by fractional operators has been long. It has required some decades and it involved various scientists. Probably, *P. G. Nutting* produced the first spark; his experimental investigation [47] has been a relevant source of inspiration for other scientists. In particular, around the 1921, Nutting focused his experimental observation on the viscoelastic behavior of materials. After several experiments, he asserted that the two equations used to describe perfectly elastic solids and perfectly viscous fluids were two special cases of a single general law. He observed that the stress and strain history during the relaxation and the creep test do not follow an exponential law, as obtained from the classic model, but they have power-law trends. In particular, Nutting proposed a power-law function to represent the force-displacement relation obtained from experimental tests. That is,

$$u = at^n F^m, \tag{21}$$

which represents the evolution in the time t of the displacement u caused by an assigned history load F. The coefficient a and the two orders n and m depend on temperature and are characteristic of the considered material, but are independent of u, t, F, and the geometry of the specimen. Note that the Hooke's law in Equation (1) is a particular case of the Equation (21) with $n = 0$ and $m = 1$ in terms of strain and stress. On the other hand, the Newton–Petroff law in Equation (2) for the perfectly-viscous fluid is another particular case of the Equation (21) when $n = m = 1$. Nutting also observed that the order n ranges from $0.2 \div 0.91$, and m from $0.75 \div 3.5$. When the order n is close to zero, the materials show a solid behavior; conversely, when n is close to 1 the materials have a behavior similar to fluids. Nutting's experience shows how the classic models, even if they are obtained as complex assembly of several classic elements, are not able to describe the viscoelastic behavior of the real materials. Indeed, by the Kelvin–Voigt model or by the Maxwell model, it is impossible to obtain such relation as in Equation (21). In this regard, in the 1936–1938, *A. N. Gemant* proposed to use the fractional derivative in the stress–strain relation to describe the mechanical behavior of real materials [29, 30]. Likewise, in the 1950s, *G. W. Scott Blair* and *J. E. Caffyn* introduced the fractional stress–strain relation in which the fractional derivative appears [62]. This new model, known as *spring-pot*, is able to reproduce the Nutting's results by a simple law. In particular, the stress–strain relation of such a model is

$$\sigma(t) = c_\beta D^\beta_{0^+}\gamma(t), \tag{22}$$

where $0 \leqslant \beta \leqslant 1$ and c_β are characteristic coefficients of the material. Later, *A. N. Gerasimov* [31] introduced a similar expression, which generalizes the stress–strain relation with the aid of the Caputo's fractional derivative. That is,

$$\sigma(t) = k_\beta\, {}^{C}D^\beta_{0^+}\gamma(t), \tag{23}$$

in this case also $0 \leqslant \beta \leqslant 1$ and k_β are characteristic coefficients of the material. Both c_β in Equation (22) and k_β in Equation (23) have the same mechanical meaning, and they can be defined as ***generalized moduli***. Scott Blair's relation in Equation (22) and Gerasimov's expression in Equation (23) coincide if the considered system is quiescent, that is, if $\gamma(t) = 0$ for $t \leqslant 0$. A formulation in terms of strain has been introduced by *G. L. Slonimsky* [64]:

$$\gamma(t) = \frac{1}{k_\beta} I^\beta_{0^+}\sigma(t), \tag{24}$$

where the Riemann–Liouville fractional integral of the stress history $I^\beta_{0^+}\sigma(t)$ appears. Being for quiescent systems, this expression also coincides with that of Scott Blair. By using the previous expressions, it is possible to summarize the fractional stress–strain relation of the spring-pot as follows:

$$\sigma(t) = c_\beta\, {}^{C}D^\beta_{0^+}\gamma(t), \tag{25}$$

and

$$\gamma(t) = c_\beta^{-1} I^\beta_{0^+}\sigma(t), \tag{26}$$

where c_β and $0 \leqslant \beta \leqslant 1$ can be obtained from a *best-fitting* of experimental data. It is useful to stress that if the order $\beta = 0$, then the fractional stress–strain relation becomes the Hooke's law in Equation (1), and when $\beta = 1$, the fractional model returns the Newtonian one in Equation (2). The order β denotes the predominant phase in the mechanical behavior of the material. In other words, if β is close to zero, then the material exhibits an elastic predominant phase; conversely, when β is close to 1, then the material is similar to a Newtonian fluid. The coefficient c_β, the proportional coefficient between the stress history and the fractional derivative of the strain history, can be defined neither as the stiffness k nor as the viscosity μ, but it must follow the dimensional relation

$$c_\beta = \mu\eta^\beta, \tag{27}$$

where k is the elastic modulus (Pa), and η is a characteristic time of the materials (sec). The relations in Equations (25) and (26) represent a mathematical model that has a perfect correspondence with the experimental results of Nutting.

2.2.2 Integral formulation of fractional viscoelasticity

Fractional stress–strain relation of the spring-pot in Equations (25) and (26) can be obtained by the Boltzmann superposition principle by choosing a proper kernel in the convolution integral. According to Nutting's experience (see Equation (21)) and more recent experimental tests [21, 24], the correct way to represent the relaxation modulus is by a power-law decaying. Therefore,

$$\phi(t) \propto t^{-\beta} \;\Rightarrow\; \phi(t) = C(\beta)t^{-\beta}, \tag{28}$$

where $0 \leqslant \beta \leqslant 1$, and the coefficient $C(\beta)$ is

$$C(\beta) = \frac{c_\beta}{\Gamma(1-\beta)}. \tag{29}$$

By taking into account Equation (27), the relaxation modulus becomes

$$\phi(t) = \frac{c_\beta t^{-\beta}}{\Gamma(1-\beta)} = \frac{k}{\Gamma(1-\beta)}\left(\frac{t}{\eta}\right)^{-\beta}. \tag{30}$$

By placing Equation (30) as kernel of the Boltzmann integral in Equation (15), the following relation holds:

$$\sigma(t) = \frac{c_\beta}{\Gamma(1-\beta)}\int_0^t \dot{\gamma}(\tau)(t-\tau)^{-\beta}d\tau = c_\beta\,{}^{C}D^{\beta}_{0^+}\gamma(t), \tag{31}$$

that is, the fractional stress–strain relation in Equation (25) of the spring-pot. In other words, if in the Boltzmann superposition integral the kernel is a power-law, then a fractional operator directly appears.

From the knowledge of the relaxation modulus, the creep compliance can be determined by using the Laplace relation in Equation (16). In particular, the Laplace transform of the power-law relaxation modulus $\phi(t)$ in Equation (30) is

$$\hat{\Phi}(s) = c_\beta s^{\beta-1}; \tag{32}$$

therefore, according to Equation (16), the creep compliance in the Laplace domain $\hat{\Psi}(s)$ must be

$$\hat{\Psi}(s) = \frac{1}{c_\beta s^{\beta+1}}, \tag{33}$$

and the inverse Laplace transform of the creep compliance yields

$$\psi(t) = \mathscr{L}^{-1}[\hat{\Psi}(s)] = \frac{t^\beta}{c_\beta \Gamma(1+\beta)} = \frac{1}{k\Gamma(1+\beta)}\left(\frac{t}{\eta}\right)^\beta, \tag{34}$$

that is, an increasing power-law function with order β.

Observe that, unlike classical models, the creep compliance obtained from decaying power-law as relaxation modulus is still able to describe the real mechanical behavior of material. By placing Equation (34) as the kernel in the Boltzmann integral in Equation (13), the strain history becomes

$$\gamma(t) = \frac{1}{c_\beta \Gamma(1+\beta)} \int_0^t \dot{\sigma}(\tau)(t-\tau)^\beta d\tau. \tag{35}$$

Integrating by parts and using the property of the Euler gamma function, we obtain the following relation:

$$\gamma(t) = \frac{\beta}{c_\beta \Gamma(1+\beta)} \int_0^t \sigma(\tau)(t-\tau)^{\beta-1} d\tau = c_\beta^{-1} I_{0^+}^\beta \sigma(t), \tag{36}$$

that is, the fractional-order relation of the spring-pot in Equation (26). Equation (31) and Equation (36) allow us to describe, with two parameters c_β and β, both creep and relaxation with the same model. This is a remarkable advantage of the spring-pot with respect to the other classical models.

2.2.3 Generalized fractional-order models

Spring-pot represents a generalized model, which contains the Hookean and Newtonian behavior as limit cases. Indeed, if the order of the involved operator is $\beta = 0$, then the spring-pot becomes a perfect spring, and when $\beta = 1$, the model yields the perfect dashpot. By using this fractional-order model, it is possible to generalize the other classical viscoelastic models. In particular, the fractional Maxwell model can be obtained by placing the spring-pot in place of the dashpot, obtaining the following fractional-order differential equation as the stress–strain relation:

$$D^{\beta}\sigma(t) + \nu_{\beta}\sigma(t) = kD^{\beta}\gamma(t), \tag{37}$$

where the characteristic parameters of the material are β, ν_{β}, and k. By performing the same substitution in the Kelvin–Voigt model, the generalized Kelvin-Voigt model is obtained. The stress–strain relation of this model is

$$D^{\beta}\gamma(t) + \nu_{\beta}\gamma(t) = \frac{\sigma(t)}{\mu}. \tag{38}$$

Also, in this case, there are three characteristic parameters of the materials (β, ν_{β}, and μ). If there are more than two fractional elements, the generic fractional stress–strain relation of the *fractional multi-elements model* is

$$\sum_{k=1}^{n} a_k D^{\alpha_k}\sigma(t) = \sum_{k=1}^{m} b_k D^{\beta_k}\gamma(t), \tag{39}$$

which includes also Equation (6) as a particular case if α_k and β_k are integers.

2.3 Mechanical model of fractional viscoelasticity

The mechanical representation of any integer-order stress–strain relations can be readily obtained as an arrangement of springs and dashpots connected to each other in various ways. However, the mechanical representation by springs and dashpots of the fractional-order stress–strain relation introduced in Equation (31) and Equation (36) is not easy to find. For this reason, in the last decades, several efforts have been devoted to the mechanical description of the fractional-order stress–strain relation [34, 55]. In particular, Bagley e Torvik observed that fractional-order stress–strain relation appear in two classical mechanics problems of fluid dynamics [68]. Thereafter, Schiessel and Blumen found a mechanical representation of the spring-pot as a hierarchical arrangement of springs and dashpots [59–61], and Heymans and Bauwens proposed a self-similar model to reproduce the fractional viscoelastic behavior [33]. Recently, another mechanical model of fractional viscoelasticity was introduced by Di Paola, Zingales and Pinnola in their works [22, 23, 49]; such an approach is described below in detail.

2.3.1 A mechanical description of fractional law

In fractional viscoelasticity, two different behaviors must be distinguished: the elasto-viscous case (EV), in which the elastic phase is predominant, and visco-elastic case

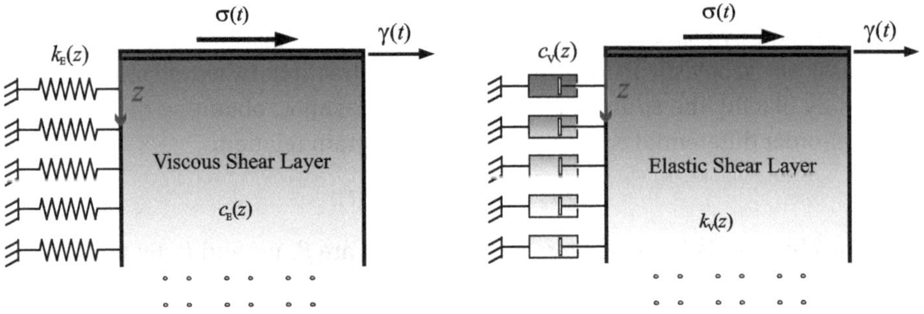

(a) Elasto-Viscous model (EV)

(b) Visco-Elastic model (VE)

Figure 2: Continuous models of fractional viscoelasticity.

(VE), where the viscous behavior prevails. For this reason, two values of orders $\beta = \beta_E \in [0, 1/2]$ and $\beta = \beta_V \in [1/2, 1]$ must be clearly identified (E and V stand for elasto-viscous and visco-elastic, respectively). In this connection, there are two ranges of β and two different mechanical models that exactly restitute the stress–strain relation expressed in Equation (31) or in Equation (36). In particular, as $0 \leqslant \beta = \beta_E \leqslant 1/2$, the mechanical model is a massless indefinite fluid column resting on a bed of independent springs as shown in Figure 2(a), whereas, if $1/2 \leqslant \beta = \beta_V \leqslant 1$, then the exact mechanical model is represented by indefinite massless shear-type column resting on a bed of independent dashpots as shown in Figure 2(b). The first mechanical description represents the EV case of fractional viscoelasticity, whereas the second one corresponds to the VE model.

The correspondence of these mechanical models and fractional-order operators has been proved by introducing a z vertical axis as shown in Figure 2 and denoting $\sigma(z, t)$ the shear stress (in the fluid or in the cantilever beam) and $\gamma(z, t)$ the normalized displacement field [55]. The fractional stress–strain relation in Equation (31) and/or Equation (36) are obtained on the top of the model between the stress $\sigma(t) = \sigma(0, t)$ and its correspondent transverse displacement $\gamma(t) = \gamma(0, t)$. All the mechanical characteristics, viscosity of fluid $c_E(z)$ and external stiffness $k_E(z)$ for the model in Figure 2(a), and the shear modulus $k_V(z)$ and external viscous coefficient of external dashpots $c_V(z)$ for the model in Figure 2(b), vary along the z according to particular functions, and precisely define k_0 and η_0, the reference values of the shear modulus and viscosity coefficient. For the EV materials ($\beta \in [0, 1/2]$), the stiffness and the viscous coefficients decay along z following a power-law trend. That is,

$$k_E(z) = \frac{k_0}{\Gamma(1+\alpha)} z^{-\alpha}, \quad c_E(z) = \frac{\mu_0}{\Gamma(1-\alpha)} z^{-\alpha}, \tag{40}$$

where $0 \leqslant \alpha \leqslant 1$. For the VE materials ($\beta \in [1/2, 1]$), the mechanical characteristics of the model in Figure 2(b) are

$$k_V(z) = \frac{k_0}{\Gamma(1-\alpha)} z^{-\alpha}, \quad c_V(z) = \frac{\mu_0}{\Gamma(1+\alpha)} z^{-\alpha}. \tag{41}$$

Then, the governing equation of the continuous model depicted in Figure 2(a) is

$$\frac{\partial}{\partial z}\left[c_E(z)\frac{\partial \dot{\gamma}(z,t)}{\partial z}\right] = k_E(z)\gamma(z,t), \tag{42}$$

whereas the equilibrium equation of the continuous model depicted in Figure 2(b) is

$$\frac{\partial}{\partial z}\left[k_V(z)\frac{\partial \gamma(z,t)}{\partial z}\right] = c_V(z)\dot{\gamma}(z,t). \tag{43}$$

The solution of the differential equations in Equation (42) and (43) can be solved by Laplace transform. In this way the solution $\hat{\gamma}(z,s)$ in Laplace domain involves the modified first- and second-kind Bessel functions, denoted respectively by $I_\alpha(\cdot)$ and $K_\alpha(\cdot)$. In particular, for the EV case, the relation is

$$\hat{\gamma}(z,s) = z^\alpha\left[B_1^E I_\alpha\left(\frac{z}{\sqrt{\eta_\alpha^E s}}\right) + B_2^E K_\alpha\left(\frac{z}{\sqrt{\eta_\alpha^E s}}\right)\right], \tag{44}$$

where $\beta = (1-\beta)/2$, and

$$\eta_\alpha^E = -\frac{\mu_0\Gamma(\alpha)}{\Gamma(-\alpha)k_0}, \tag{45}$$

whereas, for the VE case, the model leads to

$$\hat{\gamma}(z,s) = z^\alpha\left[B_1^V I_\alpha\left(z\sqrt{\eta_\alpha^V s}\right) + B_2^V K_\alpha\left(z\sqrt{\eta_\alpha^V s}\right)\right], \tag{46}$$

where $\beta = (1+\alpha)/2$, and

$$\eta_\alpha^V = -\frac{\mu_0\Gamma(-\alpha)}{\Gamma(\alpha)k_0}. \tag{47}$$

The integration constants B_j^E and B_j^V with $j = 1, 2$ are obtained by imposing the following pairs of boundary conditions:

$$(\mathrm{EV})\begin{cases}\lim_{z\to 0} c_E(z)\frac{\partial \dot{\gamma}(z,t)}{\partial z} = \sigma(0,t) = \sigma(t),\\ \lim_{z\to\infty}\gamma(z,t) = 0,\end{cases} \tag{48a}$$

$$(\mathrm{VE})\begin{cases}\lim_{z\to 0} k_V(z)\frac{\partial \gamma(z,t)}{\partial z} = \sigma(0,t) = \sigma(t),\\ \lim_{z\to\infty}\gamma(z,t) = 0.\end{cases} \tag{48b}$$

Taking into account the boundary conditions in Equation (48a) and Equation (48b), the inverse Laplace transform of Equation (44) and Equation (46) revert to the fractional stress–strain relation in Equation (36). In particular, for the EV and VE models, the strain $\gamma(t)$ on the top lamina is related to the stress as follows:

$$\begin{aligned}\gamma(t) &= \frac{1}{c_\beta^E} I_{0^+}^\beta \sigma(t),\\ \gamma(t) &= \frac{1}{c_\beta^V} I_{0^+}^\beta \sigma(t),\end{aligned} \tag{49}$$

where the coefficients c_β^E and c_β^V are

$$\begin{cases} c_\beta^E = \frac{k_0\,\Gamma(\beta)2^{2\beta-1}}{\Gamma(2-2\beta)\Gamma(1-\beta)}(\eta_\alpha^E)^\beta, & 0 \leqslant \beta \leqslant 1/2, \\ c_\beta^V = \frac{k_0\,\Gamma(1-\beta)\,2^{2\beta-1}}{\Gamma(2-2\beta)\Gamma(\beta)}(\eta_\alpha^V)^\beta, & 1/2 \leqslant \beta \leqslant 1. \end{cases} \tag{50}$$

Equations (49) are obtained from the continuous models in Figure 2, and they are the same in Equation (36). Obviously, the other relation in terms of stress, Equation (31), is obtained as the stress at the top lamina in terms of the transverse displacement. That is,

$$\begin{aligned} \sigma(t) &= c_\beta^E\,{}^{\mathrm{C}}D_{0^+}^\beta \gamma(t), \\ \sigma(t) &= c_\beta^V\,{}^{\mathrm{C}}D_{0^+}^\beta \gamma(t). \end{aligned} \tag{51}$$

Equations (49) and (51) show how it is possible to obtain the stress–stain relation with fractional-order operators from a proper physical mechanical model.

Figure 2 shows the mechanical description of the fractional-order stress–strain relation. Such a continuous model can be used to describe also the stress–strain relation of fractional multi-element model in Equation (39). In this regard, consider the case in which there are various fractional-order models connected in parallel. In this case, the stress–strain relation involves several fractional-derivative of the strain. That is,

$$\sigma(t) = c_{\beta_1} D^{\beta_1}\gamma(t) + \cdots + c_{\beta_r} D^{\beta_r}\gamma(t) + \cdots + c_{\beta_m} D^{\beta_m}\gamma(t). \tag{52}$$

Such an equation represents the stress–strain relation of m fractional-order elements, where r elements are EV, and $m - r$ elements are VE. The mechanical model of such a kind of fractional differential equation is shown in Figure 3. It is composed by a massless lamina sustained by r columns of massless Newtonian fluid resting on a bed of

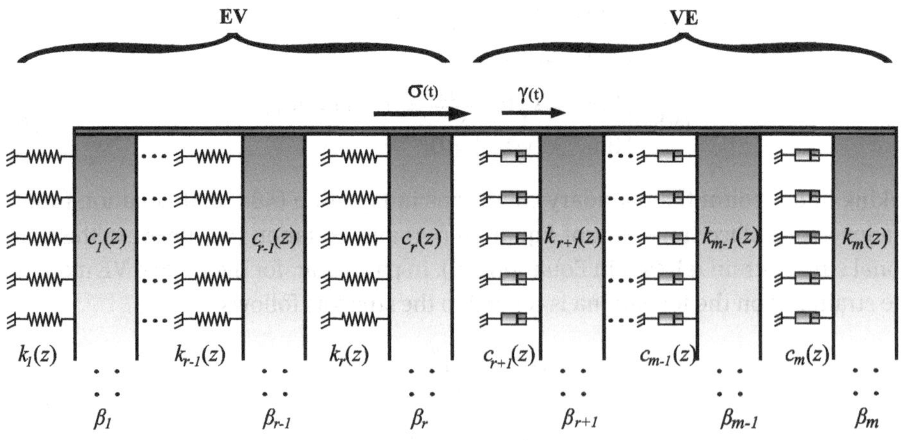

Figure 3: Continuous model of fractional multi-elements.

independent springs, and $m-r$ shear-type elastic columns resting on a bed of independent dashpots. Obviously, the parameters of each part of the model are related to all the involved fractional-order β_j with $j = 1, 2, \ldots, m$, by the relations in Equations (40) and (41).

2.3.2 Free energy of fractional-order hereditariness

The mechanical equivalence discussed in the previous section may be used to introduce an explicit expression of the free energy function associated with the fractional-order hereditariness. Indeed, as we introduce the entropy production rate due to irreversible transformations, namely $\dot{s}^{(i)} \geq 0$, we may express the second principle of thermodynamics as $\dot{s} = \dot{q}/T + \dot{s}^i$, yielding the expression of the free energy as

$$\dot{H} = \dot{W}_{\text{ext}} - T\dot{s}^{(i)} = \dot{W}_{\text{ext}}(t) - D(t), \tag{53}$$

where we used the balance equation of the first principle, and we introduced the specific dissipation rate $D(t) = T\dot{s}^{(i)} \geq 0$ due to irreversible thermodynamical transformations in the material. Assuming isothermal processes, the free-energy function H for simple hereditary material depends on the current value of the strain and on the past history undergone by the material $H(\gamma(t), \gamma^t(\tau))$. The free-energy function must be a state function, and it must fulfill the well-known restrictions on the state functions, such as:

- For any time instant, the following thermodynamic restriction holds true:
$$\sigma(t)\dot{\gamma}(t) \geqslant \dot{H}(\gamma(t), \gamma^t(\tau)) \tag{54}$$
- For the entire set of strain histories $\gamma(t-\tau)$ sharing the same value at the time instant t as $\gamma(0) = \gamma_0$, the minimum free energy must be achieved for the uniform strain history as $\gamma(t-\tau) = \text{const} = \gamma_0$ for all t.
- The derivative of the free energy with respect to the actual value of the strain must correspond to the measured stress as
$$\sigma(t) = \frac{\partial H(\gamma(t), \gamma^t(\tau))}{\partial \gamma(t)}. \tag{55}$$
- For any strain γ_0,
$$H(\gamma_0, \gamma_0^t) - H(0, 0^t) = \frac{1}{2} G_\infty \gamma_0^2, \tag{56}$$
where γ_0^t denotes a constant history of value γ_0.
- Any free energy is a state function, and therefore the following equality must be fulfilled for any pair of strain histories $\gamma_A^t(\tau)$ and $\gamma_B^t(\tau)$ corresponding to the same material state:
$$H(\gamma(t), \gamma_A^t(\tau)) = H(\gamma(t), \gamma_B^t(\tau)). \tag{57}$$

The dependence of the free energy on the state variables, that is, the current value of the strain and of the strain histories may be expressed in quadratic form as

$$H(\gamma,\gamma^t) = R(0,0)\gamma(t)^2 + 2\gamma(t)\int_{-\infty}^{t} \dot{R}(0,t-\tau)\gamma(\tau)d\tau + \int_{-\infty}^{t}\int_{-\infty}^{t} \ddot{R}(t-\tau_1,t-\tau_2)\gamma(\tau_1)\gamma(\tau_2)d\tau_1 d\tau_2, \tag{58}$$

where we used the symmetry of $R(\circ,\circ) \geq 0$. The kernel function $R(\circ,\circ)$ may be related to the relaxation function obtained by stress measures as $2R(0,0) = G(0)$ and $R(0,t) = R(t,0) = G(t)$. A more specific expression of the free-energy functional is obtained as the kernel function is related to the relaxation function as follows:

$$R(\tau_1,\tau_2) = \frac{G(t-\tau_1+t-\tau_2)}{2} = \frac{G(2t-\tau_1-\tau_2)}{2}. \tag{59}$$

Then the free-energy expression in Equation (58) is obtained in the form

$$H_{SS}(t) = \frac{1}{2}\int_{-\infty}^{t}\int_{-\infty}^{t} G(2t-\tau_1-\tau_2)\dot{\gamma}(\tau_1)\dot{\gamma}(\tau_2)d\tau_1 d\tau_2, \tag{60}$$

where the suffix *SS* stands for Stavermann and Schwartzl free energy. In Equation (60) we made the replacement of the strain variable in terms of shear strain to be consistent with subsequent derivations. The introduction of the first principle of thermodynamics yields that the specific mechanical work must equate the rate of increment of specific free energy added to the dissipation rate of the material as

$$\sigma(t)\dot{\gamma}(t) = \dot{\gamma}(t)\int_{-\infty}^{t} G(t-\tau)\dot{\gamma}(\tau)d\tau + \frac{1}{2}\int_{-\infty}^{t}\int_{-\infty}^{t} \dot{G}(2t-\tau_1-\tau_2)\dot{\gamma}(\tau_1)\dot{\gamma}(\tau_2)d\tau_1 d\tau_2 + D(t), \tag{61}$$

where we neglected nonessential dependence on the actual value of the strain only $\gamma(t)$. The observation of the energy rate reported in Equation (61) shows that the dissipation rate associated with the free energy in Equation (60) reads:

$$D(t) = -\frac{1}{2}\int_{-\infty}^{t}\int_{-\infty}^{t} \dot{G}(2t-\tau_1-\tau_2)\dot{\gamma}(\tau_1)\dot{\gamma}(\tau_2)d\tau_1 d\tau_2. \tag{62}$$

The foregoing is a quadratic form of the strain histories with measure provided by the relaxation function $G(t)$. Different choices of the kernel $R(\tau_1,\tau_2)$ in Equation (58) yield different choices for the dissipation rates, each corresponding to the same stress at time t. The explicit expressions of the free energy $\psi(t)$ and of the dissipation rate of $D(t)$ in Equations (60), (62) for the power-law relaxation function is obtained as

$$H_{ss}(t) = \frac{1}{2C_\beta \Gamma(\beta)} \int_{-\infty}^{t} \int_{-\infty}^{t} (2t - \tau_1 - \tau_2)^{-\beta} \dot{\gamma}(\tau_1)\dot{\gamma}(\tau_2) d\tau_1 d\tau_2, \tag{63a}$$

$$D(t) = \frac{\beta}{2C_\beta \Gamma(\beta)} \int_{-\infty}^{t} \int_{-\infty}^{t} (2t - \tau_1 - \tau_2)^{-(1+\beta)} \dot{\gamma}(\tau_1)\dot{\gamma}(\tau_2) d\tau_1 d\tau_2. \tag{63b}$$

The expression of the free energy in terms of the Stavermann–Schwartz functional represents, in the general context of phenomenological constitutive equations, only one of the possible representation of the more general state function describing the material behaviour (see, for example, [12]). The mechanical model reported in previous section, however, may be used to evaluate the free energy function on mechanical basis. In this regard, direct evaluation of the dissipation of the dashpots in the EV and VE mechanical and the material free energy yields Equations (63a) and (63b).

3 Fractional-order Fourier transport equation

The presence of fractional-order calculus has been also used in the theory of thermoelasticity to generalize the classical Fourier and Cattaneo transport equations [13, 28, 37, 46]. However, no physical ground in the formulation of either anomalous heat transfer or thermoelasticity theory has been provided, leading to a nonphysical representation of the thermoelastic phenomena reported in such studies.

In this section the authors obtain a fractional-order Fourier diffusion law from a multi-scale rheological model. This is done through an across-observation scales methodology, allowing for an explanation of the anomalous behavior of inhomogeneous conductors with overall thermal properties varying in time like power-laws t^β with $0 \leqslant \beta \leqslant 1$ [40]. Such consideration is used in the paper to provide a physical exact description of the fractional-order Fourier diffusion equation that is also thermodynamically consistent.

3.1 Exact thermodynamical model of power-law temperature evolution

Time-raising of temperature in the form of power-laws is obtained in this section, introducing a one-dimensional nonhomogeneous perfect conductor reported in Figure 4(a) connecting a sequence of masses. Let us assume that the $n+1$ masses may be expressed as $m_j = A_j \Delta z$ with $j = 1, 2, \ldots, n+1$, where A_j represents the cross-sectional area of the jth mass, and $\Delta z = l/(n+1)$ its length, $l = (n+1)\Delta z$ being the overall length of the system. The masses are separated from the outer systems by adiabatic walls so that thermal energy exchange may occur only along the z direction. The thermodynamic state variables describing the system are assumed as the macroscopic temperatures $T_j(t)$ of the masses m_j for $j - 1, 2, \dots, n+1$.

Energy balance of the jth mass m_j involves the rate of the internal energy U_j, and the energy flux along the conductors m_j, namely, $q_j(t)$ and $q_{j-1}(t)$, which can be written as

$$\frac{dU_j(t)}{dt} = m_j\frac{du_j(t)}{dt} = m_jC_j^{(V)}\frac{dT_j(t)}{dt} = A_{j-1}q_{j-1}(t) - A_jq_j(t), \tag{64}$$

where the mass specific heat at constant volume that is assumed to be uniform for the considered temperature interval is denoted as $C_j^{(V)} = (\frac{\partial u_j}{\partial T})_{T_0}$; $u_j(t)$ is the internal energy function density of the mass m_j.

Given the assumption that only diffusive phonon–phonon interaction yields thermal energy transport, the jth flux may be expressed as

$$q_j(t) = -\chi_j^{(T)}\frac{T_{j+1}(t) - T_j(t)}{z_{j+1} - z_j} = -\chi_j^{(T)}\frac{T_{j+1}(t) - T_j(t)}{\Delta z}, \tag{65}$$

where the thermal conductivity of the jth conductor is denoted as $\chi_j^{(T)}$. Substitution of Equation (65) into Equation (64) yields the thermal energy balance as an explicit differential equation system in the temperatures $T_j(t)$:

$$\rho\Delta z C_j^{(V)}\dot{T}_j(t) = \frac{1}{\Delta z}\left[\chi_{j+1}^{(T)}T_{j+1}(t) - (\chi_j^{(V)} + \chi_{j+1}^{(V)})T_j(t) + \chi_{j-1}^{(T)}T_{j-1}(t)\right], \tag{66}$$

where it is assumed $A = A_j$ for $j = 1, 2, \dots, n+1$ and that the masses m_j occupy the volume $A\Delta z$ so that, introducing the mass density ρ, they may be expressed as $m_j = \rho A\Delta z$ (see Figure 4). The energy balance equation reported in Equation (66) involves masses m_j with $j = 2, 3, \dots, n$ as the temperature of the m_{n+1} mass of the system has been set to the value $T_{n+1} = 0$ without loss of generality. Energy balance of mass m_1 of the thermodynamical system in Figure 4 involves an external thermal energy flux, denoted in the following formula as $\bar{q}(t)$, yielding

$$C_1^{(V)}\Delta z\rho\dot{T}_1(t) + \chi_1^{(T)}\frac{T_2(t) - T_1(t)}{\Delta z} = \bar{q}(t). \tag{67}$$

The anomalous time-scaling of the temperature field is achieved assuming that the spatial distribution of the thermal conductivity $\chi_j^{(V)}$ and the specific heat coefficient $C_j^{(V)}$ vary along the masses m_j with the relations

$$C_j^{(V)} = \frac{C_\alpha^{(V)}(j\Delta z)^{-\alpha}}{\Gamma(1-\alpha)}, \tag{68a}$$

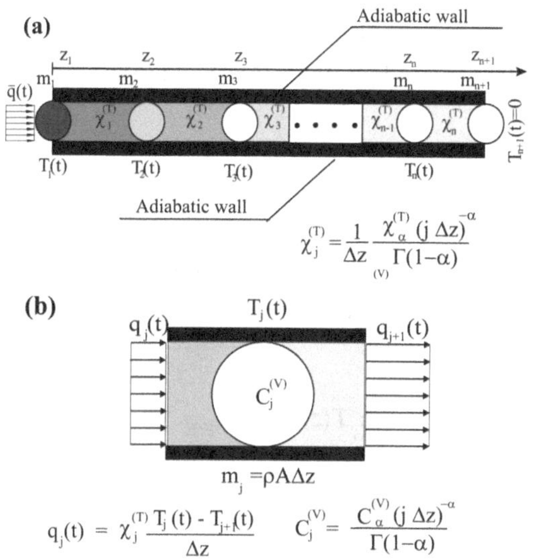

Figure 4: Thermodynamical model of anomalous temperature rising: (a) The concentrated mass system; (b) Thermal energy balance of the jth mass.

$$\chi_j^{(T)} = \frac{\chi_\alpha^{(T)}(j\Delta z)^{-\alpha}\Gamma(\frac{1+\alpha}{2})}{\Gamma(1-\alpha)}, \tag{68b}$$

where the real exponent α belongs to the interval $-1 \leqslant \alpha < 1$ for diffusion-type phenomena. The terms $C_\alpha^{(V)}$ and $\chi_\alpha^{(T)}$ are the specific heat and the thermal conductivity, respectively, with anomalous physical dimensions in the International System of Units (SI) as

$$\left[C_\alpha^{(V)}\right] = \mathrm{m}^{2+\alpha}\mathrm{K}^{-1}\mathrm{s}^{-2}, \quad \left[\chi_\alpha^{(T)}\right] = \mathrm{kg\,m}^{1+\alpha}\mathrm{K}^{-1}\mathrm{s}^{-3}. \tag{69}$$

The discretized model of the heat transfer in the nonhomogeneous conductor may be reverted in a continuous representation as we set $n \to \infty$, $\Delta z \to 0$, and $l \to \infty$. To this end, let us introduce a temperature field so that $T_j(t) \to T(z_j, t)$ and a thermal energy flux field in order to have $q_j(t) \to q(z_j, t)$. Given such assumptions, the balance equation reported in Equation (64) becomes

$$\rho C^{(V)}(z)\frac{\partial T(z,t)}{\partial t} = -\frac{\partial q(z,t)}{\partial z}. \tag{70}$$

Relation (70) describes the balance at location z between the rate of the thermal energy $\dot{U} = \rho\frac{\partial u}{\partial t}$ and the difference of the outgoing thermal energy $q(z + dz, t)$ and the incoming one $q(z, t)$ in unit time. Introducing the Fourier transport equation, obtained for $\Delta z \to 0$,

$$q(z,t) = -\chi^{(T)}(z)\frac{\partial T(z,t)}{\partial z}, \tag{71}$$

in Equation (70), the heat equation is obtained as

$$\rho C^{(V)}(z)\frac{\partial T(z,t)}{\partial t} = \frac{\partial}{\partial z}\left[\chi^{(T)}(z)\frac{\partial T(z,t)}{\partial z}\right]. \tag{72}$$

In Equation (72) the thermodynamical properties of the distributed mass system are described through the continuous counterparts of Equations (68a) and (68b), that is, $C_j^{(V)} \to C^{(V)}(z_j)$ and $\chi_j^{(T)} \to \chi^{(T)}(z_j)$:

$$C^{(V)}(z) = \frac{C_\alpha^{(V)} z^{-\alpha}}{\Gamma(1-\alpha)}, \quad \chi^{(T)}(z) = \frac{\chi_\alpha^{(T)} z^{-\alpha} \Gamma\left(\frac{1+\alpha}{2}\right)}{\Gamma(1-\alpha)}. \tag{73}$$

Accordingly, the boundary conditions associated with the heat Equation (72) are obtained as the continuous conditions on the first mass m_1 and the last mass m_{n+1} of the discrete system (see Figure 4) under consideration as

$$\bar{q}(t) = \lim_{z\to 0} -\chi^{(T)}(z) \frac{\partial T(z,t)}{\partial z}, \quad \lim_{z\to\infty} T(z,t) = 0. \tag{74}$$

The general solution of (72) is

$$\hat{T}(z,s) = B_1 z^\beta K_\beta\left(z\sqrt{\tau s}\right), \tag{75}$$

where we introduced the thermal relaxation time $[\tau] = \mathrm{s\,m}^{-2}$, and we accounted for the radiation condition for $z \to \infty$. The exponent β is related to the scaling exponent α as

$$\beta = \frac{1+\alpha}{2}. \tag{76}$$

The integration constants B_1 in Equation (75) are obtained imposing the relevant boundary conditions that are defined in the Laplace domain as

$$\lim_{z\to 0} -\chi^{(T)}(z) \frac{\partial \hat{T}(z,s)}{\partial z} = \hat{\bar{q}}(s), \tag{77}$$

which gives

$$B_1 = \frac{2^\beta \Gamma(2-2\beta)\sin(\pi\beta)(s\tau)^{-\frac{\beta}{2}}}{\pi \chi_\alpha^{(T)}} \hat{\bar{q}}(s) \tag{78}$$

with $[B_1] = \mathrm{K\,s\,m}^{-\beta}$, whereas the temperature field of the distributed mass systems is in the form

$$\hat{T}(z,s) = \frac{2^\beta \Gamma(2-2\beta)\sin(\pi\beta)(s\tau)^{-\frac{\beta}{2}}}{\pi \chi_\alpha^{(T)}} \hat{\bar{q}}(s) z^\beta K_\beta(z\sqrt{\tau s}). \tag{79}$$

As soon as the temperature $T(z,t)$ has been obtained in the whole distributed system, the explicit relation among the measured temperature at $z = 0$ as a consequence of the ingoing flux of thermal energy is obtained:

$$\hat{T}_0(s) = \lim_{z\to 0} \hat{T}(z,s) = \frac{1}{R_\beta} s^{-\beta} \hat{\bar{q}}(s), \tag{80}$$

where the anomalous thermal diffusivity coefficient R_β is expressed in terms of the physical properties of the conductor as

$$R_\beta = \frac{2^{1-2\beta}\pi\chi_\alpha^{(T)}\csc(\pi\beta)\tau^\beta}{\Gamma(2-2\beta)\Gamma(\beta)} \tag{81}$$

with $[R_\beta] = \text{kg}\,\text{K}^{-1}\,\text{s}^{\beta-3}$. Special cases of Equation (81) can be obtained by looking at some representative values of β and α as follows:

$$\lim_{\substack{\beta\to 0\\ \alpha\to -1}} R_\beta = \frac{2\chi_{-1}^{(T)}}{\pi}, \tag{82a}$$

$$\lim_{\substack{\beta\to \frac{1}{2}\\ \alpha\to 0}} R_\beta = \sqrt[4]{\pi}\sqrt{\rho\chi_0^{(T)}C_0^{(V)}}, \tag{82b}$$

$$\lim_{\substack{\beta\to 1\\ \alpha\to 1}} R_\beta = \rho C_1^{(V)}. \tag{82c}$$

Under the assumption of stationary thermal energy flux $\bar{q}(t) = \bar{q}_0 U(t)$, the time-varying temperature function $T_0(t)$ is obtained applying the inverse Laplace transform to Equation (80), yielding

$$T_0(t) = \frac{\bar{q}_0}{R_\beta\Gamma(1+\beta)} t^\beta \propto t^\beta, \tag{83}$$

that is, the power-law temperature time scaling observed in Figure 4 for the discretized mass system considered in the analysis with $\beta \in [0,1]$ (see, for example, [73] for details).

3.2 Fractional-order generalization of Fourier heat transport

Power-law rising of temperature field described in previous sections corresponds, in the context of a linear-order heat transport, to a fractional-order Fourier diffusion equation. Indeed, as we assume that the thermal energy flux across the $x = 0$ cross-section is a time-dependent function, the inverse Laplace transform of (80) yields

$$T_0(t) = \frac{1}{R_\beta}\frac{1}{\Gamma(\beta)}\int_0^t (t-\tau)^{\beta-1}\bar{q}(\tau)d\tau = \frac{1}{R_\beta} I_{0^+}^\beta \bar{q}(t), \tag{84}$$

that is, a Riemann–Liouville fractional-order integral of order $\beta \in [0,1]$.

Relation (84) is the generalization of the Fourier-Cattaneo transport equation in terms of fractional-order integrals obtained with an exact thermodynamical model of thermal energy transport. It may be observed that a generalization of the Fourier-Cattaneo equation involving fractional-order derivatives of the heat fluxes [42, 57] may be obtained in the framework of the proposed model, including in the analysis values of the decaying exponent α in the interval $-3 \leqslant \alpha \leqslant -1$ (see, for example, [15]). The inverse relation of Equation (84) could be obtained by introducing the β-order fractional derivative of both sides of (84) yielding

$$\bar{q}(t) = R_\beta D^\beta_{0^+} T_0(t), \tag{85}$$

that is, a fractional-order generalization of the transport equation analogous to fractional-order generalisation of the Darcy filtration equation as reported in [15]. The thermodynamic assessment of the transport equation may be obtained assuming that, in a three-dimensional domain V of the Euclidean space, R is referred to a three-dimensional coordinate system (O, x_1, x_2, x_3). The thermal energy flux depends on the spatial gradient $\nabla[\cdot] = \frac{\partial}{\partial x_k}[\cdot] i_k$ yielding

$$\mathbf{q}(\mathbf{x}, t) = \mathbf{K}_\beta \cdot D^\beta_{0^+} \nabla T(\mathbf{x}, t), \tag{86}$$

where i_k are the unit vectors of the coordinate system, $\mathbf{K}_\beta = l\,\mathbf{R}_\beta$ is the symmetric second-order tensor of the anomalous conductivities $k^{(\beta)}_{ij}$ with $i, j = 1, 2, 3$, and l is a characteristic length of the medium. The transport equation reported in Equation (86) yields an isotropic description of the thermal energy transport across the conductor, since the time-dependence of the temperature field involves a single-order fractional derivative.

The use of a new Fourier transport equation in the context of heat transfer must be carefully considered, since the second principle of thermodynamics, in terms of the irreversible entropy production rate $\dot{s}_u(\mathbf{x}, t)$ for unit volume, must be satisfied for any thermodynamical process $T(\mathbf{x}, t)$. In this circumstance, the Gibbs inequality yields $\dot{s}_u(\mathbf{x}, t) \geqslant 0$, which must be fulfilled for any $t \geqslant 0$ and at any location of the conductor $\mathbf{x} \in V$. In order to fulfill this requirement, we report some thermodynamical aspects of the fractional-order Fourier transport equation in Equation (86). The analysis is performed, without loss of generality, for a scalar force-flux relation, namely $q(x, t) = k_\beta D^\beta_{0^+} \nabla T x, t$, and a proper generalization can be found in a forthcoming paper.

The thermodynamical assessment of the fractional-order equation is obtained by means of the self-similar mass distribution that we used to generalize the Fourier equation as shown in the previous section. To this aim, the second principle of thermodynamics, written for the observation scale z, reads

$$\rho(z)\dot{s}(z, t) \geqslant -\frac{1}{T(z, t)} \frac{\partial q(z, t)}{\partial z}, \tag{87}$$

where $\dot{s}$ represents the entropy rate. Introducing the irreversible specific entropy rate $\dot{s}_u(z, t)$, relation (87) is rewritten as

$$\rho(z)\dot{s}(z, t) + \frac{1}{T(z, t)} \frac{\partial q(z, t)}{\partial z} = \rho(x)\dot{s}_u(z, t) \geqslant 0. \tag{88}$$

The entropy rate $\dot{s}(z,t)$ could be cast in terms of the balance among the incoming and outcoming entropy flux, namely $F(z,t)$, as

$$\dot{s}(z,t) = -\frac{1}{\rho(x)}\frac{\partial F(z,t)}{\partial z} + \dot{s}_u(z,t). \tag{89}$$

Introducing Equation (89) into Equation (88), the relevant inequality among the balance of the entropy flux and the balance of the heat flux at location z is obtained in the form

$$\frac{\partial F(z,t)}{\partial z} \geqslant \frac{1}{T(z)}\frac{\partial q(z,t)}{\partial z}. \tag{90}$$

In the context of classical irreversible thermodynamics, it is assumed that the entropy flux is a function of a state variable $u(z,t)$ that corresponds to the specific internal energy of the conductor at location z as

$$F(z,t) = \varphi(u)q(z,t), \tag{91}$$

which, after substitution into Equation (90) (omitting arguments), leads to

$$\left[\varphi(u) - \frac{1}{T}\right]\frac{\partial q}{\partial z} + \frac{\partial \varphi}{\partial z}q(z,t) \geqslant 0. \tag{92}$$

Since relation (92) must be fulfilled for any thermodynamic transformation, for the linear term $\varphi(u)$ equal to $\frac{1}{T}$, we obtain:

$$\left(\frac{\partial \varphi(u)}{\partial z}\right)q = \frac{1}{T^2}\frac{\partial T}{\partial z}q \leqslant 0. \tag{93}$$

Introducing the Fourier relation in (93), we have

$$\dot{s}_u(z,t) = \frac{\lambda(z)}{T^2}\left(\frac{\partial T(z,t)}{\partial z}\right)^2 \geqslant 0. \tag{94}$$

Relation (94) must be verified for any thermodynamical process, for any temperature field, and for any location along the conductor; this yields the thermodynamical restriction on the thermal conductivity $\lambda(z) \geqslant 0$.

It may be shown that the fractional-order generalization of the Fourier transport equation, reported in Equation (86), involves a state function of the form

$$S(x,t) = \int_{-\infty}^{t}\int_{-\infty}^{t} L(t-\tau_1, t-\tau_2)\frac{\partial[\nabla T(\tau_1,x)]}{\partial \tau_1}\frac{\partial[\nabla T(\tau_2,x)]}{\partial \tau_2}d\tau_1 d\tau_2, \tag{95}$$

where the kernel function $L(t-\tau_1, t-\tau_2)$ may be written in the form

$$L(t-\tau_1, t-\tau_2) = \frac{1}{2}G(2t-\tau_1-\tau_2) = \frac{1}{2}\frac{R_\beta l}{\Gamma(\beta)}\frac{1}{(2t-\tau_1-\tau_2)^\beta}, \tag{96}$$

which, after Frechet differentiation, takes back the fractional-order Fourier equation reported in Equation (86). It may be observed that the expression for the free energy function in Equation (95) is also obtained from the evaluation of the overall dissipation rate associated with the inhomogeneous conductor in Figure 4 (see, for example, [3, 50] for details).

4 Concluding remarks

In this chapter we showed that fractional-order generalization of some transport equations used un mechanics (constitutive hereditariness) and thermodynamics (force–flux relations) corresponds to physical models with inhomogeneous properties. The proposed macroscopic constitutive relation has been obtained through a series ordinary models, across infinite observation scales, with distribution characterized by a graded hierarchy of physical and mechanical properties ranging across the scales. The power-law exponent α of the scaling is related, in a one-to-one correspondence, to the exponent β of the power-law time evolution of the observed field. With more details, it has been shown that in presence of rigid thermal conductors, the relationship between the macroscopic heat flux and the corresponding temperature gradient involves a power-law memory in time for the thermodynamical applications. The case of mechanical hereditariness has been faced by resorting to a mechanical hierarchy that involves an inhomogeneous elastic column resting on external newtonian dashpots. From a mathematical point of view, this kind of relation is represented by fractional differential operators. In the present study the Caputo fractional derivative has been chosen as it allows the formulation of initial conditions, involving only the integer-order time-derivatives of the unknown solution, evaluated at the initial time. The proposed model may also be used to yield a proper form of the free-energy functional associated with the fractional-order generalization of the Fourier equation, which will be extensively reported in a forthcoming paper.

Bibliography

[1] E. Aigner, R. Lackner, and C. Pichler, Multiscale prediction of viscoelastic properties of asphalt concrete, *Journal of Materials in Civil Engineering*, **21**(12) (2009), 771–780.
[2] G. Ala, M. Di Paola, E. Francomano, Y. Li, and F. P. Pinnola, Electrical analogous in viscoelasticity, *Communications in Nonlinear Science and Numerical Simulation*, **19**(7) (2014), 2513–2527.
[3] G. Alaimo, V. Piccolo, A. Chiappini, M. Ferrari, D. Zonta, L. Deseri, and M. Zingales, Fractional-order theory of thermoelasticicty. i: Generalization of the Fourier equation, *Journal of Engineering Mechanics*, **144**(2) (2018).

[4] G. Alotta, M. Di Paola, and A. Pirrotta, Fractional Tajimi–Kanai model for simulating earthquake ground motion, *Bulletin of Earthquake Engineering*, **12**(6) (2014), 2495–2506.

[5] G. Alotta and M. Di Paola, Probabilistic characterization of nonlinear systems under α-stable white noise via complex fractional moments, *Physica A: Statistical Mechanics and its Applications*, **420** (2015), 265–276.

[6] G. Alotta, O. Barrera, A. C. F. Cocks, and M. Di Paola, On the behavior of a three-dimensional fractional viscoelastic constitutive model, *Meccanica*, **52**(9) (2017), 2127–2142.

[7] G. Alotta, G. Failla, and F. P. Pinnola, Stochastic analysis of a nonlocal fractional viscoelastic bar forced by Gaussian white noise, *ASCE-ASME Journal of Risk and Uncertainty in Engineering Systems, Part B: Mechanical Engineering*, **3**(3),030904 (2017).

[8] G. Alotta, O. Barrera, and E. C. Pegg, Viscoelastic material models for more accurate polyethylene wear estimation, *Journal of Strain Analysis for Engineering Design*, **53**(5) (2018), 302–312.

[9] G. Alotta, M. Di Paola, G. Failla, and F. P. Pinnola, On the dynamics of non-local fractional viscoelastic beams under stochastic agencies, *Composites Part B: Engineering*, **137** (2018), 102–110.

[10] E. Bologna and M. Zingales, Stability analysis of Beck's column over a fractional-order hereditary foundation, *Proceedings of the Royal Society A*, **474**:20180315r (2018), 1–17.

[11] F. Biagini, B. Øksendal, A. Sulem, and N. Wallner, An introduction to white-noise theory and Malliavin calculus for fractional Brownian motion, *Proceedings of the Royal Society A: Mathematical, Physical and Engineering Sciences*, **460**(2041) (2004), 347–372.

[12] S. Breuer and E. Onat, On the determination of the free energies in linear viscoelastic solids, *Zeitschrift für Angewandte Mathematik und Physik*, **15** (1964), 184–191.

[13] C. Cattaneo, Sulla conduzione del calore, *Atti Del Seminario Matematico E Fisico Del'Università Di Modena*, **3** (1948), 83–101.

[14] R. Christensen, *Theory of Viscoelasticity: an Introduction*, Elsevier, 1982.

[15] A. Compte and R. Metzler, The generalized Cattaneo equation for the description of anomalous transport processes, *Journal of Physics A: Mathematical and General*, **30**(21) (1997), 7277–7289.

[16] G. Cottone, Statistics of nonlinear stochastic dynamical systems under Lévy noises by a convolution quadrature approach, *Journal of Physics A: Mathematical and Theoretical*, **44**(18), 185001 (2011).

[17] L. Deseri, M. Di Paola, M. Zingales, and P. Pollaci, Power-law hereditariness of hierarchical fractal bones, *International Journal for Numerical Methods in Biomedical Engineering*, **29**(12) (2013), 1338–1360.

[18] L. Deseri and M. Zingales, A mechanical picture of fractional-order Darcy equation, *Communications in Nonlinear Science and Numerical Simulation*, **20**(3) (2015), 940–949.

[19] S. Di Lorenzo, M. Di Paola, F. P. Pinnola, and A. Pirrotta, Stochastic response of fractionally damped beams, *Probabilistic Engineering Mechanics*, **35** (2014), 37–43.

[20] M. Di Paola, Fokker–Planck equation solved in terms of complex fractional moments, *Probabilistic Engineering Mechanics*, **38** (2014), 70–76.

[21] M. Di Paola, V. Fiore, F. P. Pinnola, and A. Valenza, On the influence of the initial ramp for a correct definition of the parameters of fractional viscoelastic materials, *Mechanics of Materials*, **69** (2014), 63–70.

[22] M. Di Paola, F. P. Pinnola, and M. Zingales, A discrete mechanical model of fractional hereditary materials, *Meccanica*, **48**(7) (2013), 1573–1586.

[23] M. Di Paola, F. P. Pinnola, and M. Zingales, Fractional differential equations and related exact mechanical models, *Computers & Mathematics with Applications*, **66**(5) (2013), 608–620.

[24] M. Di Paola, A. Pirrotta, and A. Valenza, Visco-elastic behavior through fractional calculus: an easier method for best fitting experimental results, *Mechanics of Materials*, **43**(12) (2011), 799–806.

[25] G. Failla, A. Santini, and M. Zingales, Solution strategies for 1d elastic continuum with long-range interactions: smooth and fractional decay, *Mechanics Research Communications*, **37**(1) (2010), 13–21.

[26] G. Failla, A. Santini, and M. Zingales, A non-local two-dimensional foundation model, *Archive of Applied Mechanics*, **83**(2) (2013), 253–272.

[27] W. Flügge, *Viscoelasticity*, Blaisdell Publishing Company, Waltham, 1967.

[28] J. B. Fourier, *Theorie Analitiqué de le Chaleur*, Dover Publ., New York, 1882, English translation by Freeman.

[29] A. Gemant, A method of analyzing experimental results obtained from elasto-viscous bodies, *Journal of Applied Physics*, **7**(8) (1936), 311–317.

[30] A. Gemant, On fractional differentials, *The London, Edinburgh, and Dublin Philosophical Magazine and Journal of Science*, **25**(168) (1938), 540–549.

[31] A. N. Gerasimov, A generalization of linear laws of deformation and its application to inner friction problems, *Prikladnaâ Matematika I Mehanika*, **12** (1949), 255–259 (in Russian).

[32] E. Gerolymatou, I. Vardoulakis, and R. Hilfer, Modelling infiltration by means of a nonlinear fractional diffusion model, *Journal of Physics D: Applied Physics*, **39**(18) (2006), 4104–4110.

[33] N. Heymans and J.-C. Bauwens, Fractal rheological models and fractional differential equations for viscoelastic behavior, *Rheologica Acta*, **33**(3) (1994), 210–219.

[34] R. Hilfer, *Application of Fractional Calculus in Physics*, World Scientific, Singapore, 2000.

[35] I. S. Jesus and J. A. Tenreiro MacHado, Development of fractional order capacitors based on electrolyte processes, *Nonlinear Dynamics*, **56**(1–2) (2009), 45–55.

[36] R. M. Jones, *Mechanics of Composite Materials*, Taylor & Francis, 1999.

[37] D. Jou, G. Lebon, M. S. Mongiovì, and R. A. Peruzza, Entropy flux in non-equilibrium thermodynamics, *Physica A: Statistical Mechanics and its Applications*, **338**(3-4) (2004), 445–457.

[38] S. S. Kelley, T. G. Rials, and W. G. Glasser, Relaxation behaviour of the amorphous components of wood, *Journal of Materials Science*, **22**(2) (1987), 617–624.

[39] F. Mainardi, *Fractional Calculus and Waves in Linear Viscoelasticity*, Imperial College Press-World Scientific Publishing Co. Pte. Ltd., London, 2010.

[40] F. Mainardi, Yu. Luchko, and G. Pagnini, The fundamental solution of the space-time fractional diffusion equation, *Fractional Calculus & Applied Analysis. An International Journal for Theory and Applications.*, **4**(2) (2001), 153–192.

[41] N. Makris, G. F. Dargush, and M. C. Constantinou, Dynamic analysis of generalized viscoelastic fluids, *Journal of Engineering Mechanics*, **119**(8) (1993), 1663–1679.

[42] R. Metzler and T. F. Nonnenmacher, Fractional diffusion, waiting-time distributions, and Cattaneo-type equations, *Physical Review. E, Statistical Physics, Plasmas, Fluids, and Related Interdisciplinary Topics*, **57**(6) (1998), 6409–6414.

[43] G. D. Mino, G. Airey, M. Di Paola, F. P. Pinnola, G. D'Angelo, and D. L. Presti, Linear and nonlinear fractional hereditary constitutive laws of asphalt mixtures, *Journal of Civil Engineering and Management*, **22**(7) (2016), 882–889.

[44] M. S. Mongiovì and M. Zingales, A non-local model of thermal energy transport: the fractional temperature equation, *International Journal of Heat and Mass Transfer*, **67** (2013), 593–601.

[45] S. Müller, M. Kästner, J. Brummund, and V. Ulbricht, A nonlinear fractional viscoelastic material model for polymers, *Computational Materials Science*, **50**(10) (2011), 2938–2949.

[46] F. R. Norwood, Transient thermal waves in the general theory of heat conduction with finite wave speeds, *Journal of Applied Mechanics, Transactions ASME*, **39**(3) (1972), 673–676.

[47] P. G. Nutting, A new general law of deformation, *Journal of the Franklin Institute*, **191**(5) (1921), 679–685.

[48] M. Di Paola and M. Zingales, Long-range cohesive interactions of non-local continuum faced by fractional calculus, *International Journal of Solids and Structures*, **45**(21) (2008), 5642–5659.

[49] M. Di Paola and M. Zingales, Exact mechanical models of fractional hereditary materials, *Journal of Rheology*, **56**(5) (2012), 983–1004.

[50] V. Piccolo, G. Alaimo, A. Chiappini, M. Ferrari, D. Zonta, M. Zingales, and L. Deseri, Fractional-order theory of thermoelasticity. ii: Quasi-static behavior of bars, *Journal of Engineering Mechanics*, **144**(2) (2018).

[51] F. P. Pinnola, Statistical correlation of fractional oscillator response by complex spectral moments and state variable expansion, *Communications in Nonlinear Science and Numerical Simulation*, **39** (2016), 343–359.

[52] F. P. Pinnola, On the appearance of fractional operators in non-linear stress-strain relation of metals, *International Journal of Non-Linear Mechanics*, **105** (2018), 1–8.

[53] D. P. Pioletti and L. R. Rakotomanana, Non-linear viscoelastic laws for soft biological tissues, *European Journal of Mechanics, A/Solids*, **19**(5) (2000), 749–759.

[54] D. P. Pioletti, L. R. Rakotomanana, J.-F. Benvenuti, and P.-F. Leyvraz, Viscoelastic constitutive law in large deformations: application to human knee ligaments and tendons, *Journal of Biomechanics*, **31**(8) (1998), 753–757.

[55] I. Podlubny, *Fractional Differential Equations*, Academic Press, San Diego, 1999.

[56] V. N. Pokrovkii, *The mesoscopic Theory of Polymer Dynamics*, Springer Series in Chemical Physics, 2010.

[57] Y. Povstenko, Fractional heat conduction in a space with a source varying harmonically in time and associated thermal stresses, *Journal of Thermal Stresses*, **39**(11) (2016), 1442–1450.

[58] A. G. Radwan, A. S. Elwakil, and A. M. Soliman, On the generalization of second-order filters to the fractional-order domain, *Journal of Circuits, Systems and Computers*, **18**(2) (2009), 361–386.

[59] H. Schiessel and A. Blumen, Hierarchical analogues to fractional relaxation equations, *Journal of Physics A: Mathematical and General*, **26**(19) (1993), 5057–5069.

[60] H. Schiessel and A. Blumen, Mesoscopic pictures of the sol-gel transition: ladder models and fractal networks, *Macromolecules*, **28**(11) (1995), 4013–4019.

[61] H. Schiessel, R. Metzler, A. Blumen, and T. F. Nonnenmacher, Generalized viscoelastic models: their fractional equations with solutions, *Journal of Physics A: General Physics*, **28**(23) (1995), 6567–6584.

[62] G. W. Scott-Blair and J. E. Caffyn, An application of the theory of quasi-properties to the treatment of anomalous strain-stress relations, *The London, Edinburgh, and Dublin Philosophical Magazine and Journal of Science*, **40**(300) (1949), 80–94.

[63] M. T. Shaw and W. J. MacKnight, *Introduction to Polymer Viscoelasticity*, Wiley, 2005.

[64] G. L. Slonimsky, On the law of deformation of highly elastic polymeric bodies, *Doklady Akademii Nauk SSSR*, **140**(2) (1961), 343–346.

[65] P. D. Spanos and G. Malara, Nonlinear random vibrations of beams with fractional derivative elements, *Journal of Engineering Mechanics*, **140**(9) (2014).

[66] V. E. Tarasov, Fractional hydrodynamic equations for fractal media, *Annals of Physics*, **318**(2) (2005), 286–307.

[67] V. E. Tarasov, Large lattice fractional Fokker–Planck equation, *Journal of Statistical Mechanics: Theory and Experiment*, **2014**(9), P09036 (2014).

[68] P. J. Torvik and R. L. Bagley, On the appearance of the fractional derivative in the behavior of real materials, *Journal of Applied Mechanics, Transactions ASME*, **51**(2) (1984), 294–298.

[69] F.-Z. Wang, Z.-C. Liu, and S.-G. Hu, Influence of loading rate on compressive strength of CA mortar, *Beijing Gongye Daxue Xuebao / Journal of Beijing University of Technology*, **34**(10) (2008), 1059–1065.

[70] M. Zhang, P. Nigwekar, B. Castaneda, K. Hoyt, J. V. Joseph, A. di Sant'Agnese, E. M. Messing, J. G. Strang, D. J. Rubens, and K. J. Parker, Quantitative characterization of viscoelastic properties of human prostate correlated with histology, *Ultrasound in Medicine & Biology*, **34**(7) (2008), 1033–1042.

[71] H. W. Zhou, C. P. Wang, B. B. Han, and Z. Q. Duan, A creep constitutive model for salt rock based on fractional derivatives, *International Journal of Rock Mechanics and Mining Sciences*, **48**(1) (2011), 116–121.

[72] M. Zingales, Fractional-order theory of heat transport in rigid bodies, *Communications in Nonlinear Science and Numerical Simulation*, **19**(11) (2014), 3938–3953.

[73] M. Zingales, An exact thermodynamical model of power-law temperature time scaling, *Annals of Physics*, **365** (2016), 24–37.

Index

https://doi.org/10.1515/9783110571707-013